Advanced CMOS Devices and Applications

Advanced CMOS Devices and Applications

Editors

Yi Zhao
Choonghyun Lee

Basel • Beijing • Wuhan • Barcelona • Belgrade • Novi Sad • Cluj • Manchester

Editors
Yi Zhao
Zhejiang University
Hangzhou, China

Choonghyun Lee
Zhejiang University
Hangzhou, China

Editorial Office
MDPI
St. Alban-Anlage 66
4052 Basel, Switzerland

This is a reprint of articles from the Special Issue published online in the open access journal *Electronics* (ISSN 2079-9292) (available at: https://www.mdpi.com/journal/electronics/special_issues/CMOS_devices).

For citation purposes, cite each article independently as indicated on the article page online and as indicated below:

Lastname, A.A.; Lastname, B.B. Article Title. *Journal Name* **Year**, *Volume Number*, Page Range.

ISBN 978-3-03928-593-8 (Hbk)
ISBN 978-3-03928-594-5 (PDF)
doi.org/10.3390/books978-3-03928-594-5

© 2024 by the authors. Articles in this book are Open Access and distributed under the Creative Commons Attribution (CC BY) license. The book as a whole is distributed by MDPI under the terms and conditions of the Creative Commons Attribution-NonCommercial-NoDerivs (CC BY-NC-ND) license.

Contents

Preface

We are pleased to present this reprint of our Special Issue, a compilation of cutting-edge research and advancements in semiconductor technology. The scope of this collection encompasses diverse topics, including 3D power scaling, advanced transistor processing, high-mobility channels, and novel simulation results. Our intention is to provide a comprehensive exploration of the transformative landscape in semiconductor evolution.

The aim of this Special Issue is to contribute to the ongoing dialogue within the scientific community, addressing challenges and opportunities in semiconductor research. The scope extends to emerging non-volatile memories, gate-all-around 3D transistors, and the integration of different devices in 3D architecture.

We extend our gratitude to the esteemed contributors who have shared their expertise and insights, making this collection a valuable resource for researchers, engineers, and enthusiasts in the field. This reprint is dedicated to those at the forefront of semiconductor innovation, pushing the boundaries of technology.

We acknowledge the collaborative efforts of all involved authors, whose dedication has enriched this collection. Their commitment to advancing semiconductor science is evident in the quality and diversity of the contributions.

As Guest Editors, we are excited to present this Special Issue reprint, confident that it will serve as a reference and inspiration for future endeavors in semiconductor research. May this compilation foster new ideas and collaborations in the dynamic landscape of semiconductor evolution.

Yi Zhao and Choonghyun Lee
Editors

Editorial

Advanced CMOS Devices and Applications

Choonghyun Lee * and Yi Zhao *

College of Information Science and Electronic Engineering, Zhejiang University, Hangzhou 310058, China
* Correspondence: chlee01@zju.edu.cn (C.L.); yizhao@zju.edu.cn (Y.Z.)

Citation: Lee, C.; Zhao, Y. Advanced CMOS Devices and Applications. *Electronics* **2024**, *13*, 134. https://doi.org/10.3390/electronics13010134

Received: 26 December 2023
Accepted: 27 December 2023
Published: 28 December 2023

Copyright: © 2023 by the authors. Licensee MDPI, Basel, Switzerland. This article is an open access article distributed under the terms and conditions of the Creative Commons Attribution (CC BY) license (https://creativecommons.org/licenses/by/4.0/).

The persistent scaling of transistor dimensions has marked an era characterized by a fourfold increase in transistor density and a twofold boost in electrical performance every 2–3 years, effectively reducing their cost per function. This has been propelled by the scalability and ubiquity of complementary metal–oxide–semiconductor (CMOS) technology in silicon, a cornerstone of the semiconductor industry. However, as conventional silicon devices approach fundamental physical limits, researchers have begun fervently exploring new device architectures and novel channel materials with superior electrical properties. Fin-shaped field-effect transistors (FinFETs) have undeniably become a mainstream CMOS logic technology. However, their scalability challenges at 14/10/7/5 nm underscore the need for alternative architectures. This quest mandates the development of innovative structural, material, and process technologies that align with both current and projected silicon platform advancements.

A pivotal shift is on the horizon, foretold by the International Roadmap for Devices and Systems (IRDS)—the transition from FinFET technology to the intriguing realm of "three-dimensional (3D) power scaling". This transformative leap necessitates the creation of a new breed of high-performance, low-power transistors fashioned into 3D structures. Unlike traditional lithography-driven density increases, this technology vertically constructs each transistor with the same footprint, leading to a heightened transistor density per unit area. While 3D-stacked transistors introduce an additional degree of freedom for increasing device densities, their potential is fully harnessed when integrated with high-mobility channel materials and channel strain engineering. As standalone units, 3D-stacked transistors do not inherently enhance device performance. The infusion of non-silicon (Si) channel materials like SiGe, Ge, and III-V compounds becomes paramount for optimal integration compatibility with 3D-stacked transistors.

In tandem with 3D power scaling, state-of-the-art component research into advanced transistor processing has emerged as a cornerstone. The pursuit of high-performance, low-power technology mandates a meticulous exploration of innovative transistor processing techniques, ensuring that each step in its fabrication contributes to the overall efficiency and reliability of the device. High-mobility channels, the lifeblood of advanced semiconductor technology, come under intense scrutiny. Researchers are exploring materials and designs that maximize electron and hole mobilities, striving for optimal performance and efficiency. The marriage of 3D transistors with high-mobility channels promises a synergy that can redefine the landscape of semiconductor technology. The heterogeneous and monolithic 3D integration of different devices, chips, or wafers stands as another frontier, presenting an opportunity not only for increased device density but also for the integration of diverse functionalities on a single platform. Challenges in ensuring seamless compatibility and performance optimization across disparate components drive researchers to explore innovative solutions. In this dynamic landscape, the significance of accurate and efficient simulation cannot be overstated. Novel simulation results provide a crucial roadmap, guiding researchers through the intricate design space of emerging technologies. These simulations, often based on advanced methodologies such as artificial neural networks, offer insights into device behavior, aiding circuit designers and process engineers in optimizing their performance and functionality.

This special issue will unfold numerous contributions, each addressing a specific facet of this transformative landscape—from emerging non-volatile memories to gate-all-around 3D transistors. It will cover state-of-the-art component research into advanced transistor processing, high-mobility channels, and heterogeneous and monolithic 3D integration for novel device simulation results. Together, these endeavors chart the course for the next chapter in semiconductor evolution.

Highlighting Key Contributions

Developing advanced semiconductor technology for 5 nm logic transistor nodes and beyond is essential for future high-performance and low-power technology to be used in high-end servers, gaming chips, mobile phones, laptops, etc. To meet transistor performance targets, the innovation of device architecture, along with the introduction of high-mobility channel material and strain engineering, is mandatory.

Sagarika Mukesh and Jingyun Zhang from IBM discuss the integration scheme and challenges of full bottom isolation and multiple threshold voltages in gate-all-around nanosheet FETs (Contributor 1). Additionally, they explore the mobility of electrons and holes as a function of channel geometry, a crucial consideration in designing high-performance gate-all-around nanosheet FETs. To assist circuit designers and process engineers working on advanced semiconductor devices, SangMin Woo et al. developed a compact and accurate simulation model based on artificial neural networks (ANNs). This ANN-based compact model is approximately two times faster than the SPICE simulations of the existing compact model, and its accuracy is demonstrated through simulations of XOR, ring oscillators, and SRAM circuits (Contributor 2).

Emerging nanoscale logic and non-volatile memory (NVM) devices such as resistive, magnetic, and ferroelectric memories are vital to realize neuromorphic computation, which mimics the human brain's architecture in order to significantly increase a computer's thinking and responding power. Since the discovery of ferroelectric properties in Hf-based thin films in 2011, they have attracted much attention for their CMOS compatibility and scalability for applications in logic and memory devices. Defects and surface energy are arguably two main categories of impact factors that help stabilize the metastable orthorhombic phase in Hf-based oxides. Tianning Cui et al. discuss the intrinsic parameters used to stabilize the ferroelectric phase of HfO_2 thin films by investigating the separate effects of dopant, oxygen vacancy, and specific surface area on the crystal phase of the films (Contributor 3). Jaewook Yoo et al. describe several types of ferroelectric-based memory devices, including two-terminal-based FTJ, three-terminal-based FeFET, and FeRAM, focusing on their operational mechanisms, features, and potential applications (Contributor 4). Seonjun Choi et al. use ferroelectric materials for practical applications of 3D NAND structures, replacing the charge trapping layer in conventional flash memory with ferroelectric materials. Their TCAD simulation results show that the silicon pillar structure can maximize the operating performance of ferroelectric memory while achieving the advantages of 3D NAND structures (Contributor 5).

In addition, resistive random-access memory based on transition metal oxides has attracted increasing attention as one of the most promising candidates for the next generation of eNVM due to its prominent advantages, including low cost, high integration density, fast switching speed, high endurance, and good CMOS process compatibility. Huikai He et al. propose a simple strategy for regulating the leakage current, forming a voltage, a memory window, and uniformity by varying the thickness of a Ti buffer layer between the resistive switching layer and metal electrode. The Ti buffer layer plays a vital role in engineering the interfacial oxidation reaction, acting as an oxygen-scavenging layer to enhance resistive uniformity (Contributor 6). Wei Na et al. study RRAM devices with a metal–insulator–semiconductor structure, including an HfOx switching layer and Ge or Si bottom electrodes, to achieve a higher memory window with good endurance. A memory window over 10^5 was achieved with Pd/HfOx/p-Ge RRAM devices, and the conductance mechanism was analyzed in detail. These results suggest that HfOx/Ge RRAM devices are promising candidates for applications of FPGA and neuromorphic computation (Contributor 7).

Regarding Si CMOS technologies, germanium is an attractive choice of channel material for further advanced nodes since it has smaller effective masses for both electrons and holes than Si and SiGe and, as a group XIV element, has high process compatibility with Si. A typical cause of performance degradation in practically scaled devices is an increase in parasitic resistance. The parasitic resistance is composed of interconnecting and contact metal resistance, semiconductor resistance around the source/drain, and contact resistance just at the metal/semiconductor interface. Tomonori Nishimura describes the metal/Ge interface, including the origin of strong Fermi-level pinning (FLP) at the valence band edge of Ge, and proposes a possible method of reducing the Schottky barrier height at this interface. FLP alleviation is achieved by weakening the intrinsic metal-induced gap states at the metal/Ge interface and may be pivotal in designing scaled Ge n-FETs (Contributor 8).

A 3D-stacked device architecture alleviates the challenges involved in traditionally planar scaling transistors while continuously improving the effective chip density and performance. Here, a "monolithic" or "sequential" approach is discussed but not a "packaging" one, where interlayer connectivity is dominated by chip bonding alignment accuracy. This further enables additional functionalities to be implemented in CMOS devices, called "More Than Moore". Toshiyuki Tabata et al. illustrate the recent progress of ns and μs ultra-violet laser annealing technology, one of the most critical component technologies to realize 3D-integrated CMOS devices, where a new electrically functional Si (or other semiconductor material) layer must be fabricated directly on the underlayer components, either by wafer bonding or deposition. For 3D integration, the thermal budget should be below 500 °C to avoid the performance degradation of underlayer components (Contributor 9). Jaeyong Jeong et al. introduce heterogeneous and monolithic 3D (M3D) integration to maximize the benefits of 3D integration in terms of low power consumption, interconnection delay, and via densities. In particular, they share recent research on the M3D integration of RF devices on Si CMOS circuits and InGaAs photodetectors on Si bottom FETs for realizing future M3D-based mixed-signal systems (Contributor 10). Hyungwoo Kim delivers a comprehensive review of recent trends in metallization, including traditional barrier/liner thickness scaling and new materials and integration schemes. The innovative approaches proposed to date can contribute to scaling requirements not through direct scaling but with architectural innovations such as super vias and buried power rails (Contributor 11).

A transistor (1T) dynamic random-access memory (DRAM) without capacitors has garnered great attention due to its scalability, where 1T-DRAM utilizes the floating body effect of a partially depleted silicon-on-insulator (SOI) to retain memory performance. However, SOI wafers are significantly costly and difficult to fabricate. Geon Uk Kim et al. report a low-cost method for forming SOI-like structures using polycrystalline silicon in a TCAD simulation. By decoupling the channel and the storage layer with separation oxide, the 1T-DRAM achieves a high retention time (Contributor 12). The insulated gate bipolar transistor (IGBT) is widely used as a switching device in inverter circuits for driving motors. To improve the performance of the IGBT, Xiaodong Zhang et al. propose an IGBT with an injection-enhanced p-floating layer in a TCAD simulation, acting as a current amplification stage and suppressing the snapback effect during the turn-on period (Contributor 13).

In conclusion, the diverse array of contributions presented in this special issue represents a collective stride toward the future of semiconductor technology. From innovative device architectures and novel materials to advanced simulation techniques, each paper contributes to the next chapter in semiconductor evolution. As we delve deeper into the intricacies of 3D power scaling, high-mobility channels, and cutting-edge transistor processing, we are not just witnessing incremental advancements; we are shaping the landscape of technology for years to come. The collaborative efforts showcased here pave the way for enhanced performance, efficiency, and functionality in high-end servers, gaming chips, mobile phones, laptops, and beyond.

Funding: This work was supported in part by the Key Research and Development Program of Zhejiang Province under Grant 2021C01039.

Conflicts of Interest: The authors declare no conflicts of interest.

List of Contributions

1. Mukesh, S.; Zhang, J. A Review of the Gate-All-Around Nanosheet FET Process Opportunities. *Electronics* **2022**, *11*, 3589.
2. Woo, S.; Jeong, H.; Choi, J.; Cho, H.; Kong, J.; Kim, S. Machine-Learning-Based Compact Modeling for Sub-3-nm-Node Emerging Transistors. *Electronics* **2022**, *11*, 2761.
3. Cui, T.; Zhu, L.; Chen, D.; Fan, Y.; Liu, J.; Li, X. Independent Effects of Dopant, Oxygen Vacancy, and Specific Surface Area on Crystal Phase of HfO_2 Thin Films towards General Parameters to Engineer the Ferroelectricity. *Electronics* **2022**, *11*, 2369.
4. Yoo, J.; Song, H.; Lee, H.; Lim, S.; Kim, S.; Heo, K.; Bae, H. Recent Research for HZO-Based Ferroelectric Memory towards In-Memory Computing Applications. *Electronics* **2023**, *12*, 2297.
5. Choi, S.; Jeong, J.; Kang, M.; Song, Y. A Novel Structure to Improve the Erase Speed in 3D NAND Flash Memory to Which a Cell-On-Peri (COP) Structure and a Ferroelectric Memory Device Are Applied. *Electronics* **2022**, *11*, 2038.
6. He, H.; Tan, Y.; Lee, C.; Zhao, Y. Ti/HfO_2-Based RRAM with Superior Thermal Stability Based on Self-Limited TiOx. *Electronics* **2023**, *12*, 2426.
7. Wei, N.; Ding, X.; Gao, S.; Wu, W.; Zhao, Y. HfO_x/Ge RRAM with High ON/OFF Ratio and Good Endurance. *Electronics* **2022**, *11*, 3820.
8. Nishimura, T. Understanding and Controlling Band Alignment at the Metal/Germanium Interface for Future Electric Devices. *Electronics* **2022**, *11*, 2419.
9. Tabata, T.; Rozé, F.; Thuries, L.; Halty, S.; Raynal, P.; Karmous, I.; Huet, K. Recent Progresses and Perspectives of UV Laser Annealing Technologies for Advanced CMOS Devices. *Electronics* **2022**, *11*, 2636.
10. Jeong, J.; Geum, D.; Kim, S. Heterogeneous and Monolithic 3D Integration Technology for Mixed-Signal ICs. *Electronics* 2022, 11.
11. Kim, H. Recent Trends in Copper Metallization. *Electronics* **2022**, *11*, 2914.
12. Kim, G.; Yoon, Y.; Seo, J.; Lee, S.; Park, J.; Kang, G.; Heo, J.; Jang, J.; Bae, J.; Lee, S.; et al. Design of a Capacitorless DRAM Based on Storage Layer Separated Using Separation Oxide and Polycrystalline Silicon. *Electronics* **2022**, *11*, 3365.
13. Zhang, X.; Gong, M.; Pan, J.; Song, M.; Zhang, H.; Zhang, L. Simulation Study of Low Turn-Off Loss and Snapback-Free SA-IGBT with Injection-Enhanced p-Floating Layer. *Electronics* **2022**, *11*, 2351.

Disclaimer/Publisher's Note: The statements, opinions and data contained in all publications are solely those of the individual author(s) and contributor(s) and not of MDPI and/or the editor(s). MDPI and/or the editor(s) disclaim responsibility for any injury to people or property resulting from any ideas, methods, instructions or products referred to in the content.

Article

A Review of the Gate-All-Around Nanosheet FET Process Opportunities

Sagarika Mukesh [†,‡] and Jingyun Zhang [*,‡]

IBM Research Albany, Albany, NY 12203, USA
* Correspondence: zhangji@us.ibm.com
† Current address: 257 Fuller Road, Suite 3100, Albany, NY 12203, USA.
‡ These authors contributed equally to this work.

Abstract: In this paper, the innovations in device design of the gate-all-around (GAA) nanosheet FET are reviewed. These innovations span enablement of multiple threshold voltages and bottom dielectric isolation in addition to impact of channel geometry on the overall device performance. Current scaling challenges for GAA nanosheet FETs are reviewed and discussed. Finally, an analysis of future innovations required to continue scaling nanosheet FETs and future technologies is discussed.

Keywords: gate-all-around nanosheet FETs; multi-Vt offerings; bottom dielectric isolation; power-performance improvement; transistor scaling; Moore's Law

Citation: Mukesh, S.; Zhang, J. A Review of the Gate-All-Around Nanosheet FET Process Opportunities. *Electronics* **2022**, *11*, 3589. https://doi.org/10.3390/electronics11213589

Academic Editors: Yi Zhao and ChoongHyun Lee

Received: 3 October 2022
Accepted: 2 November 2022
Published: 3 November 2022

Publisher's Note: MDPI stays neutral with regard to jurisdictional claims in published maps and institutional affiliations.

Copyright: © 2022 by the authors. Licensee MDPI, Basel, Switzerland. This article is an open access article distributed under the terms and conditions of the Creative Commons Attribution (CC BY) license (https://creativecommons.org/licenses/by/4.0/).

1. Introduction

Gate-all-around (GAA) nanosheet field effect transistors (FETs) are an innovative next-generation transistor device that have been widely adopted by the industry to continue logic scaling beyond 5 nm technology node, and beyond FinFETs [1]. Although gate-all-around transistors have been researched for many years, the first performance bench marking on scaled pitch of 44/48 nm CPP (contact-poly-pitch) was presented less than five years ago [2–8]. To fully appreciate the advantages provided by stacked nanosheet gate-all-around transistors, it is important to understand some of the challenges faced by the state-of-the-art FinFETs, and, in general, the trends that have motivated industry wide innovations over the years. Historically, device architecture innovations have been driven by short channel effects (SCEs) that come into play while achieving power performance area (PPA) scaling. SCEs occur when the channel length is on the same order of magnitude as the source-drain depletion layers [9]. Over the years, several innovations, such as the stress technology and high-k metal gate, have enabled scaling [10,11]. FinFETs were the first-ever change in architecture in the history of transistor devices to enable scaling by introducing the trigate control, thereby giving the gate-length scaling another few generations of run-time [12,13]. The gate-all-around nanosheet FETs are only the second time in the history of transistor devices, that a completely different architecture is adopted by the industry.

Scaling FinFETs beyond 7 nm node results in exacerbated SCEs, motivating a move from a tri-gate architecture to a gate-all-around architecture [14]. Among the gate-all-around architectures explored by the semiconductor industry, while the nanowires provided best electrostatic control, wider nanosheets are the ones that provide higher "on" current and improved electrostatic control over FinFETs [15,16]. Figure 1 shows a schematic of a FinFET and a GAA nanosheet FET, where the key components of the two technologies are highlighted. The components that are common between the two technologies include the shallow trench isolation, source/drain epitaxies, and the high-k metal gate; whereas the structural differences include a tri-gate for FinFETs vs. gate-all-around for nanosheets. To achieve an advantage in performance, multiple nanosheets must be stacked on top of each other, unlike FinFETs, where one fin constitutes one device. The channel thickness is lithographically defined for FinFETs, which puts a limit on scaling due to patterning resolution, whereas this channel thickness (also referred to as T_{Si}, thickness of silicon) is defined through epitaxially grown layers of Si on top of epitaxially grown layers of low concentration germanium SiGe, providing superior channel uniformity across wafer, and eliminating process complications.

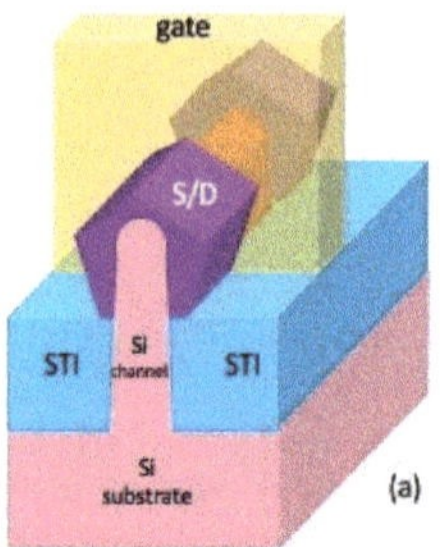

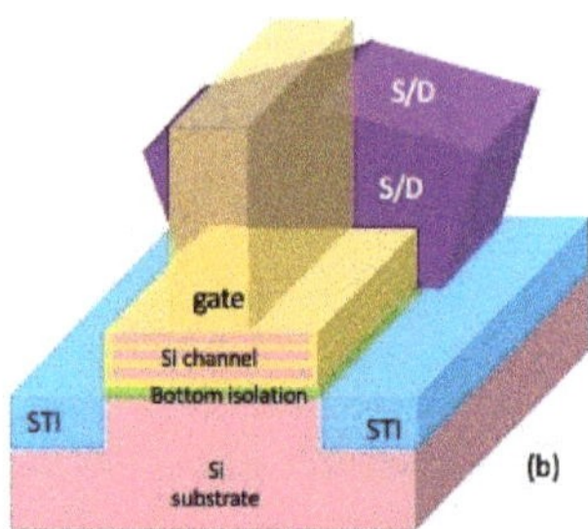

Figure 1. This figure shows a FinFET and a GAA nanosheet FET side-by-side. (**a**) A FinFET with shallow trench isolation (STI), source/drain (S/D) epitaxy, and a high-k metal trigate is depicted schematically. (**b**) A GAA nanosheet FET with STI, S/D epitaxy, bottom dielectric isolation (BDI), and high-k all-around metal is pictured. Some features, such as BDI and isolation between gate and S/D, are unique to the GAA nanosheet FETs.

Figure 2 shows a GAA-FET and highlights some of its key features that have been carefully engineered and extensively studied over the last few years. These features include discrete silicon sheets horizontally stacked to form one device, a high-k metal gate filling the space between the silicon channels, bottom dielectric isolation from bulk substrate, lithographically defined sheet width, process controlled gate-length, and inner spacer formation for gate to source-drain isolation. Some aspects of these GAA nanosheet FETs, such as inducing strain to increase hole mobility, have been a hot topic to improve overall device performance, but will not be covered in this paper [17–19]. Other aspects, such as multiple threshold voltage (Multi-V_T) options for high power and low power devices, impact of channel geometry on device performance, and integration and impact of full dielectric isolation, are reviewed in this paper [20–24].

The structure of the remaining paper is as follows: Section 2 highlights the key integration modules and shows a high-level process flow; Section 3 covers bottom dielectric isolation—its need, integration, and impact on device performance; Section 4 explores the impact of channel geometry on device performance, especially the impact of channel geometry on the hole mobility; Section 5 discusses different integration approaches for enabling multiple threshold voltages (multi-V_T) in GAA nanosheet FETs; Section 6 briefly discusses innovations in interconnects and power delivery networks that are needed to extract the value from scaled nanosheet architecture; and, finally, Section 7 discusses the direction of the transistor industry beyond GAA nanosheet FETs.

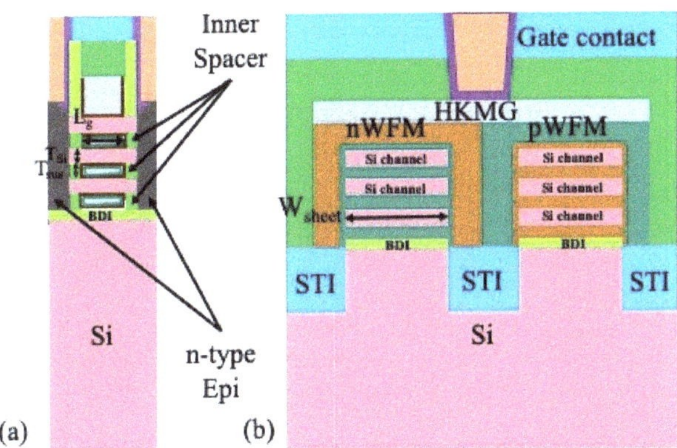

Figure 2. This figure shows a schematic for a gate-all-around nanosheet FET, with its key features highlighted. (**a**) shows a cut across the source-drain region where the key features highlighted are the bottom dielectric isolation (BDI), the thickness of the silicon channels (T_{Si}), the distance between the silicon channels (T_{sus}), and the gate length (L_g). The inner spacer, and the n-type epi are also highlighted here. (**b**) shows a cut across the gate region where the key features highlighted are shallow trench isolation (STI), n-type work function metal (WFM), p-type WFM, the high-k metal gate (HKMG), and the sheet width (W_{sheet}).

2. Integration of Gate-All-Around Nanosheet FETs

The integration of GAA nanosheet FETs involves several novel steps requiring a series of innovations to enable this technology. The key integration modules are listed below: [21]:

1. Stacked nanosheet formation: a stack of SiGe and Si are epitaxially grown on the Si substrate; the thickness of each layer can be controlled with high precision.
2. Fin reveal and STI: the devices are lithographically defined and shallow trench isolation is performed to isolate neighboring devices.
3. Dummy gate formation: a poly silicon dummy gate is formed to enable downstream processing.
4. Inner Spacer and Junction formation: n-type or p-type source/drain epitaxial layers are selectively formed on either sides of the exposed nanosheet ends [25].
5. Replacement metal gate formation:
   - Dummy gate pull: the dummy gate is etched out to reveal a cavity, at the bottom of which nanosheets are located,
   - Sacrificial SiGe channel release: the SiGe channels in between the nanosheets are etched out to enable filling up with high-k metal gate,
   - High-k Metal Gate (HKMG) formation: an interfacial oxide, a high-k dielectric layer, and the n-type or p-type work functions are selectively deposited.

3. Full Bottom Dielectric Isolation

In this section, we highlight the comparison between a full bottom dielectric isolation (BDI) and punch through stopper (PTS) scheme as examined by [24]. To introduce the problem, we first introduce the "fat-fin" effect that is unique to GAA nanosheets, where process non-idealities can result in a structure that causes an increased capacitance in the bulk region below the nanosheets, as shown in Figure 3. Although this structure is unique to GAA nanosheets, the effect, also known as sub-fin leakage, exists for FinFETs and is dealt with using the punch through stopper scheme. So, a comparison between this established PTS scheme and novel BDI scheme is performed on the basis of off-state leakage current, short channel effects, and effective capacitance (C_{eff}); it is shown that BDI could potentially provide improved C_{eff} and power-performance co-optimization.

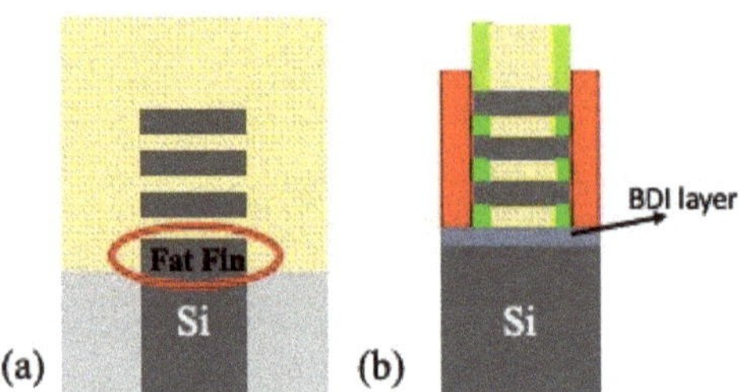

Figure 3. (**a**) A figure depicting a cross-fin cut showing high-k metal gate extension beyond the bottom sheet due to poor process control. As the metal depth below the bottom device increases, the performance penalty due to increased $C_{effective}$ also increases. (**b**) A figure showing improved process control due to full bottom dielectric isolation (FBDI) in the source/drain region.

3.1. Integration

Integrating a full bottom dielectric isolation entails adding a high-concentration SiGe layer at the bottom of the Si, SiGe nanosheet stack. Adding this layer, and later selectively etching it, requires lowering of Ge concentration in the SiGe layers used for the nanosheet stack. This introduces lower selectivity between Si and SiGe, resulting in Si loss during SiGe channel removal—requiring careful consideration of the stack thicknesses to ensure the T_{Si} is not too thin at the end of the entire process flow. We can see the BDI sitting under the S/D region in Figure 3b.

3.2. Experiments

Two splits of the PTS scheme of varying doping concentration are studied along with a full BDI scheme at V_{ds} = 0.7 V for L_{metal} of 12 nm in a 44 CPP device, where their short channel characteristics and power vs. performance is analysed.

3.3. Results and Discussion

As seen in Figure 4, full bottom dielectric isolation reduces the off-state leakage current and the DIBL, thereby improving performance and decreasing power. An 18% decrease in power is observed with a 4% improvement in performance between split with and without BDI. The devices with BDI perform better, and they also show better immunity to process variations with respect to sub-channel leakage control. So, full bottom dielectric isolation may be considered as a critical element for enabling a well-performing GAA nanosheet FET.

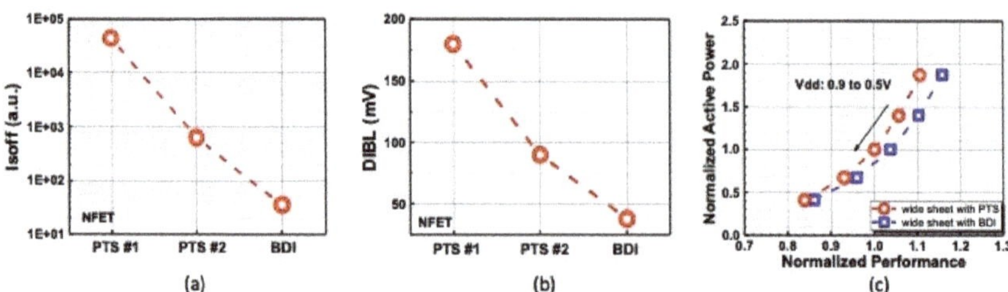

Figure 4. This figure captures key performance metrics for GAA FETs using PTS scheme and full BDI. (**a**) I_{soff} extracted from L_g =12 nm devices on PTS and BDI splits. (**b**) DIBL extracted from L_g =12 nm devices on BDI and PTS splits. (**c**) Power vs. performance correlation chart of wide sheet devices for both with and without BDI layer [24].

4. Channel Geometry Impact

In this section, the mobility of electrons and holes as a function of channel geometry are studied and 'narrow sheet effect' on carrier transport is observed [23]. T_{Si} is one of the knobs that can enable future L_g scaling needs by improving electrostatic control. Moreover, the quantization effects for $T_{Si} < 5$ nm becomes severe in SOI and FinFETs, so, it is important to study the same for GAA FETs.

4.1. Experiments

Since the mobility of holes (μ_h) is lower for the <100> plane, this plane will dominate hole transport characteristics for pFETs in GAA nanosheet FETs. To study the impact of <100> plane on hole transport, nanosheet devices are fabricated on <100> substrate with <110> transport direction. Figure 5 shows a TEM from the experiments performed, a channel length of 100 nm was chosen for this study. To study the impact of T_{Si} on hole mobility, silicon sheets of different thicknesses were epitaxially grown, and the T_{Si} was measured using TEMs.

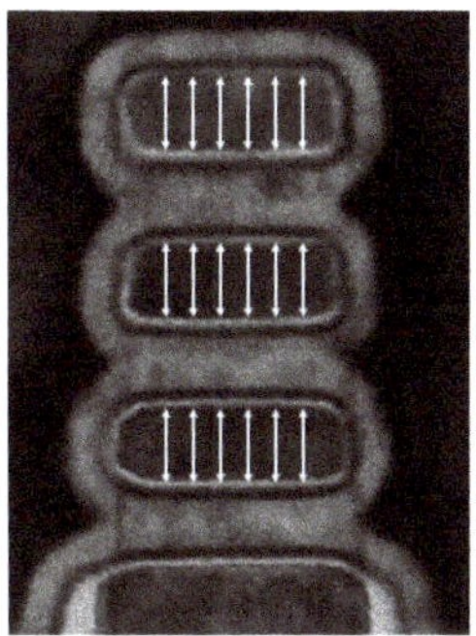

Figure 5. A TEM cross-section of GAA nanosheet FETs. The T_{Si} is uniform in thickness along the W_{sheet} direction [23].

4.2. Results and Discussion

As seen in Figure 6a, the degradation of μ_h is attributed to increased phonon scattering with thinner T_{Si}. At high fields, as in the case of N_{inv} at 10^{13}, the mobility is dominated by surface roughness, whereas the peak mobility is primarily impacted by phonon scattering. So, the impact of mobility degradation is more profound for the peak mobility case. However, this degradation of mobility is offset by sheet width W_{sheet} as seen in Figure 6b, which is primarily influenced by the contribution of <100> vs. <110> planes. Wider sheets have more contribution from the <110> plane, resulting in improved mobility, suggesting that both phonon scattering and sheet geometry effects impact hole mobility. Moreover, this dependence on W_{sheet} provides an additional knob for power and performance co-optimization in GAA nanosheet FETs.

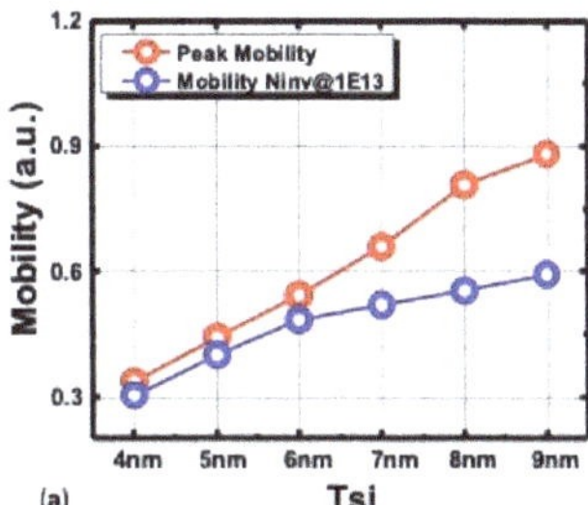

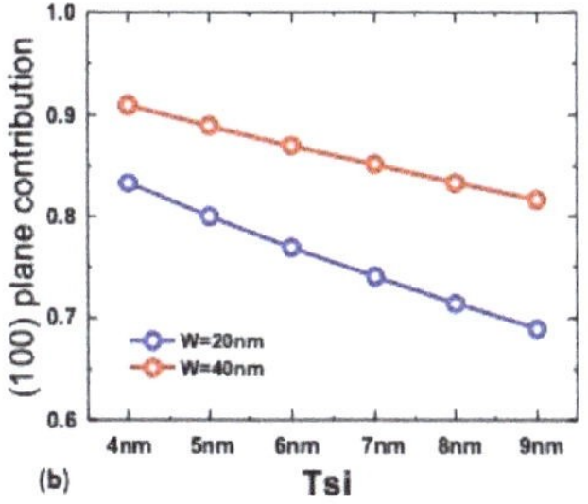

Figure 6. (a) This plot shows the extracted peak hole mobility and hole mobility for $N_{inv}@10^{13}/\text{cm}^2$ as a function of silicon channel thickness. Degradation in hole mobility is evident for thin sheet values; (b) calculated <100> plane contribution to total W_{eff} as a function of T_{Si} (a pure geometrical percentage of the whole nanosheet perimeter) [23].

5. Enabling Multiple Threshold Voltages

The ability to co-integrate multiple threshold voltages (V_T) is a key requirement for a technology to become an industry standard. Given the unique architecture of GAA FETs, the space for depositing the work function metals is limited as depicted in Figure 7. The replacement metal gate process only leaves the space between the Si channels and inner spacers open—to be filled with the work function metals as per technology requirements. This space, also known as T_{sus} (refer Figure 2), can be controlled by controlling the thickness of SiGe layer grown during the nanosheet stack development module, but is nevertheless highly constrained, and must be engineered carefully to meet the industry standards for device offerings.

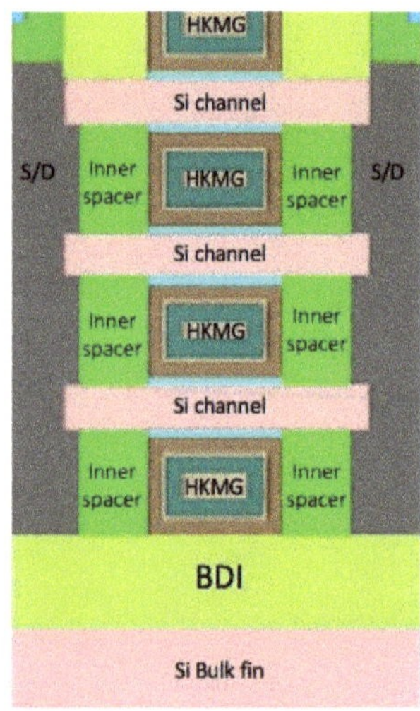

Figure 7. This figure shows a close up view of the S/D cross-section. The width of the high-k metal gate here is the gate length L_g, and the vertical thickness of this metal gate is determined by T_{sus}. Additionally, the inner spacers and bottom dielectric isolation are highlighted.

5.1. Integration

Two different approaches are proposed to accommodate multi-V_T offerings in GAA FETs—(1) WFM modification and (2) T_{sus} modification [20]. A process flow overview for WFM modification is presented in Figure 8. One of the challenges highlighted by the integration sequence for V_T modulation is that large W_{sheet} adds process challenges for WFM etch back when WFM is pinched-off between Si channels. To overcome this, ref. [21] proposed filling the space between sheets with a sacrificial material that is easy to etch, selectively opening one of the FETs, and etching away the already deposited work-function metal. This scheme is agnostic of p-type or n-type WFM, and enables both PG (p-FET first) and MY (n-FET first) schemes. This same process can be repeated to achieve different sets of work function metals or to achieve a different stack with more than two WFMs.

The second approach requires changing T_{sus} by changing the channel stack epitaxy thickness during nanosheet formation. A larger space between sheets allows the deposition of a larger volume of work function metal in this space, thereby modulating V_T. This design knob is unique to GAA nanosheet FETs when compared with FinFETs, thereby, allowing more design space for Multi-V_T options in these nanosheet FETs.

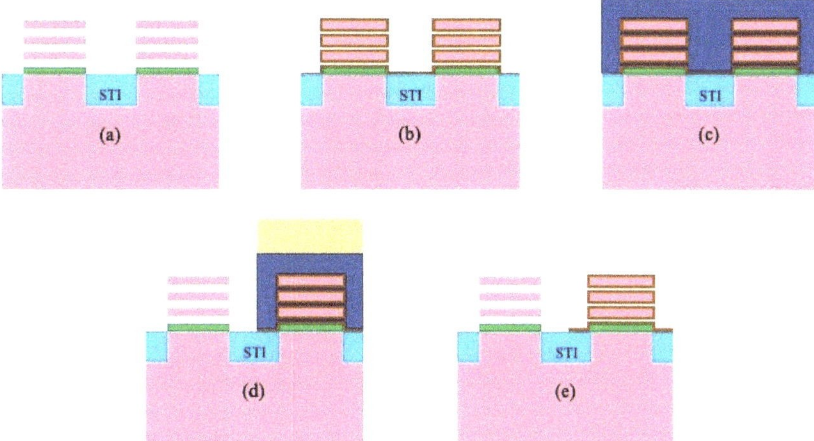

Figure 8. An example of V_T modulation as presented in [21] is shown here. (**a**) The gate region post SiGe channel release; (**b**) WFM1 deposition; (**c**) sacrificial material deposition; (**d**) selective patterning and etch of the deposited WFM; and (**e**) removal of patterning stack resulting in a structure with WFM1 along one set of sheets.

Volumeless Multiple Threshold Voltages

A volumeless multi-V_T is a term defined to represent a dipole-based V_T option where a dipole of thickness less than 5 Åis formed, followed by the base work function metals [26,27]. This innovative scheme provides space and gate resistance benefits as shown in the cited references. However, this approach does not directly translate from FinFETs to GAA nanosheet FETs, so dedicated integration of volumeless V_T is proposed in [22]. Moreover, a volumeless V_T also helps with V_T uniformity, which is important for uniform switching of transistors.

5.2. Results and Discussions

Several different flavours of V_T are created using novel integration sequence and using the unique design knobs for GAA nanosheet FETs -(a) T_{sus} design; and (b) WFM pinch-off. A dipole-based V_T scheme for nanosheet FETs is also proposed. In addition to these knobs, the T_{Si} design as discussed in Section 4, can be modulated to provide a trade-off between mobility and short-channel effects. So, overall, the GAA nanosheet FETs provide several opportunities for application-based optimization, hence they are suitable for high-power and low-power applications alike.

6. Current Challenges

This paper discusses some of the cutting-edge advances in the gate-all-around nanosheet transistor technology over the last five years, and consolidates some of pioneering work in the field. In this section, some of the processing challenges of this technology are covered as reported in the literature. These processing challenges may be broadly categorized into four areas: self-heating, mechanical stability during fabrication, device variability, and Si–SiGe intermixing.

Self-heating effects (SHE) in nanoscale devices result in significant thermal cross talk resulting in device performance degradation [28,29]. Studies have explored novel substrates, such as diamond on silicon to provide improved SHE, but such a scheme is less likely to be adopted in high-volume manufacturing. As such, this problem is open to exploration and solution [30].

An aspect of nanosheet fabrication to carefully consider is the mechanical stability of these sheets during the channel release process. Although nanosheets do allow design flexibility, aspect ratio of the sheets, and mechanical integrity of the inner spacer play an important role in overall stability of these sheets [31]. Another aspect to optimize is the device variability, which can result from several sources including, but not limited to, line-edge roughness, gate-edge roughness, non-uniform work function metal deposition, and random dopant fluctuations. A recent study analyzes these variability and proposes solution for a complementary GAA nanosheet FET structure [32].

Finally, the initial Si-SiGe stack for nanosheets itself is susceptible to thermal intermixing when going through numerous thermal cycles before the channel release step. There have been several studies examining the extent of this intermixing and the mechanism of such diffusion [33–35]. As long as the SiGe channels can etch selective to Si channel sheets, and the Si sheets are not over-etched due to Si–SiGe intermixing, this effect is tolerable.

7. Future Outlook

Although the transistor-level innovation is sufficient to drive the industry forward to the next technology nodes, this section briefly touches upon some innovations in the fields of interconnects and power-delivery for completeness.

An interesting proposal in the field of power delivery is the buried power rail (BPR), which proposes moving the power rails to be located below the transistor devices, thereby, providing area on the front-side for routing flexibility, and to reduce conductor crowding [36,37]. However, such a scheme has a short run-path as the requirement of patterning between the devices will limit contact poly pitch (CPP) scaling. To overcome this limitation, the concept of backside power delivery network (BSPDN) has been proposed, with a recent hardware demonstration of its feasibility [38]. However, this new paradigm brings several technical challenges with it, such as back-side patterning, alignment between the structures on the front-side to those on the back-side, and wafer thinning on the back-side of the wafer. If the industry as a whole decides this is the correct direction, there are tremendous opportunities of innovation for tool vendors and equipment manufacturers to enable this technology at a large scale.

8. On the Horizon

Although the industry navigates current challenges to bring the GAA nanosheet FETs to market, researchers are already thinking about what lies beyond nanosheet FETs. The top contenders to continue Moore's law scaling are the Vertical Transport FETs (VTFETs) [39] and stacked transistors [40]. VTFETs change the carrier transport direction from the traditional horizontal direction to vertical direction, thereby relaxing the contraints on scaling barriers, such as gate-length (L_g), spacer thickness, and contact size; all of which can be optimized for power or performance, based on the application. Stacked transistors offer a more conventional scaling path by stacking the nFET and pFET transistor over each other, thereby providing area benefit. However, both these technologies present several novel integration and manufacturing challenges, which may be subject of a later review.

Looking beyond the immediate future, there is a large body work on novel materials to enable 2-D transistors [41]. Molybdenum disulfide (MoS_2) is one of the top contenders for such technologies with ever improving performance based on mobility, contact resistance, and doping [42]. Graphene is another strong contender for a long time, and literature has been reporting ever improved performance for such transistors over the last decade [43]. Indium oxide is another contender for a wide-gap semiconductor material [44]. Although these technologies are promising, there is an inherent barrier to entry for them due to the large overhead cost of novel equipment for the foundries to enable large scale manufacturing of such transistors. So, silicon based transistors will continue scaling for the coming decades with the growing needs for transistors in existing and new industries.

Author Contributions: S.M. prepared the manuscript, J.Z. plotted the results, both authors reviewed the paper. All authors have read and agreed to the published version of the manuscript.

Funding: This research received no external funding.

Data Availability Statement: Not applicable.

Acknowledgments: The authors want to thank their colleagues at IBM Research, Albany for thoughtful discussions.

Conflicts of Interest: The authors declare no conflicts of interest.

Abbreviations

The following abbreviations are used in this manuscript:

GAA FETs	Gate-All-Around Field Effect Transistors
BDI	Bottom Dielectric Isolation
STI	Shallow Trench Isolation
WFM	Work Function Metal
HKMG	High-k Metal Gate
SCE	Short Channel Effects
RMG	Replacement Metal Gate
PTS	Punch Through Stopper
MOL	Middle of Line
BEOL	Back End of Line
S/D	Source/Drain
DIBL	Drain Induced Barrier Lowering
TEM	Transmission Electron Microscopy
VTFET	Vertical Transport Field Effect Transistors
PPA	Power, Performance, and Area
BPR	Buried Power Rail
BSPDN	Back-Side Power Delivery Network
CPP	Contact Poly Pitch

References

1. The International Roadmap for Devices and Systems: 2021. Available online: https://irds.ieee.org/images/files/pdf/2021/2021IRDS_MM.pdf (accessed on 2 October 2022).
2. Lee, S.-Y.; Kim, S.-M.; Yoon, E.-J.; Oh, C.-W.; Chung, I.; Park, D.; Kim, K. Three-dimensional MBCFET as an ultimate transistor. *IEEE Electron Device Lett.* **2004**, *25*, 217–219. [CrossRef]
3. Bangsaruntip, S.; Cohen, G.M.; Majumdar, A.; Zhang, Y.; Engelmann, S.U.; Fuller, N.C.M.; Gignac, L.M.; Mittal, S.; Newbury, J.S.; Guillorn, M.; et al. High performance and highly uniform gate-all-around silicon nanowire MOSFETs with wire size dependent scaling. In Proceedings of the 2009 IEEE International Electron Devices Meeting (IEDM), Baltimore, MD, USA, 7–9 December 2009; pp. 1–4. [CrossRef]
4. Kuhn, K.J. Considerations for Ultimate CMOS Scaling. *IEEE Trans. Electron Devices* **2012**, *59*, 1813–1828. [CrossRef]
5. Lauer, I.; Loubet, N.; Kim, S.D.; Ott, J.A.; Mignot, S.; Venigalla, R.; Yamashita, T.; Standaert, T.; Faltermeier, J.; Basker, V.; et al. Si nanowire CMOS fabricated with minimal deviation from RMG FinFET technology showing record performance. In Proceedings of the 2015 Symposium on VLSI Technology (VLSI Technology), Kyoto, Japan, 16–18 June 2015; pp. 140–141.
6. Mertens, H.; Ritzenthaler, R.; Hikavyy, A.; Kim, M.S.; Tao, Z.; Wostyn, K.; Chew, S.A.; De Keersgieter, A.; Mannaert, G.; Rosseel, E.; et al. Gate-all-around MOSFETs based on vertically stacked horizontal Si nanowires in a replacement metal gate process on bulk Si substrates. In Proceedings of the 2016 IEEE Symposium on VLSI Technology, Honolulu, HI, USA, 14–16 June 2016; pp. 1–2. [CrossRef]
7. Mertens, H.; Ritzenthaler, R.; Chasin, A.; Schram, T.; Kunnen, E.; Hikavyy, A.; Ragnarsson, L.-Å.; Dekkers, H.; Hopf, T.; Wostyn, K.; et al. Vertically stacked gate-all-around Si nanowire CMOS transistors with dual work function metal gates. In Proceedings of the 2016 IEEE International Electron Devices Meeting (IEDM), San Francisco, CA, USA, 3–7 December 2016; pp. 19.7.1–19.7.4. [CrossRef]
8. Loubet, N.; Hook, T.; Montanini, P.; Yeung, C.-W.; Kanakasabapathy, S.; Guillom, M.; Yamashita, T.; Zhang, J.; Miao, X.; Wang, J.; et al. Stacked nanosheet gate-all-around transistor to enable scaling beyond FinFET. In Proceedings of the 2017 Symposium on VLSI Technology, Kyoto, Japan, 5–8 June 2017; pp. T230–T231. [CrossRef]
9. Veeraraghavan, S.; Fossum, J.G. Short-channel effects in SOI MOSFETs. *IEEE Trans. Electron Devices* **1989**, *36*, 522–528. [CrossRef]
10. Ghani, T.; Armstrong, M.; Auth, C.; Bost, M.; Charvat, P.; Glass, G.; Hoffmann, T.; Johnson, K.; Kenyon, C.; Klaus, J.; et al. A 90nm high volume manufacturing logic technology featuring novel 45nm gate length strained silicon CMOS transistors. In Proceedings of the IEEE International Electron Devices Meeting 2003, Washington, DC, USA, 8–10 December 2003; pp. 11.6.1–11.6.3. [CrossRef]

11. Mistry, K.; Allen, C.; Auth, C.; Beattie, B.; Bergstrom, D.; Bost, M.; Brazier, M.; Buehler, M.; Cappellani, A.; Chau, R.; et al. A 45nm Logic Technology with High-k+Metal Gate Transistors, Strained Silicon, 9 Cu Interconnect Layers, 193nm Dry Patterning, and 100% Pb-free Packaging. In Proceedings of the 2007 IEEE International Electron Devices Meeting, Washington, DC, USA, 10–12 December 2007; pp. 247–250. [CrossRef]

12. Auth, C.; Allen, C.; Blattner, A.; Bergstrom, D.; Brazier, M.; Bost, M.; Buehler, M.; Chikarmane, V.; Ghani, T.; Glassman, T.; et al. A 22 nm high performance and low-power CMOS technology featuring fully-depleted tri-gate transistors, self-aligned contacts and high density MIM capacitors. In Proceedings of the 2012 Symposium on VLSI Technology (VLSIT), Honolulu, HI, USA, 12–14 June 2012; pp. 131–132. [CrossRef]

13. Xie, R.; Montanini, P.; Akarvardar, K.; Tripathi, M.; Haran, B.; Johnson, S.; Hook, T.; Hamieh, B.; Corliss, D.; Wang, J.; et al. A 7 nm FinFET technology featuring EUV patterning and dual strained high mobility channels. In Proceedings of the 2016 IEEE International Electron Devices Meeting (IEDM), San Francisco, CA, USA, 3–7 December 2016; pp. 2.7.1–2.7.4. [CrossRef]

14. Narendar, V.; Mishra, R.A. Analytical modeling and simulation of multigate FinFET devices and the impact of high-k dielectrics on short channel effects (SCEs). *Superlattices Microstruct.* **2015**, *85*, 357–369. [CrossRef]

15. Kim, S.-D.; Guillorn, M.; Lauer, I.; Oldiges, P.; Hook, T.; Na, M.-H. Performance trade-offs in FinFET and gate-all-around device architectures for 7 nm-node and beyond. In Proceedings of the 2015 IEEE SOI-3D-Subthreshold Microelectronics Technology Unified Conference (S3S), Rohnert Park, CA, USA, 5–8 October 2015; pp. 1–3. [CrossRef]

16. Bae, G.; Bae, B.-I.; Kang, M.; Hwang, S.M.; Kim, S.S.; Seo, B.; Kwon, T.Y.; Lee, T.J.; Moon, C.; Choi, Y.M.; et al. 3nm GAA Technology featuring Multi-Bridge-Channel FET for Low Power and High Performance Applications. In Proceedings of the 2018 IEEE International Electron Devices Meeting (IEDM), San Francisco, CA, USA, 1–5 December 2018; pp. 28.7.1–28.7.4. [CrossRef]

17. Mochizuki, S.; Bhuiyan, M.; Zhou, H.; Zhang, J.; Stuckert, E.; Li, J.; Zhao, K.; Wang, M.; Basker, V.; Loubet, N.; et al. Stacked Gate-All-Around Nanosheet pFET with Highly Compressive Strained Si1-xGex Channel. In Proceedings of the 2020 IEEE International Electron Devices Meeting (IEDM), San Francisco, CA, USA, 12–18 December 2020; pp. 2.3.1–2.3.4. [CrossRef]

18. Tsutsui, G.; Mochizuki, S.; Loubet, N.; Bedell, S.W.; Sadana, D.K. Strain engineering in functional materials. *AIP Adv.* **2019**, *9*, 030701. [CrossRef]

19. Murray, C.E.; Yan, H.; Lavoie, C.; Jordan-Sweet, J.; Pattammattel, A.; Reuter, K.; Hasanuzzaman, M.; Lanzillo, N.; Robison, R.; Loubet, N. Mapping of the mechanical response in Si/SiGe nanosheet device geometries. *Commun. Eng.* **2022**, *1*, 11. [CrossRef]

20. Zhang, J.; Ando, T.; Yeung, C.W.; Wang, M.; Kwon, O.; Galatage, R.; Chao, R.; Loubet, N.; Moon, B.K.; Bao, R.; et al. High-k metal gate fundamental learning and multi-Vt options for stacked nanosheet gate-all-around transistor. In Proceedings of the 2017 IEEE International Electron Devices Meeting (IEDM), San Francisco, CA, USA, 2–6 December 2017; pp. 22.1.1–22.1.4. [CrossRef]

21. Bao, R.; Watanabe, K.; Zhang, J.; Guo, J.; Zhou, H.; Gaul, A.; Sankarapandian, M.; Li, J.; Hubbard, A.R.; Vega, R.; et al. Multiple-Vt Solutions in Nanosheet Technology for High Performance and Low Power Applications. In Proceedings of the 2019 IEEE International Electron Devices Meeting (IEDM), San Francisco, CA, USA, 7–11 December 2019; pp. 11.2.1–11.2.4. [CrossRef]

22. Bao, R.; Durfee, C.; Zhang, J.; Qin, L.; Rozen, J.; Zhou, H.; Li, J.; Mukesh, S.; Pancharatnam, S.; Zhao, K.; et al. Critical Elements for Next Generation High Performance Computing Nanosheet Technology. In Proceedings of the 2021 IEEE International Electron Devices Meeting (IEDM), San Francisco, CA, USA, 13–15 December 2021; pp. 26.3.1–26.3.4. [CrossRef]

23. Yeung, C.W.; Zhang, J.; Chao, R.; Kwon, O.; Vega, R.; Tsutsui, G.; Miao, X.; Zhang, C.; Sohn, C.-W.; Moon, B.K.; et al. Channel Geometry Impact and Narrow Sheet Effect of Stacked Nanosheet. In Proceedings of the 2018 IEEE International Electron Devices Meeting (IEDM), San Francisco, CA, USA, 1–5 December 2018; pp. 28.6.1–28.6.4. [CrossRef]

24. Zhang, J.; Frougier, J.; Greene, A.; Miao, X.; Yu, L.; Vega, R.; Montanini, P.; Durfee, C.; Gaul, A.; Pancharatnam, S.; et al. Full Bottom Dielectric Isolation to Enable Stacked Nanosheet Transistor for Low Power and High Performance Applications. In Proceedings of the 2019 IEEE International Electron Devices Meeting (IEDM), San Francisco, CA, USA, 7–11 December 2019; pp. 11.6.1–11.6.4. [CrossRef]

25. Loubet, N.; Kal, S.; Alix, C.; Pancharatnam, S.; Zhou, H.; Durfee, C.; Belyansky, M.; Haller, N.; Watanabe, K.; Devarajan, T.; et al. A Novel Dry Selective Etch of SiGe for the Enablement of High Performance Logic Stacked Gate-All-Around NanoSheet Devices. In Proceedings of the 2019 IEEE International Electron Devices Meeting (IEDM), San Francisco, CA, USA, 7–11 December 2019; pp. 11.4.1–11.4.4. [CrossRef]

26. Bao, R.; Zhou, H.; Wang, M.; Guo, D.; Haran, B.S.; Narayanan, V.; Divakaruni, R. Extendable and Manufacturable Volume-less Multi-Vt Solution for 7 nm Technology Node and Beyond. In Proceedings of the 2018 IEEE International Electron Devices Meeting (IEDM), San Francisco, CA, USA, 1–5 December 2018; pp. 28.5.1–28.5.4. [CrossRef]

27. Bao, R.; Greene, B.; Kwon, U.; Lee, S.; Bruley, J.; Wang, W.; Zhao, K.; DeHaven, P.W.; Li, Z.; Wong, K.; et al. Replacement metal gate resistance in FinFET architecture modelling, validation and extendibility. In Proceedings of the 2015 IEEE International Electron Devices Meeting (IEDM), Washington, DC, USA, 7–9 December 2015; pp. 34.3.1–34.3.4. [CrossRef]

28. Cai, L.; Chen, W.; Du, G.; Kang, J.; Zhang, X.; Liu, X. Investigation of self-heating effect on stacked nanosheet GAA transistors. In Proceedings of the 2018 International Symposium on VLSI Technology, Systems and Application (VLSI-TSA), Hsinchu, Taiwan, 16–19 April 2018; pp. 1–2. [CrossRef]

29. Bury, E. Kaczer, B.; Linten, D.; Witters, L.; Mertens, H.; Waldron, N.; Zhou, X.; Collaert, N.; Horiguchi, N.; Spessot, A.; et al. Self-heating in FinFET and GAA-NW using Si, Ge and III/V channels. In Proceedings of the 2016 IEEE International Electron Devices Meeting (IEDM), San Francisco, CA, USA, 3–7 December 2016; pp. 15.6.1–15.6.4.

30. Rathore, S.; Jaisawal, R.K.; Kondekar, P.N.; Bagga, N. Design Optimization of Three-Stacked Nanosheet FET From Self-Heating Effects Perspective. *IEEE Trans. Device Mater. Reliab.* **2022**, *22*, 396–402. [CrossRef]

31. Lee, K.-S.; Park, J.-Y. Inner Spacer Engineering to Improve Mechanical Stability in Channel-Release Process of Nanosheet FETs. *Electronics* **2021**, *10*, 1395. [CrossRef]

32. Yang, X.; Li, X.; Liu, Z.; Sun, Y.; Liu, Y.; Li, X.; Shi, Y. Impact of Process Variation on Nanosheet Gate-All-Around Complementary FET (CFET). *IEEE Trans. Electron Devices* **2022**, *69*, 4029–4036. [CrossRef]

33. Xia, G.; Hoyt, J.L.; Canonico, M. Si–Ge Interdiffusion in Strained Si/Strained SiGe Heterostructures and Implications for Enhanced Mobility Metal-Oxide-Semiconductor Field-Effect Transistors. *J. Appl. Phys.* **2007**, *101*, 044901. [CrossRef]

34. Dong, Y. *A Systematic Study of Silicon Germanium Interdiffusion for Next Generation Semiconductor Devices (T)*; University of British Columbia: Vancouver, BC, Canada, 2014. Available online: https://open.library.ubc.ca/collections/ubctheses/24/items/1.0167516 (accessed on 2 October 2022).

35. Thornton, C.; Tuttle, B.; Turner, E.; Law, M.; Pantelides, S.; Wang, G.; Jones, K.The Diffusion Mechanism of Ge During Oxidation of Si/SiGe Nanofins. *ACS Appl. Mater. Interfaces* **2022**, *14*, 29422–29430. [CrossRef]

36. Prasad, D.; Teja Nibhanupudi, S.S.; Das, S.; Zografos, O.; Chehab, B.; Sarkar, S.; Baert, R.; Robinson, A.; Gupta, A.; Spessot, A.; et al. Buried Power Rails and Back-side Power Grids: Arm® CPU Power Delivery Network Design Beyond 5 nm. In Proceedings of the 2019 IEEE International Electron Devices Meeting (IEDM), San Francisco, CA, USA, 7–11 December 2019; pp. 19.1.1–19.1.4. [CrossRef]

37. Salahuddin, S.; Perumkunnil, M.; Dentoni Litta, E.; Gupta, A.; Weckx, P.; Ryckaert, J.; Na, M.H.; Spessot, A. Buried Power SRAM DTCO and System-Level Benchmarking in N3. In Proceedings of the 2020 IEEE Symposium on VLSI Technology, Honolulu, HI, USA, 16–19 June 2020; pp. 1–2. [CrossRef]

38. Chen, R.; Sisto, G.; Jourdain, A.; Hiblot, G.; Stucchi, M.; Kakarla, N.; Chehab, B.; Salahuddin, S.M.; Schleicher, F.; Veloso, A.; et al. Design and Optimization of SRAM Macro and Logic Using Backside Interconnects at 2 nm node. In Proceedings of the 2021 IEEE International Electron Devices Meeting (IEDM), San Francisco, CA, USA, 13–15 December 2021; pp. 22.4.1–22.4.4. [CrossRef]

39. Jagannathan, H.; Anderson, B.; Sohn, C.-W.; Tsutsui, G.; Strane, J.; Xie, R.; Fan, S.; Kim, K.-I.; Song, S.; Sieg, S.; et al. Vertical-Transport Nanosheet Technology for CMOS Scaling beyond Lateral-Transport Devices. In Proceedings of the 2021 IEEE International Electron Devices Meeting (IEDM), San Francisco, CA, USA, 13–15 December 2021; pp. 26.1.1–26.1.4. [CrossRef]

40. Wang, J.; Suk, S.D.; Chu, A.; Hook, T.; Young, A.; Krishnan, R.; Bao, R.; Seshadri, I.; Senapati, B.; Zalani, V.; et al. Challenges and Opportunities for Stacked Transistor: DTCO and Device. In Proceedings of the 2021 Symposium on VLSI Technology, Kyoto, Japan, 13–19 June 2021; pp. 1–2.

41. Das, S.; Sebastian, A.; Pop, E.; McClellan, C.J.; Franklin, A.D.; Grasser, T.; Knobloch, T.; Illarionov, Y.; Penumatcha, A.V.; Appenzeller, J.; et al. Transistors based on two-dimensional materials for future integrated circuits. *Nat. Electron.* **2021**, *4*, 786–799. [CrossRef]

42. Rai, A.; Movva, H.C.P.; Roy, A.; Taneja, D.; Chowdhury, S.; Banerjee, S.K. Progress in Contact, Doping and Mobility Engineering of MoS2: An Atomically Thin 2D Semiconductor. *Crystals* **2018**, *8*, 316. [CrossRef]

43. Akinwande, D.; Huyghebaert, C.; Wang, C.-H.; Serna, M.I.; Goossens, S.; Li, L.-J.; Wong, H.-S.P.; Koppens, F.H.L. Graphene and two-dimensional materials for silicon technology. *Nature* **2019**, *573*, 507–518. [CrossRef]

44. Bierwagen, O. Indium oxide—A transparent, wide-band gap semiconductor for (opto)electronic applications. *Semicond. Sci. Technol.* **2015**, *30*, 024001. [CrossRef]

 *electronics*

MDPI

Article

Machine-Learning-Based Compact Modeling for Sub-3-nm-Node Emerging Transistors

SangMin Woo, HyunJoon Jeong, JinYoung Choi, HyungMin Cho, Jeong-Taek Kong * and SoYoung Kim *

College of Information and Communication Engineering, SungKyunKwan University, Suwon 16419, Korea
* Correspondence: jtkong@skku.edu (J.-T.K.); ksyoung@skku.edu (S.K.); Tel.: +82-31-299-4598

Abstract: In this paper, we present an artificial neural network (ANN)-based compact model to evaluate the characteristics of a nanosheet field-effect transistor (NSFET), which has been highlighted as a next-generation nano-device. To extract data reflecting the accurate physical characteristics of NSFETs, the Sentaurus TCAD (technology computer-aided design) simulator was used. The proposed ANN model accurately and efficiently predicts currents and capacitances of devices using the five proposed key geometric parameters and two voltage biases. A variety of experiments were carried out in order to create a powerful ANN-based compact model using a large amount of data up to the sub-3-nm node. In addition, the activation function, physics-augmented loss function, ANN structure, and preprocessing methods were used for effective and efficient ANN learning. The proposed model was implemented in Verilog-A. Both a global device model and a single-device model were developed, and their accuracy and speed were compared to those of the existing compact model. The proposed ANN-based compact model simulates device characteristics and circuit performances with high accuracy and speed. This is the first time that a machine learning (ML)-based compact model has been demonstrated to be several times faster than the existing compact model.

Keywords: artificial neural network; compact model; nanosheet FETs; TCAD/SPICE simulation

Citation: Woo, S.; Jeong, H.; Choi, J.; Cho, H.; Kong, J.-T.; Kim, S. Machine-Learning-Based Compact Modeling for Sub-3-nm-Node Emerging Transistors. *Electronics* **2022**, *11*, 2761. https://doi.org/10.3390/electronics11172761

Academic Editor: Yi Zhao

Received: 27 July 2022
Accepted: 29 August 2022
Published: 1 September 2022

Publisher's Note: MDPI stays neutral with regard to jurisdictional claims in published maps and institutional affiliations.

Copyright: © 2022 by the authors. Licensee MDPI, Basel, Switzerland. This article is an open access article distributed under the terms and conditions of the Creative Commons Attribution (CC BY) license (https://creativecommons.org/licenses/by/4.0/).

1. Introduction

According to Moore's Law, integrated circuits (ICs) have advanced rapidly over the last few decades in the semiconductor industry [1]. The difficulty and cost of improving performance are increasing as the transistors in ICs continue to scale down. To address these issues, new technologies have emerged for transistors (e.g., FinFETs, nanowire FETs (NWFETs), nanosheet FETs (NSFETs), and negative-capacitance FETs (NCFETs)). The iterative real fabrications and evaluations of new nano-transistors require a lot of time and money. Before producing the best transistor, it is crucial to quickly complete transistor modeling and IC simulation in an effort to save time and costs. In circuit simulation, the performance (e.g., power, delay) of a circuit created for a particular technology is assessed. Circuit simulation requires a compact model, which is crucial for effective design and analysis. The existing compact models (e.g., BSIM) are composed of mathematically and physically based device-characteristic equations [2]. However, developing a new compact model suitable for next-generation devices is very complicated, necessitating the involvement of numerous experts, as the development of a sophisticated compact model typically takes several years [3]. To address these shortcomings, researchers are working hard to develop new modeling methodologies for predicting the performance of new devices by using ML methods.

1.1. Related Work

J. Y. Lim et al. developed a model for predicting line edge roughness (LER)-induced performance variation of FinFETs using the ANN technique [4]. Q. Chen et al. employed an ANN technique with three hidden layers. As inputs, the terminal voltages (V_{gs}, V_{ds}), gate width (W), and gate length (L) of a thin-film transistor were used to build a model that predicted I_{ds} and C_{gs} [5]. K. Ko et al. predicted process variation effects using a gate-all-around (GAA) vertical FET with three hidden layers and four key electrical parameters (work function variation (WFV), gate length (L_g), channel thickness (T_{ch}), and equivalent oxide thickness (EOT)) [6]. Prior studies have used ANN models to simulate an inverter after predicting the I-V characteristics. A simple NMOS resistive-load inverter was simulated or a CMOS inverter was simulated applying a static or simple voltage as input data [7,8]. Z. Zhang et al. used the ANN technique to predict the I-V curve of a tunnel FET and performed simulations for an inverter, 6T-SRAM, and 2-NAND. However, the simulation speed was not discussed [9]. K. Mehta et al. used ML with an autoencoder algorithm to predict the device characteristics of a small dataset. However, when the accuracy was measured using the R2 score, it was found to be inadequate [10]. F. Klemme et al. modeled the I-V characteristics of a negative-capacitance FinFET, an emerging device, using ML. The number of input data points was used to calculate simulation time and accuracy [3]. The number of neurons in each of the two hidden layers could reach 500, but the modeling error was around 5%. Although the I-V characteristics and process prediction of semiconductor devices have been actively pursued using ML techniques, there are very few studies that model the C-V characteristics properly. Y. Wang, et al. discussed the overall accuracy of ML models and circuit simulations in SPICE [11]. The SPICE simulation speed of the ML-based model was much slower than that of the existing compact model. The overall ML learning speed and SPICE simulation speed were improved using local fitting (a separate model for each device instance). However, because local fitting was learned using only one device, it had a limitation in terms of scalability. The global fitting method covered only 36 devices, but was about five times slower than the local fitting method.

1.2. Contributions

There are models that use ML algorithms, such as autoencoders and convolutional neural networks (CNNs), to reflect the I-V and C-V characteristics of devices [10,12]. An autoencoder allows modeling with less data, but requires many hidden layers and produces noisy predictions. CNN models are excellent in data generation. However, this method requires a large amount of data and is difficult to train, and it is difficult to compute convolutions. A disadvantage of these complex computational methods and many-step models is that the computation time for circuit simulation in SPICE becomes longer. The ANN method used in this paper has demonstrated validity and excellent performance when compared to other regression algorithms [13]. ANNs can accurately predict linear and nonlinear data relationships, allowing them to approximate the physical equations of devices. In addition, because the calculation process is simpler than those in other ML methods, it has the advantage of reducing circuit simulation time when used with SPICE. Therefore, in this paper, we propose a new, powerful ANN-based compact model that can predict the characteristics of an NSFET, which is spotlighted as next-generation sub-3-nm devices. We aggressively expand the device range to sub-3-nm nodes (i.e., gate length (L_g) = 11 nm, sheet thickness (T_{sheet}) = 4 nm, spacer length (L_{sp}) = 3 nm, and oxide thickness (T_{ox}) = 1 nm). Various experiments were carried out in order to create a compact model using ML. Instead of feature extraction work, five key geometric parameters that affect global variation and can be controlled by designers were carefully chosen from the existing NSFET research. The proposed modeling framework reduces the complexity of the ML-based model using smaller numbers of hidden layers and neurons, but predicts the I-V and C-V characteristics with high accuracy and speed. The accuracy and speed of the completed ANN-based compact model implemented in Verilog-A were compared to those of the existing compact model in an HSPICE simulation. To the best of our knowledge, this is the first time that a proposed ANN-based compact model has outperformed the recently developed compact model [2].

1.3. Paper Organization

This paper is organized as follows. In Section 2, we present the creation of a dataset for an NSFET device design and the training of the ANN model. Section 3 discusses the input data preprocessing and output data scaling for smooth learning, the structure and techniques of the ANN model, and the overall workflow. Section 4 describes the modeling of the device's I-V and C-V characteristics. Section 5 shows the simulation results for the XOR, ring oscillator, and 6T-SRAM circuits using HSPICE compared with BSIM-CMG as the reference compact model [2]. Finally, Section 6 summarizes the conclusions of the paper and discusses future work.

2. Device Design and Dataset Generation

2.1. Process Flow of the Nanosheet FET (NSFET)

The actual NSFET process sequence is as follows. Si and SiGe are sequentially deposited on a wafer, followed by the formation of a dummy poly gate and gate space. In the next step, a dry etch process is performed for the first source/drain (S/D) recess to selectively remove Si/SiGe. A wet etch process is used to create an internal spacer. Then, a second S/D recess is performed to create a space for the bottom oxide to be filled. Chemical vapor deposition (CVD) is used to fill the space left over after the second S/D recess process. Subsequently, S/D growth and implantation are performed. Finally, HfO_2/TiN formation is accomplished via dummy gate removal and atomic layer deposition (ALD), with stress engineering added to improve hole mobility in the case of PMOS. The contact and wiring processes are not described. In this paper, an NSFET with bottom oxide is designed with this process in mind, and the detailed process flow is described [14].

2.2. Construction of Device Datasets

Using the dataset of an NSFET, which was highlighted as a new device after the FinFET, a compact model based on ML was designed. Compared to a FinFET with a three-sided gate, the NSFET has higher gate control capabilities and a larger effective width at the same size due to its four-sided gate [15]. Figure 1 shows the three-dimensional structure and cross-sectional views of a sub-3-nm-node NSFET designed in accordance with International Technology for Devices and Systems (IRDS) 2021. A device of the sub-3-nm node was designed and simulated using Sentaurus TCAD [16].

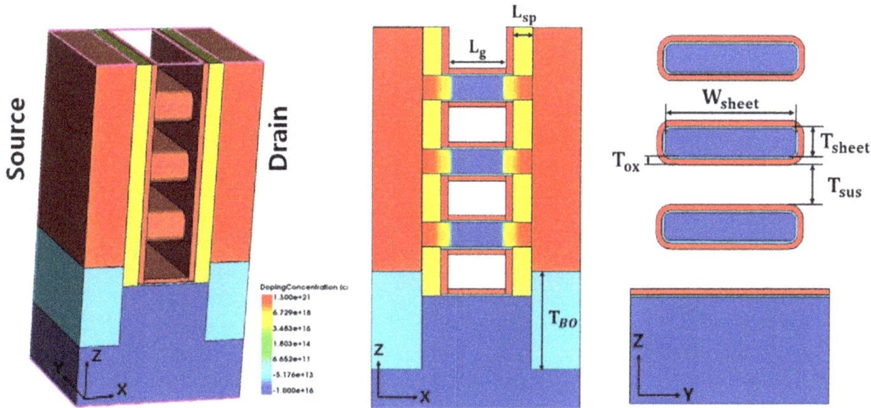

Figure 1. A schematic diagram of a 3D TCAD simulation of the NSFET.

2.3. Simulation Conditions

The quantum-potential model and Fermi model were used to study the quantum effects at the nanoscale. The movement of carriers in the low electric field along the short channel length were captured using a quasi-ballistic mobility model. For remote phonon and coulomb scattering effects, the Lombardi model was used, where the inversion and accumulation layer model was used to account for surface roughness caused by impurities and phonons in the thin layer. To account for bandgap changes caused by doping, the Slotboom bandgap narrowing model was applied to all regions of the semiconductor [17]. The Shockley–Read–Hall (SRH), Auger, and SurfaceSRH models were used as recombination models, and the band-to-band tunneling (BTBT) model was used to consider the tunneling and quantum confinement of small devices [18,19]. The stress-induced hole mobility in PMOS was studied using the h-multivalley model [16]. The sub-band model was used among the piezo models to apply the quasi-Fermi energy level based on doping, and the silicon <110> direction was used for the channel direction [20]. Each carrier valley change due to stress was calculated using the deformation potential model [21]. Calibration to the actual I-V data of IBM 3-nm NSFETs was performed to demonstrate the validity of the physical formula applied to TCAD. Figure 2 shows the calibration results [22]. Figure 3 shows the characteristics of the 3-nm NSFET generated using the physics formula after calibration. Figure 3a displays the I-V symmetry of N-type and P-type NSFETs, while Figure 3b shows the gate capacitance. Table 1 shows the electrical characteristics obtained using the TCAD simulation.

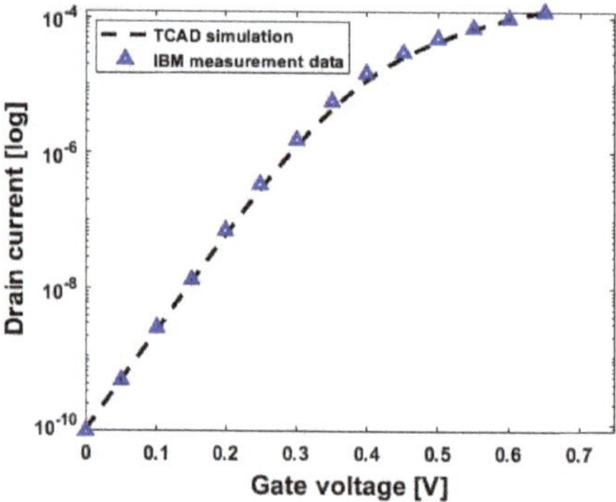

Figure 2. Calibrated I-V curve with the IBM measurement data.

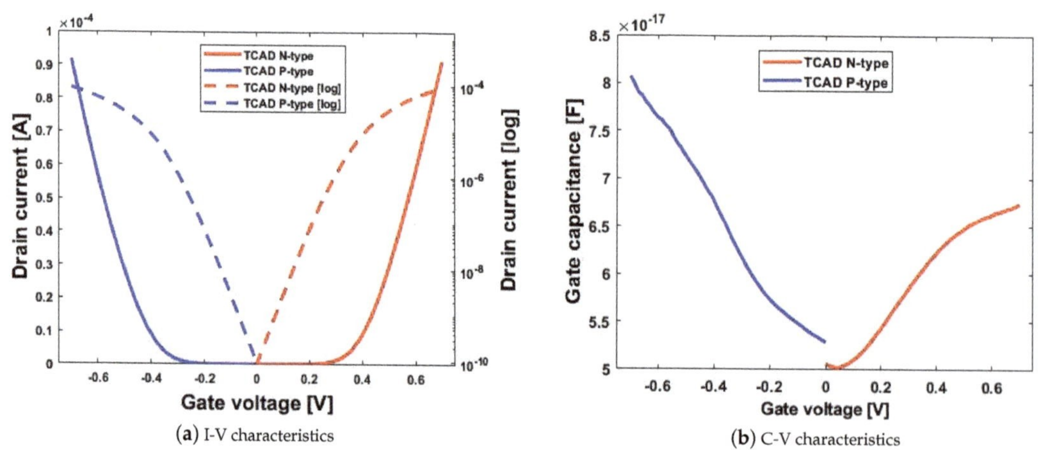

(**a**) I-V characteristics

(**b**) C-V characteristics

Figure 3. (**a**) I-V symmetry graph of an N-type NSFET and P-type NSFET. (**b**) Gate capacitance of an N-type NSFET and P-type NSFET.

Table 1. The electrical characteristics obtained using a TCAD simulation.

	N-Type NSFET	P-Type NSFET
V_{th} [mV]	0.218	0.220
I_{off} [pA]	98.01	95.03
I_{on} [uA]	91.13	90.34
SS [1] [mV/dec]	65.8	66.4

[1] Sub-threshold swing.

2.4. Construction of Device Datasets

The ranges of calibration and the data split were determined based on the IRDS roadmap organized down to the 1.5 nm node and actual data published by IBM, as shown in Table 2 [23]. Compared to previously reported 36 devices [11], 405 devices were created by selecting the five most important structural parameters in the NSFET and splitting them based on the application scope. The I-V and C-V characteristics were extracted to create datasets. Table 2 contains information about the created datasets. The TCAD simulator was used to create the dataset based on a temperature of 27 °C. Section 5 confirms that the five structural parameters chosen accurately represent the device characteristics. Existing papers were organized around typical values, but we conducted research on devices that had undergone sub-3-nm scaling (i.e., L_g = 11 nm, T_{sheet} = 4 nm, L_{sp} = 3 nm, T_{ox} = 1 nm). Because the device characteristics were nonlinearly dependent on structural parameters, it was difficult to predict the exact characteristics of the devices with different structural parameters when limited to a narrow range, as illustrated in Figure 4. The use of differently sized devices in one circuit, such as in an SRAM cell, is difficult to model. The global device modeling method presented in this paper, which covers multiple devices with a single model, is an approach that can foster collaboration between designers and process engineers. Furthermore, the proposed ANN model can reduce the use of TCAD data because it can predict with high accuracy the characteristics of devices with untrained structural parameters.

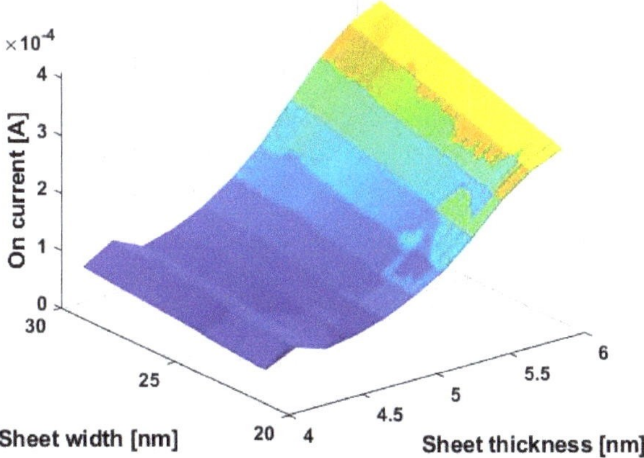

Figure 4. Nonlinear electrical characteristics of NSFETs.

Table 2. Key parameters of NSFETs for the TCAD simulation.

Tech Node	3 [nm]	2.1 [nm]	1.5 [nm]	3 [nm]	Sub-3 [nm]
		IRDS		IBM	Our Datasets
L_g [nm]	16	14	12	12	11, 12, 13
W_{sheet} [nm]	30	30	30	50	21, 23, 25, 27, 29
T_{sheet} [nm]	8	7	6	5	4, 5, 6
L_{sp} [nm]	6	5	4	5	3, 4, 5
T_{ox} [nm]	1.4	1.37	1.37	1.5	1, 1.5, 2
T_{sus} [nm]	10	10	10	10	10
S/D doping [cm^{-3}]	-	-	-	-	6.5×10^{20}
Channel doping [cm^{-3}]	-	-	-	-	1×10^{16}

3. ANN Model Architecture and Methodology

3.1. The Architecture of the Proposed ANN Model

The proposed model architecture using the ANN structure is shown in Figure 5. The calculation method of the ANN model is formulated in Equation (1). As shown in Table 2, the model of the I-V characteristics of the 405 devices composed of a combination of 5 key parameters has one input layer, two hidden layers, and one output layer. The numbers of neurons in the hidden layers are 20 and 15, respectively. The C-V characteristics of the 405 devices are less complex than that of the I-V characteristics. One input layer, two hidden layers, and one output layer comprise the model of the C-V characteristics. We reduced the complexity of the ANN model using a model with 10 and 5 neurons in the hidden layers, respectively. As input values, five important geometric parameters of the NSFET were chosen, and two terminal voltage biases were used. The I-V and C-V characteristics of the 405 devices were used as output values. The five chosen key parameters had the advantage of being able to replace the dimension reduction processes, such as PCA (principal component analysis), used for many parameters of the traditional compact model, and they still accurately represent the I-V and C-V characteristics, as verified in Section 5.

$$\mathbf{Y} = \mathbf{W}_3 \times g(\mathbf{W}_2 \times g(\mathbf{W}_1 \times \mathbf{X} + \mathbf{B}_1) + \mathbf{B}_2) + \mathbf{B}_3 \tag{1}$$

In ML, the loss function is critical. When a model is trained, it learns in the direction of minimizing the loss function. The loss function of the proposed ANN model takes the device physics into account. α, β, and γ were multiplied by each operation region of the device to determine the loss. To reduce errors in the medium region and ON region, which are important in device operation, weights of $\alpha = 1$, $\beta = 2$, and $\gamma = 3$ were multiplied (the value was adjusted so that the ON region's error was less than 1%). The physics-augmented loss function had the advantage of being easily adjusted based on the operation region where the error is to be reduced. Equation (2) represents the physics-augmented loss function.

$$Loss\,function = \frac{1}{n} \sum_{i=1}^{n} \left(\alpha \left(y_{true,off} - y_{pred,off} \right)^2 + \beta \left(y_{true,medium} - y_{pred,medium} \right)^2 + \gamma \left(y_{true,high} - y_{pred,high} \right)^2 \right) \tag{2}$$

The ADAM optimizer was used in the learning process. A continuous and smooth activation function, the hyperbolic tangent function (tanh), was used. One of the hyperparameters that has a significant impact on learning outcomes is the learning rate. As a result, one of the learning rate scheduler methods, ReduceLROnPlateau Scheduler, was used. The ReduceLROnPlateau scheduler technique is one of several methods employed to reduce the learning rate and continue learning by multiplying the learning rate by a constant factor if the value of the valid loss remains constant for a certain period of time during training. Early stopping was also used to prevent overfitting, which is a major issue in ML.

The errors of the I-V characteristics were 2.5%, 2.0%, and 1.0% for the OFF region (logarithmic scale of 0.2%, V_{gs} = 0.0 to 0.2 Volt), medium V_{gs} region (V_{gs} = 0.2 to 0.5 Volt), and high V_{gs} region (V_{gs} = 0.5 to 0.7 Volt), respectively.

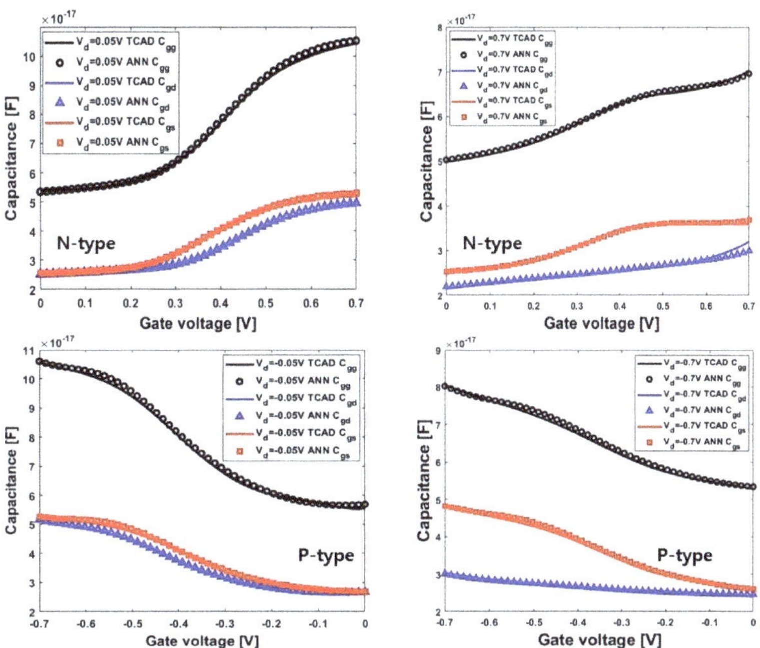

Figure 8. Comparison of the ANN model and simulated C-V characteristics using TCAD for N-type and P-type NSFETs (gate length = 12 nm, sheet width = 25 nm, sheet thickness = 5 nm, spacer length = 4 nm, oxide thickness = 1.5 nm).

The errors of the C-V characteristics were 1.3% for C_{gg}, 1.5% for C_{gd}, and 1.5% for C_{gs}. These findings indicate that the proposed ANN model is capable of high-accuracy global fitting. Existing compact models employ the binning method because it is difficult to achieve accuracy within 1–2% error using a single-model parameter set, even after parameter extraction through more than 10 complicated processes. The presented ANN model can form a global device model with a smaller error using a simpler process. Previously published ANN models of next-generation transistors have more complex structures, higher computational costs, and longer training times (hours to learn). However, the proposed ANN model reduces the computational cost and training time by using fewer hidden layers and neurons than existing ANN models while maintaining higher accuracy. Both the ANN I-V model and ANN C-V model used one million epochs, where the training times were about 1 h.

5. SPICE Simulation of Circuits Using the Developed ANN Models

The SPICE simulations using the developed ANN model are summarized in this section. Three circuits were simulated to validate the ANN-based compact model. The operation of the XOR circuit, which is one of the complex combinational logic gates, was verified, and the ring oscillator was chosen to verify the transient operation. Finally, the operation margins of the 6T-SRAM circuit with various structural parameters were simulated. HSPICE was chosen for circuit simulation using the ANN-based compact model. Verilog-A is a de facto standard modeling language that allows model developers to focus on modeling while significantly reducing the development time. There is a published example of how to efficiently write a compact model in Verilog-A [25]. After learning the ANN I-V and ANN C-V models, the ANN-based compact model was built in Verilog-A using the weights and biases of the ANN structure. The simulation time was reduced during the construction of the ANN-based compact model by removing unnecessary 'for' and 'list' statements. For accuracy verification, data extracted from BSIM-CMG were used and compared to those of the ANN-based compact model.

5.1. XOR, Ring Oscillator, and SRAM Simulation

Figure 9 shows a graph comparing XOR gate simulation between the ANN-based compact model and BSIM-CMG. For the XOR gate simulation, the first input voltage was 0.7 V with a period of 2 ns, and the second was 0.7 V with a period of 4.5 ns. Figure 10 shows a graph comparing the 17-stage ring oscillator simulation of the ANN-based compact model to that of BSIM-CMG. The initial voltage was set to 0 V, and the simulation was performed over a 1-ns transient period.

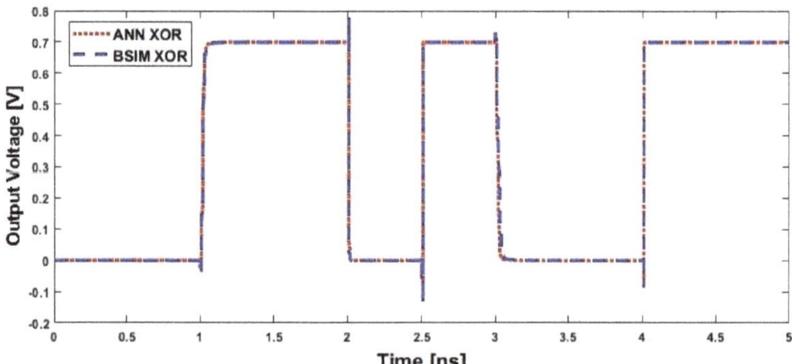

Figure 9. Comparison of the ANN model and the BSIM-CMG model for the XOR simulation.

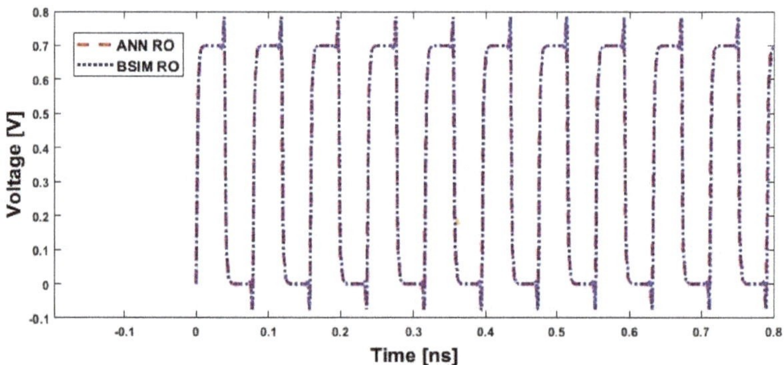

Figure 10. Comparison of the ANN model and the BSIM-CMG model for the 17-stage ring oscillator simulation.

Figure 11 shows the 6T-SRAM simulation results for the hold, read, and write operation margins [26]. In the 6T-SRAM simulation, the device width was set to 1:a:b in order to account for the static noise margin (SNM) [27]. Table 3 shows the simulation results and errors between the ANN-based compact model and BSIM-CMG for each circuit. The proposed ANN-based compact model built in Verilog-A achieved a high accuracy of about 1% in all three circuit simulations, except for the SRAM read margin.

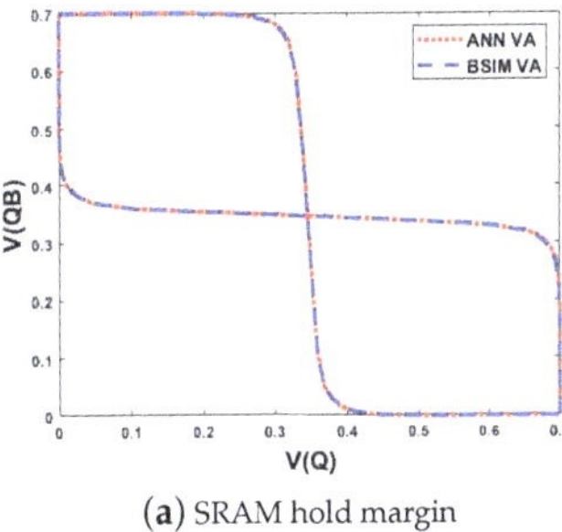

(a) SRAM hold margin

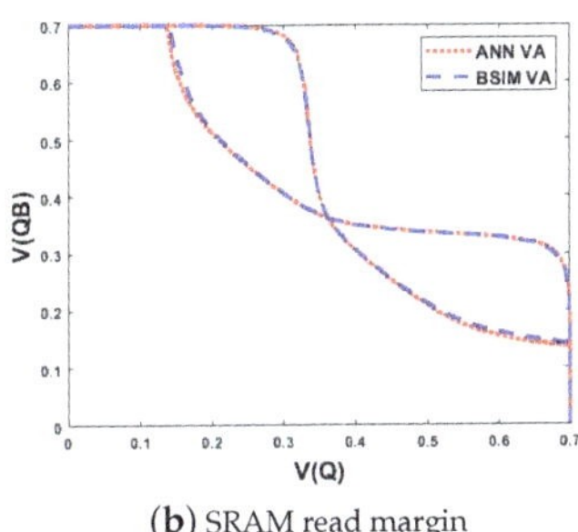

(b) SRAM read margin

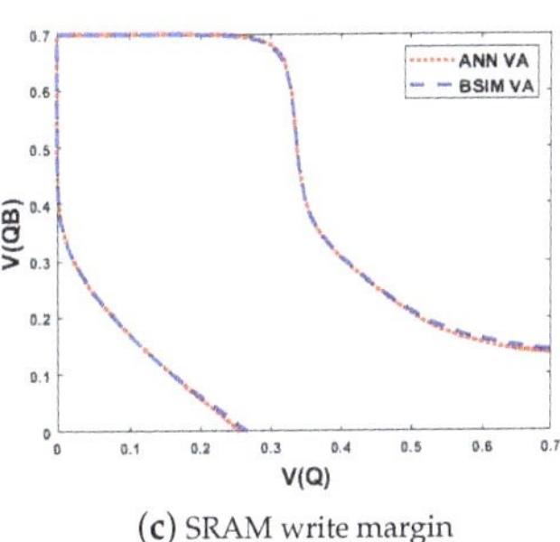

(c) SRAM write margin

Figure 11. Comparison of the ANN and BSIM models' results for the 6T-SRAM simulation: (a) hold operation margin, (b) read operation margin, and (c) write operation margin.

Table 3. Circuit performance simulation results and errors between the ANN-based compact model and BSIM-CMG.

		BSIM-CMG	ANN Model	Errors [%]
XOR	A to Y Delay [ps]	11.96	11.88	0.68
	B to Y Delay [ps]	5.94	5.87	1.13
17-RO	Delay [ps]	79.50	78.70	1.01
SRAM (Hold)	Margin [mV]	286.98	285.88	0.39
SRAM (Read)	Margin [mV]	129.68	126.01	2.91
SRAM (Write)	Margin [mV]	223.16	221.72	0.65

5.2. SPICE Simulation Performance Comparison

We compared the simulation speeds of the ANN-based compact model and BSIM-CMG in Verilog-A. Because the simulation of a small circuit was completed quickly, in order to clearly investigate the speed comparison, each simulation was completed with 1000 iterations using the Monte Carlo method in HSPICE. The simulation speed was measured while increasing the number of stages of the ring oscillator.

5.2.1. Global Device Model

Figure 12 shows the simulation time for the proposed ANN-based compact model and the BSIM-CMG Verilog-A version. The results demonstrate that the ANN-based compact model written in Verilog-A was more than twice as fast as BSIM-CMG. Because the ANN-based compact model primarily computes multiplication and activation functions and is simpler than BSIM-CMG, it is more than twice as fast [28].

5.2.2. Single-Device Model

A single-device model was built because it is much more efficient than the global device model when a circuit does not require separate device sizing. The amount of data in a single device is relatively small. The single I-V model, like the global device model, had two hidden layers, but with 15 and 10 fewer nodes, respectively. The C-V model had only one hidden layer with five nodes. The single-device model had fewer nodes than the global device model, but was faster and more accurate, with less than 0.3% prediction error. As shown in Figure 13, the speed of the single-device model was approximately twice as fast as the global device model due to its greater conciseness. Based on the experimental results shown in Figures 12 and 13, if the ANN-based compact model is embedded in HSPICE using the C programming language, it is predicted to be several times faster than BSIM-CMG (blue dashed line), as shown by the red dashed line at the bottom of Figure 13. Table 4 compares the characteristics of the single-device and the global device models.

Table 4. Characteristics of the single-device model and the global device model.

	Single-Device Model		Global Device Model	
ANN model sizes (numbers of neurons in hidden layers)	I-V	C-V	I-V	C-V
	(5, 5)	(5)	(20, 15)	(10, 5)
Simulation errors	0.3% or less	0.3% or less	Off region: 2.5% (log scale: 0.2%) Medium region: 2.0% High region: 1.0%	C_{gg}: 1.3% C_{gd}: 1.5% C_{gs}: 1.5%
Simulation complexity [1]	I-V: $(7 \times 5) + (5 \times 5) + (5 \times 1) = 65$ C-V: $(7 \times 5) + (5 \times 3) = 50$ Activation function (tanh) calculation = 15		I-V: $(7 \times 20) + (20 \times 15) + (15 \times 1) = 455$ C-V: $(7 \times 10) + (10 \times 3) = 100$ Activation function (tanh) calculation = 50	
Advantages	1. The simulation speed is fast 2. The accuracy is very high		1. Different devices can be simulated By changing the structural parameters of the circuit 2. The circuit specifications (e.g., power, delay) can be evaluated based on the structural parameters 3. TCAD usage is reduced by predicting values between structures	

[1] Complexity of the calculation of multiplication and activation functions in the ANN-based compact model.

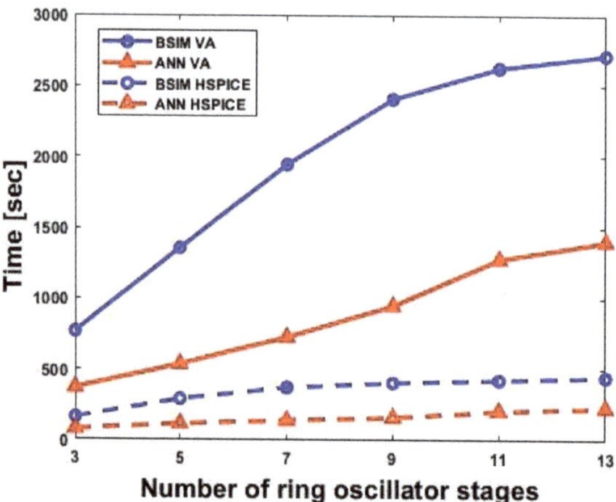

Figure 12. Time comparison of Monte Carlo simulations by stages of a ring oscillator using the global device model.

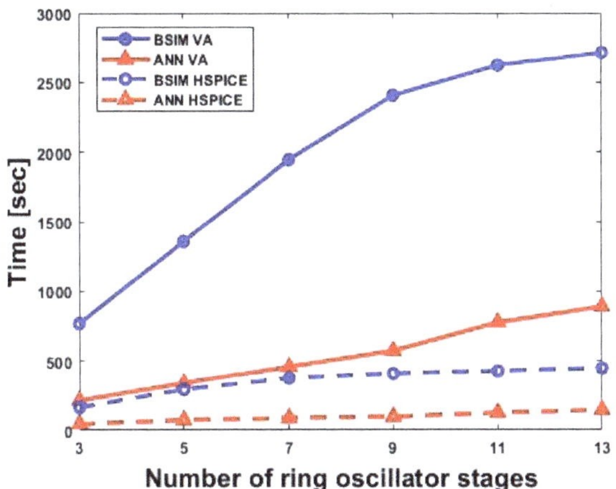

Figure 13. Time comparison of Monte Carlo simulations by stages of a ring oscillator using the single-device model.

6. Conclusions

In this paper, an ANN-based compact model was developed to predict the I-V and C-V characteristics of 405 NSFETs, including all typical devices and sub-3-nm devices. This approach can be very useful when designers and process engineers work together. The ANN model was created by selecting five key geometric parameters. Even after using a large number of global device datasets (405), fewer neurons and hidden layers were used to reduce the complexity of the ANN model and accurately perform device modeling at a high speed. It was confirmed that the five geometric parameters chosen were able to represent more than 98% of the I-V and C-V characteristics. When tested on the datasets that were not included in training, the predicted values of the ANN model and TCAD simulations matched very closely. Additionally, the TCAD data usage was effectively reduced. The ANN-based compact model was implemented in Verilog-A. The modeling flow was automated using Python. The ANN-based compact model proposed in this work is approximately two times faster than the SPICE simulation of the existing compact model, and it can be further accelerated by 2–3 times by using a single-device model. The accuracy of the proposed ANN-based compact model was also demonstrated through simulations of XOR, ring oscillators, and SRAM circuits. In addition, the physics-augmented loss function can be used to reduce the error in the desired operation region. The developed ANN-based compact modeling framework is being expanded and applied to a negative-capacitance NSFET. In addition, ANN-based statistical analyses will be performed to reflect global and local variations.

Author Contributions: S.W., H.J. and J.C. contributed to the writing of this article. S.W. and H.J. contributed to the primary idea of this research. J.C. and H.C. performed the simulations. This research was planned and executed under the supervision of J.-T.K. and S.K. All authors have read and agreed to the submitted version of the manuscript.

Funding: This work was supported by an Institute of Information and Communications Technology Planning Evaluation (IITP) grant funded by the Korean government (MSIT) (No. 2021-0-00754, Software Systems for AI Semiconductor Design) and by a National Research Foundation of Korea grant funded by the Korean government (MISP) (NRF-2020R1A2C1011831, 2020R1A5A1019649).

Acknowledgments: The EDA tools were supported by the IC Design Education Center (IDEC), Korea.

Conflicts of Interest: The authors declare no conflict of interest.

References

1. Root, D.E. Future Device Modeling Trends. *IEEE Microw. Mag.* **2012**, *13*, 45–59. [CrossRef]
2. Dunga, M.V.; Lin, C.-H.; Niknejad, A.M.; Hu, C. BSIM-CMG: A Compact Model for Multi-Gate Transistors. In *FinFETs and Other Multi-Gate Transistors*; Springer: Berlin/Heidelberg, Germany, 2008; pp. 113–153.
3. Klemme, F.; Prinz, J.; Santen, V.M.V.; Henkel, J.; Amrouch, H. Modeling Emerging Technologies Using Machine Learning: Challenges and Opportunities. In Proceedings of the 2020 International Conference on Computer-Aided Design, New York, NY, USA, 2–5 November 2020.
4. Lim, J.H.; Shin, C.H. Machine Learning (ML)-Based Model to Characterize the Line Edge Roughness (LER)-Induced Random Variation in FinFET. *IEEE Access* **2020**, *8*, 158237–158242. [CrossRef]
5. Chen, Q.; Chen, G. Artificial Neural Network Compact Model for TFTs. In Proceedings of the 2016 International Conference on Computer Aided Design for Thin-Film Transistor Technologies, Beijing, China, 26–28 October 2016.
6. Ko, K.; Lee, J.K.; Kang, M.G.; Jeon, J.W.; Shin, H.C. Prediction of Process Variation Effect for Ultrascaled GAA Vertical FET Devices Using a Machine Learning Approach. *IEEE Trans. Electron Devices* **2019**, *66*, 4474–4477. [CrossRef]
7. Lei, Y.; Huo, X.; Yan, B. Deep Neural Network for Device Modeling. In Proceedings of the 2018 IEEE Electron Devices Technology and Manufacturing Conference, Kobe, Japan, 13–16 March 2018.
8. Zhang, L.; Chan, M. Artificial Neural Network Design for Compact Modeling of Generic Transistors. *J. Comput. Electron.* **2017**, *16*, 825–832. [CrossRef]
9. Zhang, Z.; Wang, R.; Chen, C.; Huang, Q.; Wang, Y.; Hu, C.; Wu, D.; Wang, J.; Huang, R. New-Generation Design-Technology Co-Optimization (DTCO): Machine-Learning Assisted Modeling Framework. In Proceedings of the 2019 Silicon Nanoelectronics Workshop, Kyoto, Japan, 9–10 June 2019.
10. Mehta, K.; Wong, H. Prediction of FinFET Current-Voltage and Capacitance-Voltage Curves Using Machine Learning with Autoencoder. *IEEE Electron Device Lett.* **2021**, *42*, 136–139. [CrossRef]
11. Wang, J.; Kim, Y.H.; Ryu, J.S.; Jeong, C.W.; Choi, W.S.; Kim, D.S. Artificial Neural Network-Based Compact Modeling Methodology for Advanced Transistors. *IEEE Trans. Electron Devices* **2021**, *68*, 1318–1325. [CrossRef]
12. Hirtz, T.; Huurman, S.; Tian, H.; Yang, Y.; Ren, T.-L. Framework for TCAD augmented machine learning on multi-I-V characteristics using convolutional neural network and multiprocessing. *J. Semicond.* **2021**, *42*, 124101. [CrossRef]
13. Shirakawa, K.; Shimiz, M.; Okubo, N.; Daido, Y. A Large-Signal Characterization of an HEMT Using a Multilayered Neural Network. *IEEE Trans. Microw. Theory Tech.* **1997**, *45*, 1630–1633. [CrossRef]
14. Yoo, S.; Kim, S. Leakage Optimization of the Buried Oxide Substrate of Nanosheet Field-Effect Transistors. *IEEE Trans. Electron Devices* **2022**, *69*, 4109–4114. [CrossRef]
15. Kim, S.-D.; Guillorn, M.; Lauer, I.; Oldiges, P.; Hook, T.; Na, M.-H. Performance Trade-offs in FinFET and Gate-All-Around Device Architecture for 7nm-node and Beyond. In Proceedings of the 2015 SOI-3D-Subthreshold Microelectronics Technology Unified Conference, Rohnert Park, CA, USA, 5–8 October 2015.
16. Synopsys Inc. *Sentaurus Device, User Manual, Version Y-2019*; Synopsys Inc.: Mountain View, CA, USA, 2019.
17. Yoon, J.-S.; Jeong, J.; Lee, S.; Baek, R.-H. Optimization of nanosheet number and width of multi-stacked nanosheet FETs for sub-7-nm node system on chip applications. *Jpn. J. Appl. Phys.* **2019**, *58*, SBBA12. [CrossRef]
18. Ryu, D.; Kim, M.; Yu, J.; Kim, S.; Lee, J.-H.; Park, B.-G. Investigation of Sidewall High-k Interfacial Layer Effect in Gata-All-Around Structure. *IEEE Trans. Electron Devices* **2020**, *67*, 1859–1863. [CrossRef]
19. Yoon, J.-S.; Jeong, J.; Lee, S.; Baek, R.-H. Systematic DC/AC Performance Benchmarking of Sub-7-nm Node FinFETs and Nanosheet FETs. *IEEE J. Electron Devices Soc.* **2018**, *6*, 942–947. [CrossRef]
20. Sun, Y.; Thompson, E.; Nishida, T. Physics of strain effect in semiconductors and metal-oxide-semiconductor field-effect transistors. *J. Appl. Phys.* **2007**, *101*, 104503. [CrossRef]
21. Reboh, S.; Coquand, R.; Augendre, E.; Barraud, S.; Maitrejean, S.; Viner, M.; Faynot, O.; Loubet, N.; Guillorn, M.; Fetterolf, S.; et al. An analysis of stress evolution in stacked GAA transistors. In Proceedings of the IEEE Slilcon Nanoelectronics Workshop (SNW), Honolulu, HI, USA, 12–13 June 2016.
22. Loubet, N.; Hook, T.; Montanini, P.; Yeung, C.-W.; Kanakasabapathy, S.; Guillom, M.; Yamashita, T.; Zhang, J.; Miao, X.; Wang, J.; et al. Stacked nanosheet gate-all-around transistor to enable scaling beyond FinFET. In Proceedings of the 2017 IEEE Symposium on VLSI Technology, Kyoto, Japan, 5–8 June 2017.
23. Moore, M. International Roadmap for Devices and Systems (IRDS™) Edition. IEEE. 2021. Available online: https://irds.ieee.org/images/files/pdf/2021/2021IRDS_MM.pdf (accessed on 25 August 2022).
24. Paszke, A.; Gross, S.; Massa, F.; Lerer, A.; Bradbury, J.; Chanan, G.; Killeen, T.; Lin, Z.; Gimelshein, N.; Antiga, L.; et al. PyTorch: An Imperative Style, High-Performance Deep Learning Library. In Proceedings of the Advances in Neural Information Processing Systems, Vancouver, BC, Canada, 8–14 December 2019; pp. 8024–8035.
25. Mcandrew, C.C.; Coram, G.J.; Gullapalli, K.K.; Jones, J.R.; Nagel, L.W.; Roy, A.S.; Roychowdhury, J.; Scholten, A.J.; Smit G.D.J.; Wang, X.; et al. Best Practices for Compact Modeling in Verilog-A. *IEEE J. Electron Devices Soc.* **2015**, *3*, 383–396. [CrossRef]
26. Lim, W.; Chin, H.C.; Lim, C.S.; Tan, M.L.P. Performance Evaluation of 14nm FinFET-Based 6T SRAM Cell Functionality for DC and Transient Circuit Analysis. *J. Nanomater.* **2014**, *2014*, 105. [CrossRef]

27. Song, T.; Jung, H.; Yang, G.; Tang, H.; Kim, H.; Seo, D.; Kim, H.; Rim, W.; Baek, S.; Baeck, S.; et al. 3nm Gate-All-Around (GAA) Design-Technology Co-Optimization (DTCO) for succeeding PPA by Technology. In Proceedings of the Custom Integrated Circuits Conference (CICC), Newport Beach, CA, USA, 24–27 April 2022.
28. Kao, M.Y.; Kam, H.; Hu, C. Deep-Learning-Assisted Physics-Driven MOSFET Current-Voltage Modeling. *IEEE Electron Device Lett.* **2022**, *43*, 974–977. [CrossRef]

Article

Independent Effects of Dopant, Oxygen Vacancy, and Specific Surface Area on Crystal Phase of HfO$_2$ Thin Films towards General Parameters to Engineer the Ferroelectricity

Tianning Cui [1,2], Liping Zhu [3], Danyang Chen [1,2], Yuyan Fan [1,2], Jingquan Liu [1] and Xiuyan Li [1,*]

[1] National Key Laboratory of Science and Technology on Micro/Nano Fabrication, Shanghai Jiao Tong University, Shanghai 200240, China; cuicui520@sjtu.edu.cn (T.C.); c19970528@sjtu.edu.cn (D.C.); fanyuyan@sjtu.edu.cn (Y.F.); jqliu@sjtu.edu.cn (J.L.)
[2] Department of Micro/Nano Electronics, Shanghai Jiao Tong University, Shanghai 200240, China
[3] Department of Physics, Fudan University, Shanghai 200433, China; zhuliping@fudan.edu.cn
* Correspondence: xiuyanli@sjtu.edu.cn

Abstract: Many factors have been confirmed to affect ferroelectric phase formation in HfO$_2$-based thin films but there was still a lack of general view on describing them. This paper discusses the intrinsic parameters to stabilize the ferroelectric phase of HfO$_2$ thin films to approach this general view by investigating the separate effects of dopant, oxygen vacancy (V$_O$), and specific surface area on the crystal phase of the films. It is found that in addition to extensively studied dopants, the ferroelectric orthorhombic phase can also be formed in pure HfO$_2$ films by only introducing sufficient V$_O$ independently, and it is also formable by only increasing the specific surface area. By analyzing the common physics behind these factors, it is found that orthorhombic phase formation is universally related to strain in all the above cases with a given temperature. To get a general view, a physical model is established to describe how the strain influences ferroelectric phase formation during the fabrication of HfO$_2$-based films based on thermodynamic and kinetics analysis.

Keywords: ferroelectricity engineering; HfO$_2$; impactor factors; general parameters

Citation: Cui, T.; Zhu, L.; Chen, D.; Fan, Y.; Liu, J.; Li, X. Independent Effects of Dopant, Oxygen Vacancy, and Specific Surface Area on Crystal Phase of HfO$_2$ Thin Films towards General Parameters to Engineer the Ferroelectricity. *Electronics* **2022**, *11*, 2369. https://doi.org/10.3390/electronics11152369

Academic Editor: Ahmed Abu-Siada

Received: 9 June 2022
Accepted: 21 July 2022
Published: 28 July 2022

Publisher's Note: MDPI stays neutral with regard to jurisdictional claims in published maps and institutional affiliations.

Copyright: © 2022 by the authors. Licensee MDPI, Basel, Switzerland. This article is an open access article distributed under the terms and conditions of the Creative Commons Attribution (CC BY) license (https://creativecommons.org/licenses/by/4.0/).

1. Introduction

HfO$_2$-based high-k dielectrics have been applied as an alternative to thin SiO$_2$ gate insulators in Si technology since 2007 [1]. Continuously, a ferroelectric property in Si-doped HfO$_2$ thin films, which was not expected in the thermodynamic phase diagram of HfO$_2$, was observed in 2011 [2]. As soon as it was discovered, it attracted great attention because the HfO$_2$-based ferroelectrics with good CMOS compatibility and scalability are promising to overcome the critical obstacles in the application of conventional ferroelectric materials in memory and logic devices [3–5]. Clarification of fundamental material science essentially is critical to the device demonstration in the era beyond Moore's Law. In the last decade, intensive studies on understanding the physical origin of ferroelectricity and controlling the fundamental properties of HfO$_2$-based thin films have been carried out. It has been agreed that the metastable orthorhombic (O-phase) phase, which is formed during the transition from tetragonal (T-phase) to monoclinic (M-phase) phase, is the origin of ferroelectricity in HfO$_2$ [1,6]. Additionally, many experimental factors, including doping concentration, oxygen vacancy (V$_O$), film thickness, annealing conditions, and capping layers, have been confirmed to influence the stabilization of the O-phase [7–17]. However, a general model describing the intrinsic driving force to determine the stabilization of the O-phase from the T–M phase transition in HfO$_2$ thin film has not been clarified clearly up to now.

It has been argued that defects and surface energy are two main categories of impact factors to help stabilize the O-phase [7,8]. The former refers to dopants and V$_O$, while the latter is usually related to mechanical clamping of electrodes, substrate orientation, and film thickness [9–17]. However, in these studies, the above categories and factors are usually synergistic to affect the ferro-O phase formation. For example, ferroelectricity in undoped HfO_2 films was usually observed in a sandwich structure with TiN or TaN as bottom and top electrodes followed by post metalization annealing. However, the top layer would not only introduce Vo through the scavenging effect but also bring extra stress into the films [15–20]. Additionally, thinner films with larger surface energies enable more ferro-O phase formation, while this effect was mostly demonstrated in doped films [13–21]. Therefore, it is difficult to discuss the effect of every single factor and to get the intrinsic parameters behind these factors for engineering the ferroelectricity of the HfO_2-based films. It is even difficult to confirm whether the ferroelectric phase is formable or not in an undoped film.

In this study, we intend to approach a general view of phase formation in the films. For achieving reliable conclusions, we investigate the independent effect of dopants, V_O, and specific surface area of HfO_2-based thin films by carefully designing the experiment process. Our results show that all the above factors enable the formation of ferroelectric O-phase individually. By analyzing our results and the reported ones of other factors, it is found that most of the factors are related to strain universally. On the basis of this, a thermodynamic model is further proposed to give a universal description of the driving forces on the stabilization of the metastable O-phase.

2. Experimental

The 30 nm undoped and Si-doped HfO_2 (HSO) films were fabricated by plasma-enhanced atomic layer deposition (PE-ALD) at 200 °C on p-Si substrates (HfO_2/p-Si) and heavily doped p-Ge substrates (HSO/p^+-Ge), respectively. Tetrakis-ethylmethylamino-hafnium (TEMAHf) and oxygen plasma were used as precursors for HfO_2. Bis-terarybutylamino-silane (BTBAS) was used for silicon doping. Post-deposition annealing (PDA) was carried out in 0.005 atm O_2 for HSO films and in 0.005 atm O_2, 0.005 atm N_2, and ultra-high vacuum (UHV, base pressure of 10^{-7} Pa) for undoped HfO_2 films, respectively. The heating conditions were all set at 10 °C/s which began from 100 °C and all films were annealed at 650 °C for 30 s. A Premtek rapid thermal processing furnace was used for O_2 or N_2 annealing and a chamber of thermal desorption spectroscopy (EMD-WA1000S/W (ESCO, Ltd.)) was used for UHV annealing. Then, Au was deposited by thermal evaporation as the top electrode (TE). For some samples, films were etched into cuboid arrays by Ar^+ plasma whose side lengths were set to 2 μm, 5 μm, and 10 μm before PDA.

Figure 1 shows the experiment process designed to control doping, V_O, and specific surface area separately. For controlling doping singly, quadrivalent Si dopants were selected because they do not induce V_O in the film like trivalent cation, and O_2-PDA was carried out for eliminating the V_O generated in film deposition. Meanwhile, employment of Ge substrate and PDA process help to minimize the interface effect such as inducing extra V_O or stress. While for controlling Vo independently, undoped HfO_2 films were annealed in O_2, N_2, or UHV respectively. Similarly, the PDA process was employed, but the Si substrate was selected because less leakage is induced on Si than on Ge in UHV-PDA. While for the individual control of the specific surface area, undoped HfO_2 films were etched into different sizes before annealing to explore its impact, while Ge substrate and PDA were also employed for the reasons stated above.

Film thickness was measured by spectroscopic ellipsometry (Filmetrics F50). Doping concentration, defined as dop% = Si/(Si + Hf), was confirmed by X-ray photoelectron spectroscopy (XPS, AXIS-ULTRADL DLD). Grazing incidence X-ray diffraction (GIXRD) was used to detect phase composition by a Riguaku D/max 2500 V (Cu-Kα radiation, λ = 0.154 nm). The incidence angle was fixed at 1° and the emergence angle varied from 27° to 33°. An atomic force microscope (AFM, Bruker ICON, Billerica, MA, USA) was utilized for detecting surface morphologies. Polarization-electric field (P-E) properties were measured by semiconductor parameter analyzer (Keithley 4200, Cleveland, OH,

USA) and triangular voltage pulse was set at 10 KHz. Positive-up-negative-down (PUND) measurement was carried out to check the ferroelectricity of samples with UHV annealing.

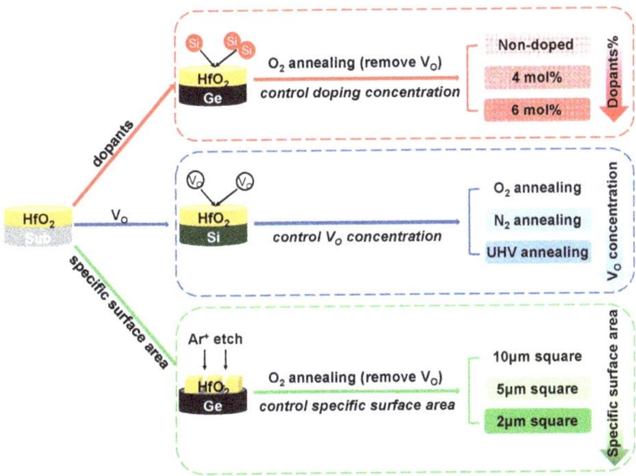

Figure 1. The process for independent control of doping concentration, V$_O$ concentration, and specific surface area in HfO$_2$ thin film.

3. Results

3.1. Independent and Common Effects of Dopants and V_O

Firstly, we consider the independent effect of doping concentration and V$_O$ on the ferroelectric phase of HfO$_2$ thin films. GIXRD patterns of doped films with different doping concentrations and undoped-HfO$_2$ films with different annealing conditions are summarized in Figure 2a,b. The M-phase appears at both 28.5° and 31.6°, while the peak around 30.5° is attributed to T-, O-, or cubic phases (C-phase) whose difference is difficult to distinguish solely. Therefore, it is referred to as the O/T/C phase hereafter [22]. In order to clarify the impact quantitatively, we extracted the O/T/C phase fraction (R$_{O/T/C}$) by Gaussian peak splitting. R$_{O/T/C}$ is defined as the ratio of O/T/C phase peak intensity to the total intensity (I), expressed as

$$R_{O/T/C} = (I_{O/T/C}/(I_{M(-111)} + I_{O/T/C} + I_{M(111)})), \tag{1}$$

Based on the literature, ferroelectricity in HfO$_2$ is observed with R$_{O/T/C}$ of 20–90% [23]. The effect of doping concentration and V$_O$ on R$_{O/T/C}$ are summarized in Figure 2c,d where the red regions are those with ferroelectricity. As for the doped films, 60% of the O/T/C phase was stabilized with a suitable doping concentration (~4 mol%) without other effects. The M-phase tends to form with lower doping concentrations while the T-phase with higher ones. The P-E characteristics in Figure 2e confirm ferroelectricity in ~4 mol% HSO films and non-ferroelectricity in ~6 mol% or undoped ones. This is consistent with reported data [1]. As for undoped HfO$_2$ films, a pure M-phase is formed with N$_2$- or O$_2$-PDA as expected. Interestingly, ~60% of the O/T/C phase is formed with UHV-PDA. This indicates the formation of ferroelectric O-phase with this condition. In Figure 2f, the ferroelectricity of the film with UHV-PDA is further confirmed by PUND measurement. These results clearly suggest that either sufficient V$_O$ or sufficient dopants enable the stabilization of the O-phase in HfO$_2$ thin film independently. Although the effect of dopants and V$_O$ on phase formation have been investigated widely, previous studies neglected to separate the effect of these two parameters, thus making it difficult to get the intrinsic and general view. Here, we clearly show their independent effect and amply confirm that the ferroelectric phase can be formed in really undoped HfO$_2$ films with an amount of Vo. However, it should be noticed that introducing too much V_O into the films will arouse large leakage which brings negative effects from the perspective of application engineering.

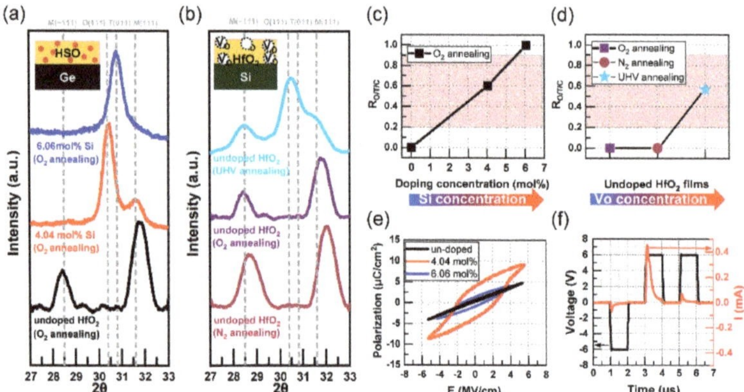

Figure 2. Summarized GIXRD patterns of (**a**) HSO films with different doping concentrations and (**b**) undoped HfO_2 films with different annealing atmospheres. (**c**,**d**) Extracted $R_{O/T/C}$ by fitting curves from (**a**,**b**). (**e**) PE curves of undoped, ~4% and ~6% HSO films with O_2 annealing. Only ~4% of doped HfO_2 films show ferroelectricity. (**f**) The difference in current measured by PUND measurement demonstrates that undoped HfO_2 films with UHV annealing are ferroelectric.

In addition to the independent effect of Si dopants and V_O on promoting O/T/C phase stabilization, the combined effect of two factors was further considered. Figure 3a shows P−E results of low doped HSO films (~2.5 mol%) with N_2- or O_2-PDA, respectively. As expected, no obvious polarization is observed in ~2.5 mol% HSO with O_2-PDA. However, the polarization of ~2.5 mol% HSO films with N_2-PDA increases rapidly which even becomes greater than that of 4 mol% HSO films. Figure 3b,c shows the surface topography of the films with N_2- and O_2-PDA respectively. Both films have excellent uniformity, while films with O_2-PDA have larger grains (RMS = 0.158) compared to that of N_2-PDA (RMS = 0.203). It is well known that the T-phase has smaller grains than the M-phase [24]. Thus, these results suggest that a certain amount of O/T phase can also be formed if the additive concentration of Vo and Si is sufficient, though the individual concentrations of them are not enough. Namely, V_O plays a similar role as a dopant.

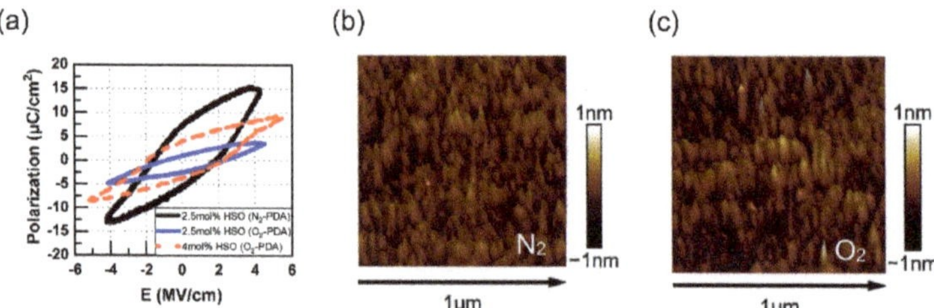

Figure 3. (**a**) P-E curves of 2.5 mol% HSO films with N_2- and O_2-PDA, and 4 mol% HSO films with O_2-PDA. AFM results of 2.5 mol% HSO films which were annealed in (**b**) N_2 and (**c**) O_2, respectively.

3.2. Effect of Specific Surface Area

Next, we investigate the independent effect of surface area. The effect of reducing the planar size of the film is very important to device performance when the technology node is within a nanometer. To get a clear view experimentally, we etched 30 nm ~4 mol% Si-doped and 10 nm undoped HfO_2 films into cuboids with different sizes, followed by O_2-PDA to study the influence of the specific surface area. Figure 4a shows the morphology diagram of the cuboid arrays and Figure 4b shows the AFM analysis of a cuboid with a length of 10 μm. The etching depth is about 30 nm, indicating that HfO_2 films are removed without residual. Figure 5a,b show GIXRD patterns of doped and undoped HfO_2 cuboid arrays with a length of 2 μm, 5 μm, and 10 μm, respectively. The signal of the undoped HfO_2 cuboid array with a length of 2 μm is too weak to analyze, so we did not show the result. Similarly, we extracted $R_{O/T/C}$ from the curves and plotted them in Figure 5c. It is shown that more O/T/C phases tend to be stabilized as the length of the cuboid decrease, namely the specific surface area increases, regardless of doping. In particular, the emergency of the O/T/C phase in the undoped HfO_2 array with a length below 10 μm suggests that the O-phase could be formable by only increasing the specific surface area largely. $R_{O/T/C}$ of 25% is reached in our results for undoped films with a length of 5 μm. Although it is still relatively small for getting enough polarization, we believe that it can be optimized by further decreasing the size of the film to the nanoscale. These results clearly show the enhanced effect of specific surface area on the ferroelectric phase formation. Moreover, decreasing film thickness is another common way to increase specific surface area and it has been intensively investigated in previous studies. It has been shown a similar result that thinner HfO_2 films tend to form T-phase while thicker films tend to form M-phase. In general, our results provide an approach to controlling the properties of HfO_2-based devices in size scaling. This is quite valuable for the demonstration of nanoscale devices.

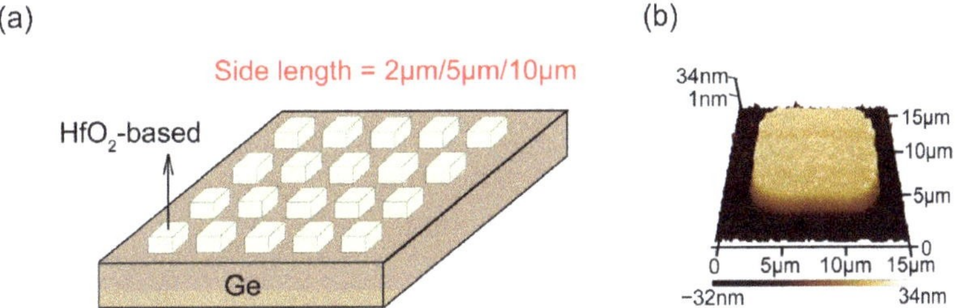

Figure 4. (**a**) Graphical presentation of cuboid array films whose side lengths were set to 2 μm, 5 μm, and 10 μm before PDA. (**b**) AFM diagram of one cuboid in a 10 μm array, showing that HfO_2-based films were all etched.

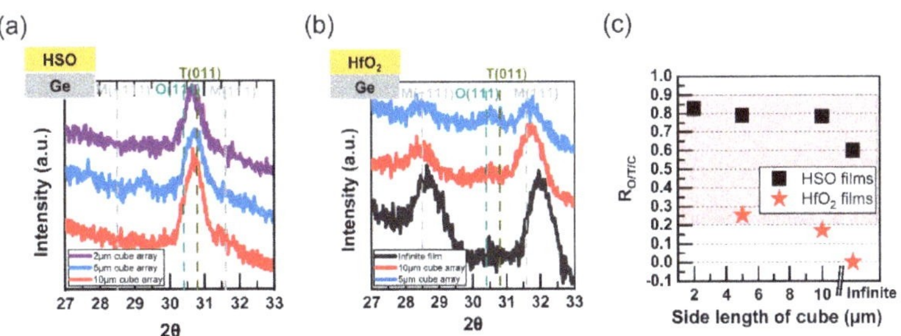

Figure 5. GIXRD results of (**a**) HSO (~4 mol% of Si) cuboid array films and (**b**) undoped HfO_2 cuboid array films. The peak] intensity of the O/T/C phase increases with the decrease of array size. (**c**) Extracted $R_{O/T/C}$ by fitting GIXRD results in (**a**,**b**). O-phase seems formable in undoped films only by decreasing the side length.

4. Discussion

Finally, we discuss the intrinsic parameters for controlling ferroelectric O-phase formation in HfO_2 films by combining the above results with those published previously. Apart from Si, which was discussed in our study, quadrivalent cation dopants, including Ge, Si, Zr, and various trivalent ones, including Y, Al, La, Gd, Sc, and Sr have been confirmed to stabilize the O-phase in HfO_2 films by sputtering or ALD [1,25,26]. However, quinquevalent cations, such as Nb, do not show such an effect [23]. To get a general view, the radii, and the concentration for the ferro-O phase stabilization of these dopants are summarized in Figure 6a,b. It is found that all the quadrivalent dopants with smaller radii than Hf (~71 pm) and lower doping concentrations are needed for smaller radii dopants. This indicates that the tensile strain caused by dopants with smaller radii than Hf is likely to make HfO_2 films ferroelectric. These quadrivalent dopants will cause lattice to shrinkage, and the smaller the dopant radius is (shorter bond length), the larger the strain is. Thus, it can be understood that a lower doping concentration is needed for dopants with smaller radii to get the same strain as dopants with larger radii. As for trivalent dopants, interestingly, all of them work with similar doping concentrations for O-phase stabilization but the radii vary from 53 pm (<Hf) to 115 pm (>Hf). It has been proposed that trivalent dopants introduce V_O into HfO_2 [25]. In addition, anion dopant N has been reported to have a similar effect [27]. Additionally, our results also confirm that even the pure V_O is able to stabilize the O-phase in HfO_2 thin film. Therefore, we consider that all the trivalent dopants, anion N dopant works in the same manner as V_O. It is well known that V_O also induces tensile strain into HfO_2 [28]. Thus, the intrinsic effect of all the dopants including V_O can be understood as inducing tensile strain. Meanwhile, we noted that the doping concentration to achieve ferroelectricity in HfO_2 by SPD was slightly lower than that by the ALD process even for the same dopants. This may be because the sputtering process induces more Vo additionally than ALD which affects the ferroelectricity with dopants commonly as mentioned above [20]. Therefore, it is worth noting that the tensile strain is related to dopants and V_O, and tensile stress is associated with the direction of the in-plane. It is different if the direction is changed.

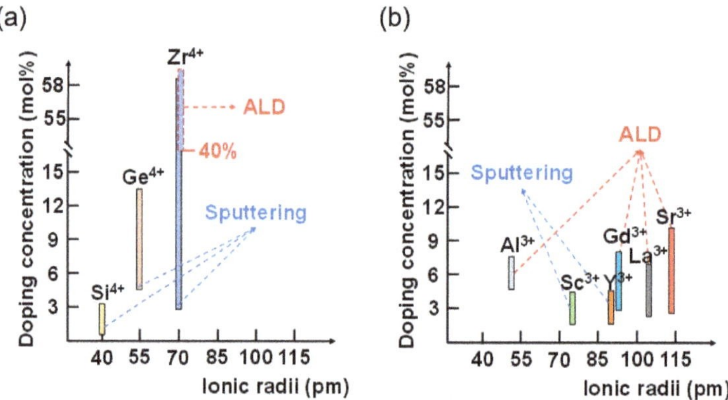

Figure 6. (a) Summarized doping concentration and ionic radii of (**a**) quadrivalent (Si, Ge, Zr) and (**b**) trivalent (Al, Sc, Y, Gd, La, Sr) dopants that make HfO_2 films ferroelectric [22,23].

Considering the effect of specific surface area, it is well known that the increase in surface area is a kind of increase in surface tension. Surface tension can be regarded as tensile stress from the surface and enables the distortion of crystalline lattice [29]. In addition, there is also other external stress acting on the surface to promote the formation of the O-phase, such as mechanical capping or selecting the substrate [2,8,9,30]. Moreover, an early paper found that ferroelectricity in HfO_2 can be observed on substrates inducing tensile stress in the film but not on those inducing compressive stress [31]. Therefore, the effect of specific surface area may also be originated from the effect of tensile strain from the surface.

To sum up, in controlling phase composition in HfO_2 thin films, all the effects of the dopants, V_O, and specific surface area, which are investigated in this work, as well as other factors reported in the literature can be generalized to be a strain effect. To understand this more deeply, we further consider it from a thermodynamic view. The change in Gibbs free energy of an elastic dielectric with fixed pressure is given by

$$dG = -SdT - \varepsilon_i d\sigma_i - D_i dE_i, \tag{2}$$

where S is entropy, T is temperature, ε is strain, σ is stress, D is electric displacement, E is the electric field, and i is the coordinates (i = 1, 2, 3). Therefore, in the process of HfO_2-based film fabrication, two parameters, temperature and strain, affect the phase formation intrinsically as E = 0. Since the O-phase is thought to be stabilized in the phase transition from T- to M-phase, we will only discuss the M-, O-, and T-phases in the following [32]. It is well known that T-phases are generally under high temperature while M is for the reverse case [2]. Based on Equation (2), it is understandable that strain has a similar effect to temperature, that is, the T-phase is more stable with larger strain and the M-phase is more stable with lower strain, which is consistent with our results. Thus, the phase formation and transition with different strains and given annealing temperature (<700 °C) are discussed as follows. For the film with relatively small strain, namely, the film with low doping concentration, no V_O, or small specific surface area, the M-phase is the most stable. Thus, the M-phase is nucleated and dominant after annealing which behaves as non-ferroelectric (first panel of Figure 7a). When sufficient strain is induced in the film by any method, the free energy of the T-phase is lowered relative to the M-phase, which makes it possible for the T-phase to be stabilized. Especially in the stage of initial nucleation, the tiny crystallite (radius ~2 nm) with larger surface energy promotes the nucleation of the T-phase [33]. As the grain grows and the thermal process continues in annealing, part of the strain may be released and the T-phase may become less stable than the M-phase. It has been studied that the activation energy of the T–O phase transition (~30 meV f.u.$^{-1}$) is lower than that of the T–M phase transition (~300 meV f.u.$^{-1}$) due to the similar crystal structure between T- and O-phases [11,34]. Thus, T–O–M phase transition may occur and a metastable O-phase is formed transitionally (second panel of Figure 7a). When the strain in the film is further increased, corresponding to high doping concentration in our experiments, the T-phase keeps being the most stable even with partial strain releasing [13,30]. In this case, there is only the T-phase in the films (third panel in Figure 7a). The schematics of phase concentration in HfO_2-based film with strain increases at a given temperature are shown in Figure 7b.

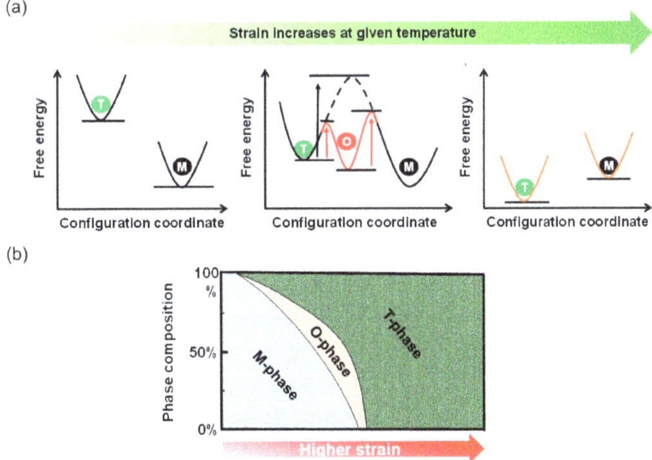

Figure 7. (**a**) Free energy diagram with different strains at a given temperature which determines phase composition in HfO_2 film. (**b**) The schematic diagram of the mechanism for general parameters influencing phase stability in HfO_2 films.

5. Conclusions

In conclusion, the separate effects of dopant, V_O, and specific surface area on O-phase formation in ferroelectric HfO_2 films have been demonstrated experimentally. It has been confirmed that in addition to sufficient dopants, only sufficient V_O or specific surface area enables us to stabilize the ferroelectric O-phase in HfO_2 films independently. By summarizing the results, a strain effect is considered to be the origin of these factors and this view is applicable for interpreting most of the results reported to date. The strain effect has also been understood from a thermodynamic view. T-phase nucleation with sufficient strain and its transition to the M-phase through a T–O–M pathway is a key to the formation of a ferroelectric O-phase. Thus, engineering the strain in HfO_2-based films is critical for controlling the ferroelectricity of HfO_2 thin films.

Author Contributions: Conceptualization, T.C. and X.L.; formal analysis, T.C. and X.L.; investigation, T.C., L.Z., D.C. and Y.F.; resources, J.L. and X.L. writing—original draft preparation, T.C.; writing—review and editing, X.L. All authors have read and agreed to the published version of the manuscript.

Funding: This research was funded by the National Natural Science Foundation of China (91964110, 61904103, 62111540163), Natural Science Foundation of Shanghai (19ZR1475300, 19JC1416700), and partly by the Interdisciplinary Program of Shanghai Jiao Tong University (project number ZH2018QNA09).

Institutional Review Board Statement: Not applicable.

Informed Consent Statement: Not applicable.

Data Availability Statement: Not applicable.

Acknowledgments: We would like to thank Emeritus Akira Toriumi, Koji Kita, and Nishimura from the University of Tokyo for discussions and thank Liying Wu from the Center of Advanced Electronic Materials and Devices (AEMD) of Shanghai Jiao Tong University for supporting plasma enhanced atomic layer deposition.

Conflicts of Interest: The authors declare no conflict of interest.

References

1. Bohr, M.; Chau, R.; Ghani, T.; Mistry, K. The High-K Solution. *IEEE Spectr.* **2007**, *44*, 29. [CrossRef]
2. Böscke, T.S.; Müller, J.; Bräuhaus, D.; Schröder, U.; Böttger, U. Ferroelectricity in hafnium oxide thin films. *Appl. Phys. Lett.* **2011**, *99*, 102903. [CrossRef]
3. Müller, J.; Polakowski, P.; Müller, S.; Mikolajick, T. Ferroelectric Hafnium Oxide Based Materials and Devices: Assessment of Current Status and Future Prospects. *ECS J. Solid State Sci. Technol.* **2015**, *4*, N30–N35. [CrossRef]
4. Park, M.H.; Lee, Y.H.; Mikolajick, T.; Schroeder, U.; Hwang, C.S. Review and perspective on ferroelectric HfO_2-based thin films for memory applications. *MRS Commun.* **2018**, *8*, 795. [CrossRef]
5. Park, M.H.; Lee, Y.H.; Kim, H.J.; Kim, Y.J.; Moon, T.; Kim, K.D.; Müller, J.; Kersch, A.; Schroeder, U.; Mikolajick, T.; et al. Ferroelectricity and Antiferroelectricity of Doped Thin HfO_2-Based Films. *Adv. Mater.* **2015**, *27*, 1811. [PubMed]
6. Sang, X.; Grimley, E.D.; Schenk, T.; Schroeder, U.; Lebeau, J.M. On the structural origins of ferroelectricity in HfO_2 thin films. *Appl. Phys. Lett.* **2015**, *106*, 162905. [CrossRef]
7. Park, M.H.; Lee, D.H.; Yang, K.; Park, J.-Y.; Yu, G.T.; Park, H.W.; Materano, M.; Mittmann, T.; Lomenzo, P.D.; Mikolajick, T.; et al. Review of defect chemistry in fluorite-structure ferroelectrics for future electronic devices. *J. Mater. Chem. C* **2020**, *8*, 10526. [CrossRef]
8. Shiraishi, T.; Katayama, K.; Yokouchi, T.; Shimizu, T.; Oikawa, T.; Sakata, O.; Uchida, H.; Imai, Y.; Kiguchi, T.; Konno, T.J. Impact of mechanical stress on ferroelectricity in $(Hf_{0.5}Zr_{0.5})O_2$ thin films. *Appl. Phys. Lett.* **2016**, *108*, 262904. [CrossRef]
9. Park, M.H.; Kim, H.J.; Kim, Y.J.; Moon, T.; Kim, K.; Hwang, C.S. Study on the degradation mechanism of the ferroelectric properties of thin $Hf_{0.5}Zr_{0.5}O_2$ films on TiN and Ir electrodes. *Appl. Phys. Lett.* **2014**, *105*, 072901.
10. Batra, R.; Tran, H.D.; Ramprasad, R. Stabilization of metastable phases in hafnia owing to surface energy effects. *Appl. Phys. Lett.* **2016**, *108*, 172902.
11. Park, M.H.; Lee, Y.H.; Kim, H.J.; Schenk, T.; Lee, W.; Kim, K.D.; Fengler, F.P.G.; Mikolajick, T.; Schroeder, U.; Hwang, C.S. Surface and grain boundary energy as the key enabler of ferroelectricity in nanoscale hafnia-zirconia: A comparison of model and experiment. *Nanoscale* **2017**, *9*, 9973. [CrossRef]

12. Kim, K.D.; Park, M.H.; Kim, H.J.; Kim, Y.J.; Moon, T.; Lee, Y.H.; Hyun, S.D.; Gwon, T.; Hwang, C.S. Ferroelectricity in undoped-HfO$_2$ thin films induced by deposition temperature control during atomic layer deposition. *J. Mater. Chem. C* **2016**, *4*, 6864. [CrossRef]
13. Materlik, R.; Künneth, C.; Kersch, A. The Origin of Ferroelectricity in Hf$_{1-x}$Zr$_x$O$_2$: A Computational Investigation and a Surface Energy Model. *J. Appl. Phys.* **2015**, *117*, 134109. [CrossRef]
14. Hamouda, W.; Pancotti, A.; Lubin, C.; Tortech, L.; Richter, C.; Mikolajick, T.; Schroeder, U.; Barrett, N. Physical chemistry of the TiN/Hf$_{0.5}$Zr$_{0.5}$O$_2$ interface. *J. Appl. Phys.* **2020**, *127*, 064105. [CrossRef]
15. Schroeder, U.; Materano, M.; Mittmann, T.; Lomenzo, P.D.; Mikolajick, T.; Toriumi, A. Recent progress for obtaining the ferroelectric phase in hafnium oxide based films: Impact of oxygen and zirconium. *Jpn. J. Appl. Phys.* **2019**, *58*, SL0801. [CrossRef]
16. Szyjka, T.; Baumgarten, L.; Mittmann, T.; Matveyev, Y.; Schlueter, C.; Mikolajick, T.; Schroeder, U.; Müller, M. Chemical Stability of IrO$_2$ Top Electrodes in Ferroelectric Hf$_{0.5}$Zr$_{0.5}$O$_2$-Based Metal–Insulator–Metal Structures: The Impact of Annealing Gas. *Phys. Stat. Solidi Rapid Res. Lett.* **2021**, *15*, 2100027. [CrossRef]
17. Mittmann, T.; Michailow, M.; Lomenzo, P.D.; Gärtner, J.; Falkowski, M.; Kersch, A.; Mikolajick, T.; Schroeder, U. Stabilizing the ferroelectric phase in HfO$_2$-based films sputtered from ceramic targets under ambient oxygen. *Nanoscale* **2021**, *13*, 912. [CrossRef] [PubMed]
18. Lomenzo, P.D.; Takmeel, Q.; Zhou, C.; Fancher, C.M.; Lambers, E.; Rudawski, N.G.; Jones, J.L.; Moghaddam, S.; Nishida, T. TaN interface properties and electric field cycling effects on ferroelectric Si-doped HfO$_2$ thin films. *J. Appl. Phys.* **2015**, *117*, 134105. [CrossRef]
19. Nishimura, T.; Xu, L.; Shibayama, S.; Yajima, T.; Migita, S.; Toriumi, A. Ferroelectricity of nondoped thin HfO$_2$ films in TiN/HfO$_2$/TiN stacks. *Jpn. J. Appl. Phys.* **2016**, *55*, 08PB01. [CrossRef]
20. Mittmann, T.; Materano, M.; Lomenzo, P.D.; Park, M.H.; Stolichnov, I.; Cavalieri, M.; Zhou, C.; Chung, C.C.; Jones, J.L.; Szyjka, T.; et al. Origin of Ferroelectric Phase in Undoped HfO$_2$ Films Deposited by Sputtering. *Adv. Mater. Interfaces* **2019**, *6*, 1900042. [CrossRef]
21. Tian, X.; Shibayama, S.; Nishimura, T.; Yajima, T.; Migita, S.; Toriumi, A. Evolution of ferroelectric HfO$_2$ in ultrathin region down to 3 nm. *Appl. Phys. Lett.* **2018**, *112*, 102902. [CrossRef]
22. Mueller, S.; Mueller, J.; Singh, A.; Riedel, S.; Sundqvist, J.; Schroeder, U.; Mikolajick, T. Incipient Ferroelectricity in Al-Doped HfO$_2$ Thin Films. *Adv. Funct. Mater.* **2012**, *22*, 2412. [CrossRef]
23. Xu, L.; Shibayama, S.; Izukashi, K.; Nishimura, T.; Yajima, T.; Migita, S.; Toriumi, A. General relationship for cation and anion doping effects on ferroelectric HfO$_2$ formation. In Proceedings of the 2016 IEEE International Electron Devices Meeting (IEDM), San Francisco, CA, USA, 3–7 December 2016; p. 25.
24. Garvie, R.C. The occurrence of metastable tetragonal zirconia as a crystallite size effect. *J. Phys. Chem.* **1965**, *69*, 1238–1243. [CrossRef]
25. Xu, L.; Nishimura, T.; Shibayama, S.; Yajima, T.; Migita, S.; Toriumi, A. Kinetic pathway of the ferroelectric phase formation in doped HfO$_2$ films. *J. Appl. Phys.* **2017**, *122*, 124104. [CrossRef]
26. Schroeder, U.; Yurchuk, E.; Muller, J.; Martin, D.; Schenk, T.; Polakowski, P.; Adelmann, C.; Popovici, M.I.; Kalinin, S.V.; Mikolajick, T. Impact of different dopants on the switching properties of ferroelectric hafnium oxide. *Jpn. J. Appl. Phys.* **2014**, *53*, 08LE02. [CrossRef]
27. McBride, J.R.; Hass, K.C.; Poindexter, B.D.; Weber, W.H. Raman and x-ray studies of Ce$_{1-x}$Re$_x$O$_{2-y}$, where Re = La, Pr, Nd, Eu, Gd, Tb. *J. Appl. Phys.* **1994**, *76*, 2435. [CrossRef]
28. Jena, S.; Tokas, R.B.; Misal, J.S.; Rao, K.D.; Udupa, D.V.; Thakur, S.; Sahoo, N.K. Effect of O$_2$/Ar gas flow ratio on the optical properties and mechanical stress of sputtered HfO$_2$ thin films. *Thin Solid Film.* **2015**, *592*, 135. [CrossRef]
29. Lukyanchuk, I.A.; Schilling, A.; Gregg, J.M.; Catalan, G.; Scott, J.F. Origin of ferroelastic domains in free-standing single-crystal ferroelectric films. *Phys. Rev. B* **2009**, *79*, 144111. [CrossRef]
30. Park, M.H.; Kim, H.J.; Kim, Y.J.; Moon, T.; Hwang, C.S. The effects of crystallographic orientation and strain of thin Hf$_{0.5}$Zr$_{0.5}$O$_2$ film on its ferroelectricity. *Appl. Phys. Lett.* **2014**, *104*, 072901. [CrossRef]
31. Katayama, K.; Shimizu, T.; Sakata, O.; Shiraishi, T.; Nakamura, S.; Kiguchi, T.; Akama, A.; Konno, T.J.; Uchida, H.; Funakubo, H. Orientation control and domain structure analysis of {100}-oriented epitaxial ferroelectric orthorhombic HfO$_2$-based thin films. *J. Appl. Phys.* **2016**, *119*, 134101. [CrossRef]
32. Kisi, E.H. Influence of Hydrostatic Pressure on the $t{\to}o$ Transformation in Mg-PSZ Studied by In Situ Neutron Diffraction. *J. Am. Ceram. Soc.* **1998**, *81*, 741–745. [CrossRef]
33. Park, M.H.; Lee, Y.H.; Kim, H.J.; Kim, Y.J.; Moon, T.; Kim, K.D.; Hyun, S.D.; Mikolajick, T.; Schroeder, U.; Hwang, C.S. Understanding the formation of the metastable ferroelectric phase in hafnia–zirconia solid solution thin films. *Nanoscale* **2018**, *10*, 716. [CrossRef] [PubMed]
34. Park, M.H.; Lee, Y.H.; Mikolajick, T.; Schroeder, U.; Hwang, C.S. Thermodynamic and Kinetic Origins of Ferroelectricity in Fluorite Structure Oxides. *Adv. Electron. Mater.* **2019**, *5*, 1800522. [CrossRef]

 *electronics*

Review

Recent Research for HZO-Based Ferroelectric Memory towards In-Memory Computing Applications

Jaewook Yoo [1], Hyeonjun Song [1], Hongseung Lee [1], Seongbin Lim [1], Soyeon Kim [1], Keun Heo [2,*,†] and Hagyoul Bae [1,*,†]

1   Department of Electronic Engineering, Jeonbuk National University, Jeonju 54896, Republic of Korea; dbwodnr9797@jbnu.ac.kr (J.Y.); shg0768@jbnu.ac.kr (H.S.); lhs0409@jbnu.ac.kr (H.L.); sincelove@jbnu.ac.kr (S.L.); sy21620@jbnu.ac.kr (S.K.)
2   Semiconductor Science and Technology, Jeonbuk National University, Jeonju 54896, Republic of Korea
*   Correspondence: kheo@jbnu.ac.kr (K.H.); hagyoul.bae@jbnu.ac.kr (H.B.)
†   These authors contributed equally to this work.

Abstract: The AI and IoT era requires software and hardware capable of efficiently processing massive amounts data quickly and at a low cost. However, there are bottlenecks in existing Von Neumann structures, including the difference in the operating speed of current-generation DRAM and Flash memory systems, the large voltage required to erase the charge of nonvolatile memory cells, and the limitations of scaled-down systems. Ferroelectric materials are one exciting means of breaking away from this structure, as Hf-based ferroelectric materials have a low operating voltage, excellent data retention qualities, and show fast switching speed, and can be used as non-volatile memory (NVM) if polarization characteristics are utilized. Moreover, adjusting their conductance enables diverse computing architectures, such as neuromorphic computing with analog characteristics or 'logic-in-memory' computing with digital characteristics, through high integration. Several types of ferroelectric memories, including two-terminal-based FTJs, three-terminal-based FeFETs using electric field effect, and FeRAMs using ferroelectric materials as capacitors, are currently being studied. In this review paper, we include these devices, as well as a Fe-diode with high on/off ratio properties, which has a similar structure to the FTJs but operate with the Schottky barrier modulation. After reviewing the operating principles and features of each structure, we conclude with a summary of recent applications that have incorporated them.

Keywords: HZO; ferroelectric memory device; NVM; in-memory computing; neuromorphic; FTJ; Fe-diode; FeFET; FeRAM

Citation: Yoo, J.; Song, H.; Lee, H.; Lim, S.; Kim, S.; Heo, K.; Bae, H. Recent Research for HZO-Based Ferroelectric Memory towards In-Memory Computing Applications. *Electronics* **2023**, *12*, 2297. https://doi.org/10.3390/electronics12102297

Academic Editors: Yi Zhao and ChoongHyun Lee

Received: 3 April 2023
Revised: 30 April 2023
Accepted: 12 May 2023
Published: 19 May 2023

Copyright: © 2023 by the authors. Licensee MDPI, Basel, Switzerland. This article is an open access article distributed under the terms and conditions of the Creative Commons Attribution (CC BY) license (https://creativecommons.org/licenses/by/4.0/).

1. Introduction

Software is being developed to develop learning algorithms for machine learning, deep learning, and artificial intelligence (AI). Hardware must also be developed that matches the performance of this software [1–3]. Structurally separated processing and memory devices have high energy costs and poor data bandwidth performance (Von Neumann bottlenecks) [4,5]. The human brain shows high accuracy and as it calculates and remembers in the analog domain [6,7]. Therefore, neuromorphic systems that imitate the human brain can be expected to overcome the Von Neumann bottleneck. Efforts are also underway to overcome the so-called "memory wall", i.e., the speed gap between logic and memory in the current CMOS stage, which occurs when hundreds of processes are performed in parallel by a graphic processing unit or custom-designed processor [8–12]. Resistive random-access memory (RRAM), phase change material (PCM), and magnetic random-access memory (MRAM) have been proposed as new non-volatile forms of memory that reduce the physical distance between computing components and data for memory and processing to overcome the fundamental limitations of existing CMOS systems [13–16]. However, these candidates are too slow, possess only a limited data bandwidth, and offer no price advantage [14]. As an alternative, in-memory computing has been proposed as a means of breaking away from the Von Neumann structure itself. In-memory computing improves latency and energy by performing calculations without physical separation from the storage where the data are located. Ferroelectric-based non-volatile memory with independent switching mechanisms and high-power efficiency has recently been recognized for its potential in neuromorphic and in-memory computing [17–22].

Ferroelectric materials are those with spontaneous two-way remanent polarization even in the absence of an electric field applied from the outside. Perovskite, for example, is a well-known ferroelectric material [23–26]. Perovskite may possess polarization properties depending on the location of Ti or Zr cations. In general, a phase transition occurs to tetragonal systems with ferroelectricity at low temperatures. When an external electric field higher than the coercive field (E_c) is applied, the position of the positive ion in the material moves to the opposite position compared to the conventional one, and polarization is reversed [26–28]. Accordingly, even when an electric field is not applied, binary states of 0 and 1 can be obtained by using reversing polarization. For this reason, these materials have begun to attract attention as an NVM [29]. However, polarization is reduced by an irregular interface layer formed between the silicon and ferroelectric material, reducing ferroelectricity, and for that reason is not compatible with the CMOS process. Moreover, since the size of the metal atom is relatively large and has a three-dimensional structure, ferroelectric properties do not appear in thin films of 50~70 nm or less [30].

In 2011, it was reported that SiO_2 doping performed on an HfO_2 thin film resulted in the expression of ferroelectricity. Since then, doping attempts with Si, Zr, Al, Y, Gd, Sr, La, etc., have been made [31–41]. Among these, Zr has the advantage of having a lower heat treatment temperature than other dopants and a similar atomic radius and lattice parameter to Hf, so HZO ($HF_{1-x}Zr_xO_2$) doped with Zr in Hf is receiving a lot of attention. HZO shows ferroelectric characteristics as its crystal structure between an electrode and a ferroelectric thin film is changed by stress from a conventional monoclinic phase to an orthorhombic phase at a temperature of 400~600 °C. HZO shows the highest ferroelectricity when the composition ratio of Hf and Zr is 1:1, and since HZO is usually deposited through atomic layer deposition (ALD), the composition ratio can be controlled and thin films can be produced easily [42,43]. In addition, HZO has ferroelectric characteristics even in thin films, so it is easy to integrate into devices and is compatible with existing CMOS processes [44]. However, studies noted its insufficient endurance, sustainability, and reliability, as well as the difficulty associated with manufacturing the back-end-of-line (BEOL) process at high temperatures. Ferroelectric memory can be classified into four types: ferroelectric tunnel junction (FTJ), Fe-diode with diode operation, FeFET with field effects, and ferroelectric random-access memory (FeRAM) [23,24,44]. This review paper introduces the operation of these four types of memory devices and some of their applications.

2. Ferroelectric Memory Operation Category

2.1. Ferroelectric Tunneling Junction (FTJ)

A FTJ is based on tunneling current conversion between the upper and lower metal electrodes and ferroelectrics. FTJs were first proposed by Esaki et al. in 1971 and only began to attract attention as it became possible to manufacture a very thin high-quality ferroelectric thin film [27,45]. FTJs have a metal-ferroelectric-metal (MFM) structure in which electrodes are formed on both sides of a nanometer thick of ferroelectric thin film. The potential barrier of the energy band between the electrode and the ferroelectrics varies depending on the polarization direction of the ferroelectric, and the amount of tunneling current changes with the change in resistance (a phenomenon called tunneling electrical resistance (TER)) [30,46]. TER occurs through three mechanisms: "direct tunneling" (when electrons pass directly through the tunneling barrier), "Fowler-Nordheim tunneling" (when electron tunneling occurs in a relatively thick tunneling barrier), and "thermionic emission" (when carriers receive thermal energy and exceed potential barrier) [47].

The bistable resistance states for the two opposite polarities can be obtained, which means that they are eventually polarized to the low resistance (LRS) and high resistance states (HRS) as shown in energy band diagrams of Figure 1a,b. FTJs can be driven at low power because the program and erase occurs in nanoseconds and do not lose data thanks to a non-destructive read process. In principle, the on/off states of an FTJ follow linear or quasi-linear I-V relationships, so an additional selector device is required to reduce the skip current in the crossbar arrangement.

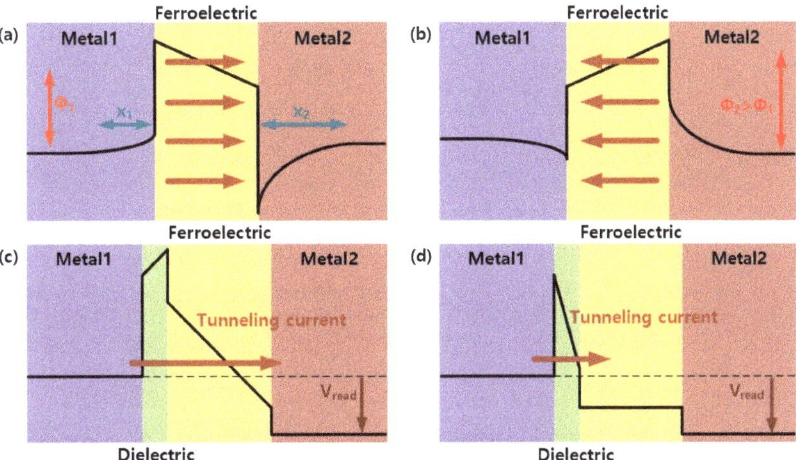

Figure 1. Conduction band schematic by polarization during FTJ operation of an MFM and MIFM structure: (**a**) Asymmetric screening length x from other effective tunneling barrier height Φ: LRS (On); (**b**) HRS (Off); (**c**) direct tunneling through tunneling barriers; (**d**) Fowler-Nordheim tunneling that passes through only part of the tunneling barrier [47].

The FTJ using tunneling current causes problems from thin films compared to hafnium-based materials and other ferroelectric devices. When polarization is reversed due to a high E_c value (~1 MV/cm), a hafnium-based ferroelectric has a small voltage margin, causing failures and poor cycling endurance. In addition, a high tunneling current may occur that traps electrons during operation. Trapping affects the barrier via remanent polarization and lowers the LRS and HRS values. Moreover, when a very thin polycrystalline layer is used, the memory window (MW) is dramatically reduced as parasitic currents are generated due to unwanted conduction from the interface. To address this problem, a structural metal-insulator-ferroelectric-metal (MIFM) with additional tunneling barriers inserted into the dielectric membranes has been explored as an alternative [46,48,49]. Figure 1c,d show that direct tunneling and Fowler-Nordheim tunneling contribute to tunneling current, respectively. Although the current density is slightly reduced from the stack structure, it can be useful for neuromorphic computing using large-scale parrel arrays.

Hwang et al. examined an FTJ with an MIFM structure by inserting a TiO_2 layer to increase the ferroelectricity of the HZO films. The insertion layer increased the TER ratio of the device (which had an asymmetric structure). Figure 2a shows that the TER was maintained up to 10^7 times when a 10 µs bipolar pulse was applied. In addition, the TER ratio of the device was measured for 28 h (Figure 2b), and their research confirmed that the ratio was maintained for more than 10 years through the fitting [49].

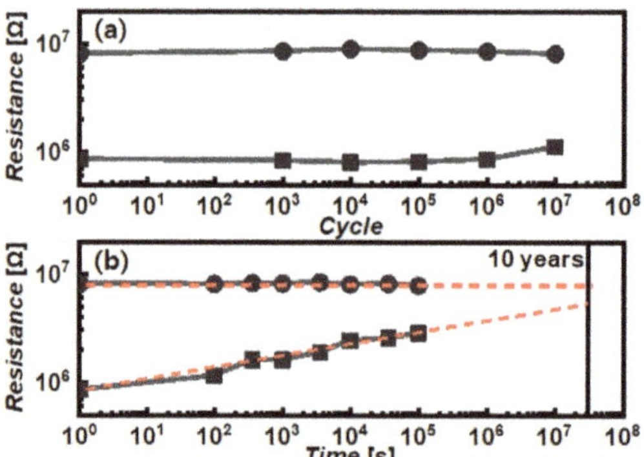

Figure 2. (a) Endurance characteristics and (b) retention characteristics of FTJ devices with a TiO_2 layer inserted in the HZO film with permission from Ref. [49], 2021, IEEE.

2.2. Ferroelectric Diode (Fe-Diode)

Fe-diodes use polarization inversion caused by Schottky barrier modulation at an interface between a metal electrode and a ferroelectric thin film. The molecular structure and energy band diagram of the TiN/HZO/TiN Fe-diode is exhibited in Figure 3a,b, respectively. In a conventional metal-semiconductor contact, a spatial charge region in the semiconductor is formed to align the Fermi level, and a charge exacts same and opposite exists on the interface of the metal. The interfacial charge of a single ferroelectric Schottky diode and the band diagram is therefore subject to continuous and sustainable variability by the application of an external electric field [50,51]. The positive and negative polarization charges result in the accumulation of electrons that, respectively reduce the barrier height and increase the band bending by the positively charged oxygen vacancy. The current-voltage characteristic has a bistable operation due to the polarization dependence of the Schottky barrier. In the ferroelectric, as the rigidity of the lattice increases, the dielectric constant decreases in inverse proportion to the increasing electric field [50]. After polarization inversion, the changed electric field of the ferroelectric, and the depletion width of the Schottky diode decrease linearly until the depletion region end is zero at the maximum value of the interface. Finally, the transmission characteristic of the ferroelectric Schottky diode is obtained from a bias exceeding the threshold voltage.

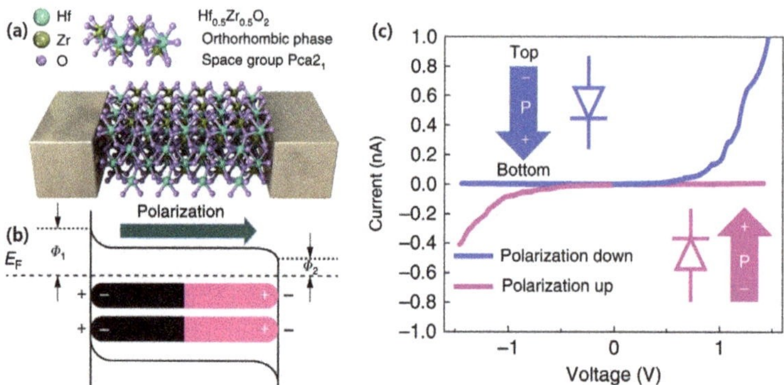

Figure 3. (a) Molecular structure of $Hf_{0.5}Zr_{0.5}O_2$ and structural mimetic diagram of a Fe-diode; (b) Energy band diagram showing Schottky to ohmic internal contact in a TiN/HZO/TiN structure modulated from polarization; (c) Nonlinear I-V curve for two polarizations of ferroelectric with permission from Ref. [52], 2020, Springer Nature.

Figure 3c shows the I-V characteristics of an Fe-diode [52]. The blue and red lines refer to positive voltage sweep and negative voltage sweep. In the former, the current becomes high and changes to a positive forward diode, while in the latter case, the current remains high and changes to a negative forward diode.

Fe-diodes and FTJs are both two-terminal devices, but the Fe-diode differs in its unique nonlinearity, which is defined as the ratio of the read current in V_r and $V_r/2$ attributable to the Schottky barrier. This means that the leakage current may be effectively reduced at the intersection of the array. Therefore, each memory cell and an external selector device need not be additionally connected in series. Moreover, if the two-terminal device repeatedly switches ferroelectric polarization, an error occurs during the read operation. The generated error can be solved by restoring the initial polarization state using the existing read pulse and inversion pulse as inputs.

2.3. Ferroelectric Field Effect Transistor (FeFET)

A FeFET incorporates a ferroelectric material into a gate oxide and is a non-destructive method that induces a change in current in a channel according to the polarization switching of the ferroelectric material. For this reason, the electric field that reaches the channel can be reduced, and the switching speed of the polarization is brisk [44,47]. Fast read/write operations are possible that use less power than static random-access memory (SRAM) and dynamic random-access memory (DRAM). Even if the gate voltage applied from the outside is stopped, the remanent polarization of the ferroelectric remains and can be applied to nonvolatile memory, and the character appears as hysteresis in the transfer curve [53]. In short, FeFET employs the same driving method as an existing FET in which ferroelectricity in the insulator site of the existing FET structure is used to generate an inverse layer of the semiconductor channel. It differs, however, in that the polarization of the ferroelectric induced by the applied gate voltage changes the threshold voltage [54]. This difference in threshold voltage is defined as MW, which can be obtained by multiplying the Ec by the thickness of the ferroelectric material and then doubling the result.

A Hf-based ferroelectric material has a larger E_c value than other materials, so it is easy to scale down and implement a high MW, though the issue of write endurance and read-after-write delay due to retention and parasitic charge trapping remains. This can be addressed by using a MIFM structure that inserts an insulating layer or by adding a layer of oxide with good conductivity to the FeFET channel. Mo et al. used an ultrathin IGZO channel in FeFET, and when the P/E voltage is higher than 2V, V_{th} is effectively shifted to record high endurance and retention [55,56]. The former characteristics can be determined through drain current measurement over time in the absence of applied voltage, and the latter characteristics can be extracted in the number of cycles by repeatedly applying positive/negative voltage pulses. When the program and erase were operated 10^8 times, 2Pr = 15 $\mu C/cm^2$ was recorded without wake-up and significant degradation. In addition, it has been confirmed that both the program and the erase state are maintained for at least 1 year. Hoffmann et al. achieved high write endurance by inserting SiN_x layers to improve the parasitic charge trapping of FeFETs [57]. It has been confirmed that the retention degradation caused by the reverse conversion of ferroelectric domains in a large demarcation theater is capable of overcoming retention instability by controlling the gate work function. Gaddam et al. reported the use of HfO_2 as a seed layer above and below HZO in a MFM structure composed of TiN-HZO-TiN (Figure 4a) [58]. Figure 4b shows that the thickness of the insert layer is inversely proportional to its ferroelectric characteristics, and a 1 nm seed layer in the HZO bottom layer of HZO yields a maximum (P_r) of 22.1 $\mu C/cm^2$, while a 1 nm seed layer inserted into the upper layer of HZO yields a maximum (P_r) of 19.6 $\mu C/cm^2$. Endurance and retention characteristics were measured and compared, which can be confirmed in Figure 4c,d. In both cases, the upper and lower layer show improved properties after the insert layer is added.

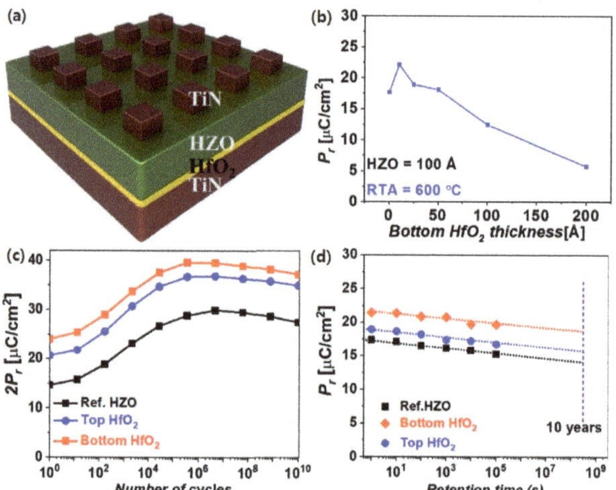

Figure 4. (**a**) Schematic of HfO_2 using bottom insert layer in an HZO MFM structure; (**b**) remanent polarization of the thickness of the HfO_2 bottom insert layer; (**c**) endurance and (**d**) retention characteristics with no HfO_2 insert layer or location with permission from Ref. [58], 2020, IEEE.

2.4. Ferroelectric RAM (FeRAM)

FeRAM is a memory with a 1T-1C structure that uses a ferroelectric as a capacitor-based material [59]. FeRAM is similar to DRAM in which the capacity layer has been replaced by a nonlinear dielectric, and the plate line of FeRAM, unlike DRAM, does not maintain a constant voltage because it requires pulses to switch the ferroelectric polarization direction [47]. The binary information is stored in the ferroelectric capacitor and the transistor allows random access for read/write operations. Because ferroelectricity is used, it has NVM characteristics, but the stored values have destructive characteristics because the polarization is detected by the read pulse as to whether it is inverted, and the cell needs to be recovered.

Unfortunately, FeRAM has the disadvantage of having a larger cell size than other memories because it requires a plate line. Accordingly, FeRAM footprints range from 15 to 20 F^2, making them larger than other ferroelectric memory devices, including FeFETs (4–8 F^2), FTJs (4 F^2), and Fe-diodes (4 F^2). Therefore, these are unlikely candidates for next-generation ferroelectric devices because integration is more difficult even if a scale-down is performed [44].

3. Ferroelectric Device Application

Recently, researchers have focused on implementing artificial neurons and synapses with memory devices using ferroelectric materials and using these in neuromorphic computing systems. [60–65]. The basic behavior of computing is based on the matrix product of the voltage generated in artificial spiking neurons and the variable resistance called synapses [66]. Conventional CMOS can also construct neurons and synapses, but it requires a great deal of unit devices, resulting in low energy efficiency and difficulty in device integration [67]. Hf-based ferroelectric materials have a bistable polarization state and thus can be used as a non-volatile memory that stores binary information via the regulation of electric field strength [68]. Hf-based ferroelectric devices are stochastically capable of unexpected switching and multi-value polarization control and have a larger band gap (5.0 to 5.5 eV) than perovskite-structured ferroelectric devices (3 to 4 eV). Their high stability against leakage current renders them suitable for high-speed and low-power storage applications even if the scale of the device is reduced in charge-based ferroelectric memory devices (three-terminal devices). Furthermore, two-terminal ferroelectric memory devices are available for high-performance, low-power, and large-scale parallel memory computing applications due to their excellent analog switching characteristics, high nonlinearity, and uniformity [69–73]. However, the two-terminal device has current density, sneak current, and retention characteristics, and the three-terminal device has endurance, and the number of states in the scaled-down device causes problems. This section highlights the benefits of ferroelectric devices as memory devices from an application perspective.

3.1. Two-Terminal Devices

3.1.1. FTJ

Neuromorphic applications perform many analog tasks such as addition, multiplication, etc., and require linear and symmetric conductance tuning, low current write-read operations, fast transition speed, and high endurance. [74,75]. In the MFM structure of a FTJ or MFIM structure with an added insertion layer, related studies are noted because the current density is low, and a crossbar arrangement structure is possible [76]. In general, when the crossbar arrangement structure is expanded through the linear I-V characteristics of FTJ, there is the problem of high current flow from high conductivity. However, research has been conducted to provide ultra-low conductivity and ultra-low current by developing nonlinear I-V characteristics [77]. In short, FTJ brings great advantages in terms of large-scale parallel computing and integration [78].

Reservoir computing (RC) refers to energy-consuming networks that compute time data. They have attracted attention for their low cost of learning-only weights among reading layers [79]. Currently, data processing power must be improved, and research is ongoing to develop a dynamic RC hardware system by combining it with an RRAM-based readout layer using nonlinear features based on FTJ devices. The RRAM used in one experiment showed stable Set/Reset operation from the I-V curve by measuring 1T-1R 100 times. When the dielectric effect is removed from the polarization current over time in the FTJ; four peak values from the beginning of the polarization current are used as RC computing nodes. This experiment confirmed the trend of the recognition accuracy of the RC hardware system simulation value and the measured value matched for digital signals.

Chen et al. have reported electron synapses for neuromorphic systems that use three-dimensional vertical HZO-based FTJs, which can be confirmed in Figure 5a [80]. Throughout the training, the FTJ synapse showed analog-like conductivity and low energy consumption characteristics when synapse weights were updated, as well as high integration performance. As shown in Figure 5b,c, implemented in hardware, the synapse was trained by applying boost/decrease pulses through a 1/2 bias to change conductivity. When the original state is maintained, caution is needed because both the word line and the cell related to the bit line can be selected if an appropriate bias is not applied. Subsequently, as shown in Figure 5d, recognition accuracy of 96% was confirmed through 50 test character patterns through a hardware neural network (HNN) application.

Ota et al. used an FTJ that enhanced the performance and reliability of in-memory reinforcement learning (RL) through structural and material engineering, which can be confirmed in Figure 6a [81]. Compared to conventional memristors, which are difficult to program for accurate conductive states, RL can achieve human-level performance on a variety of control or decision-making tasks [82]. As shown in Figure 6b, the simulation of the standard problem showed that the training speed could be optimized at the standard deviation value (σ) and at the same normalized conductance change by adjusting the pulse voltage. Even at an optimal pulse voltage, if the σ increased, the minimum convergence time also increased, which means that the smaller σ better maximizes RL performance in Figure 6c. It was confirmed that the endurance is improved as the voltage margin (V_m) increases. As shown in Figure 6d,e, to increase V_m, the thickness of HZO and the concentration of Zr should be reduced. In this case, it is advantageous to reduce the thickness of the HZO in order to secure the variability of the RL system. V_m fluctuation with respect to the scale of the device can be solved by voltage compensation.

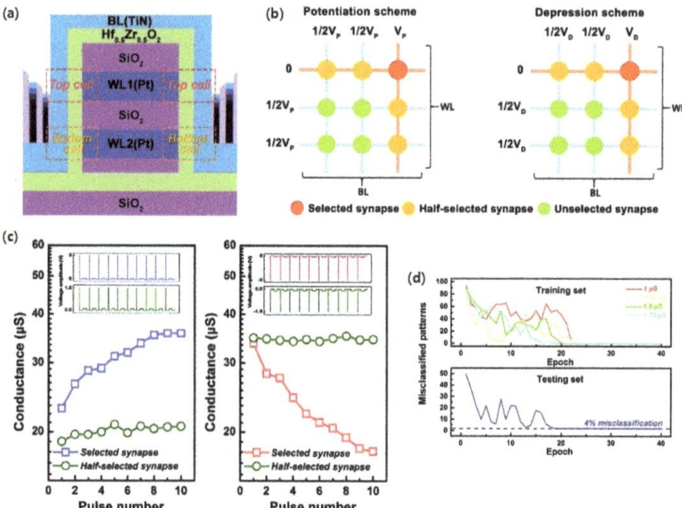

Figure 5. (**a**). Section cut of 3D vertical ferroelectric HZO−based FTJ array structure. (**b**) Diagram of 1/2 bias training of FTJ array structure. (**c**) The curve of the change in conductance according to the pulse number, the difference in conductance between the selected-synapse and the half-selected synapse effectively changed. (**d**) The test set trained 50 letter patterns at 20 epochs, and these characters recorded high accuracy of more than 96%. With permission from Ref. [80], 2018, RSC.

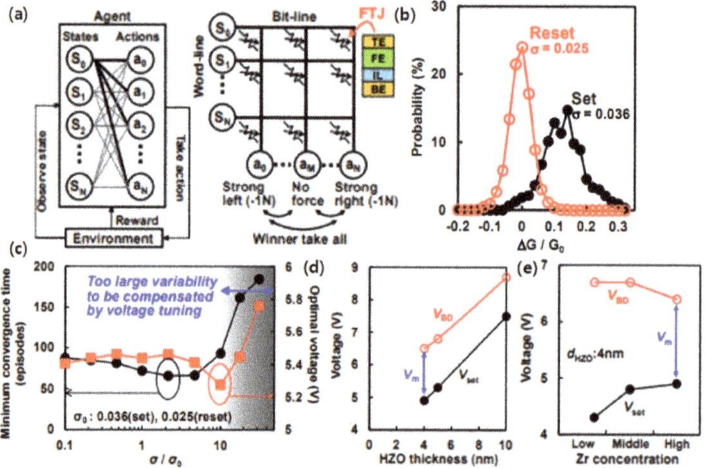

Figure 6. (**a**) The RL system in a parallel structure operates by read operation the maximum current from Word-line to Bit−line. (**b**) Changes in conductivity of FTJ by positive and negative pulses. (**c**) Minimum convergence time and optimal voltage by different conductance. (**d**) V_m in inverse proportion to HZO thickness. (**e**) V_m in inverse proportion to Zr concentration with permission from Ref. [81], 2019, IEEE.

In the crossbar array, the parasitic component between the sneak path current and the mutual is fatal to the FTJ's operation pursuant to Kirchhoff's current law, and as a result, the sum of the currents is difficult to determine by a subsequent amplifier [67,78,83]. Goh et al. implemented a crossbar array without a selector using TaN as the insertion layer to prevent sneak current and confirmed that this was possible up to 4 kbit [84]. With the addition of the insertion layer, stable intermediate states of 30 or more, linear potentiation and depression are possible, as is the possibility of synapse implementation with long-term reinforcement and suppression, spiking-timing-dependent plasticity (STDP), and low energy consumption. As a result, we looked at instances of neuromorphic computing and in-memory that used an HZO−based FTJ compatible with CMOS as a synaptic device. In addition, we identified the difficulties in implementing these in large crossbar array environments and briefly summarized the research that sought solutions to these problems.

Kim et al. implemented a physical unreachable function (PUF) by simulating a 4×4 FTJ array [85]. PUF uses unique variability as an electrical parameter, compared to storing sequential passwords in existing memory devices. It is attracting attention because it has advantages in terms of price and is strong against side-channeling attacks. To implement this PUF, RRAM was considered a promising structure, and thus research was initiated to use it as an encryption device using FTJ. This study was simulated with an FTJ array of three sizes. It has been confirmed that reliability due to deviation between devices has more influence than reliability due to the scale-down of devices. That being said, even if the FTJ is scaled down, a stable operation may be performed in the intersection array. Lim et al. proposed SR-FTJ (self-rectifying-FTJ) through simulation that allows content-addressable memory (CAM) and PUF to operate in two layers [86]. CAM is specialized in in-memory computing, specializing in high-speed search of cache controllers or routers. However, since it has a security problem, research has been conducted in giving it high area efficiency by vertically combining it with the PUF introduced above. A certain difference in the leakage current of SR-FTJ was used as CAM data, and the leakage current was used as a value of PUF. Although its performance is still low, such memory can reduce its area by 88.1 to 97.4% compared to the previous one.

3.1.2. Fe-Diode

A design of a Hf-based Fe-diode was published in 2020 by Luo et al. [52]. The current density and nonlinearity were very high compared to the widely examined perovskite. Relative to FTJ, a higher on/off ratio could be obtained using Schottky contact. The following year, a study conducted by Bae described a diode formed by stacking an HZO layer on an oxide IGZO layer [87]. When IGZO and a ferroelectric are combined, they have high mobility and ideal SS, and high retention and endurance characteristics [55,56]. Therefore, when IGZO is efficiently positive bias and the polarization of ferroelectricity is up-polar, and negative biases and polarization of the ferroelectricity is down-polar, the charges have a depletion state and an accumulation state, respectively. The single device had a high on/off ratio of 3×10^5 and can be used as a high-speed, low-power non-volatile memory operating at 800 ps and 0.8 fJ. To implement logic-in-memory, the Boolean logic 'NOT' was operated using two diodes, with a significant difference in input/output current of as much as 10^4. In addition, a Fe-diode that combines HZO and IGZO called TCAM (Ternary-Content-Addressable-Memory) was implemented using non-volatile and ambipolar characteristics. This was verified as a potential device for in-memory data processing and neuromorphic computing [88].

To implement logic-in-memory with an Fe-diode, researchers have sought to improve the retention performance through the insertion of a HfO_x layer into the HZO layer [89]. Since the retention defect in the ferroelectric was caused by the leakage current and the resulting trapping effect, retention performance was improved by adding the HfO_x insertion layer [90]. A logic 'XOR' operation was performed on these devices by three sequential cycles (write 1, write 2, and read). The logical variables p and q were used for 16 Boolean functions. For one fluorine function, for example, 0 and 1 were excluded when 0 and 1 are 0 V and +7.5 V in the write cycle, and 0 and 1 operate when each was −4 V and 4 V in the read cycle. The operation was confirmed with the pulse results, and the experiment confirmed that this can be used as a logic-in-memory device in an MFIM structure in which an insertion layer is added to the ferroelectric layer.

3.2. Three-Terminal Devices

3.2.1. FeFET

A FeFET has a similar driving method as MOSFET, therefore efforts have been made to improve memory performance using a structural approach. An HZO-based FinFET device was investigated by Bae et al. [91]. Space and energy distributions were extracted from the trap density of the ferroelectric layer and interface layer, which were stacked on the gate, through low-frequency noise measurement (a non-destructive analysis). In addition, degradation due to radiation damage was analyzed based on the interface trap and bulk trap. A 3D FeFET was developed by Florent et al. that connected three transistors in series in the vertical direction, confirming the potential for achieving CMOS process compatibility and high integration as an NVM. To significantly improve on the low endurance characteristics in the three-terminal device created by Liao et al., a 3D GAA nano-sheet was structurally introduced, and the corner dead zone of ferroelectric polarization was suppressed [92]. In Figure 7a, HZO is used as a double layer to enable multiple E_c, which can store 3 bits of data. Figure 7b shows that the relatively nonuniform vectors at the edge of the GAA offset each other for a single HZO film using TCAD, resulting in weak polarization. In Figure 7c,d, the high endurance of 10^{11} and the metal-HZO interface dead layer are improved by using HZO as a double layer. Using the ITO-IGZO heterojunction channel developed by Chen et al., one study showed that the top interface defects can be self-compensated in the top gate and dual gate structure to improve the ultra-low sub-pA leakage, disturb-free, and sneak-current-free read/write operations [93]. First, the defect self-compensation effect on the ITO-IGZO channel FeFET, as well as the FeFET using only the IGZO channel, was investigated as the top gate, and it was confirmed that the trapping/de-trapping of electrons in the IGZO channel device lost the hysteresis response. As a result, MW converged to 0, and Figure 8a shows that the deep-level trap density of the low IGZO channel device is 100 times higher than that of the ITO-IGZO channel device. XPS analysis suggests that oxygen vacancy in Figure 8b is less in ITO-IGZO devices than in IGZO devices because metals with high bond dissociation energy (BDE) suppress oxygen defects. Meanwhile, read-after-write delay tests were performed on devices with BEOL dual gate and IGZO channel devices, and delays of 200 ns and 10 s were observed (Figure 8c), respectively. As illustrated in the energy band diagram of Figure 8d, interface and bulk deep-level traps are effectively passivated by the self-compensation effect. Finally, BEOL, which was structurally and thermally disadvantageous, was developed as a BEOL top gate FeFET with a very low delay value of 200 ns and an ideal SS of 62 mV/dec using heterogeneous bonding channels.

Unlike a two-terminal device, an FET structure changes channel conductivity by the gate and determines the weight of the input/output. This system shows an advantage in online learning because the learning and option are easily separated [44,94,95]. TiN/HZO/TiN structures were created using a 10 nm HZO thin film, and remanent polarization was measured in three ways (Figure 9a–c) to investigate synaptic properties [96]. The best way for remanent polarization proved to be increasing the voltage amplitude while retaining the same pulse interval. In this way, 32 polarization states were generated and converted into channel current states by simulation, and the MNIST simulation showed a pattern recognition accuracy of 84.34%, as shown in Figure 9d,e. Storage computing systems operating based on the parallel computing of FeFETs examined have been shown by Nako et al. to be well-suited for large-scale in-memory computing, which is updated over time in real-world applications such as speech recognition via online learning [97]. Voice numbers from 0 to 9 are converted into 78 frequency channels and 48 time steps, and the voltage pulse of a single frequency channel can be used as input as the gate voltage of a single FeFET. A number is determined in one device through the drain current obtained from there, according to the majority vote of the finally determined FeFET connected in parallel. In this study, several measures were proposed to increase detection accuracy. The time step used to make virtual nodes was adjusted, the voltage conversion and frequency channel selection methods were optimized, and drain, source, and substrate currents were measured and applied in detection to improve accuracy.

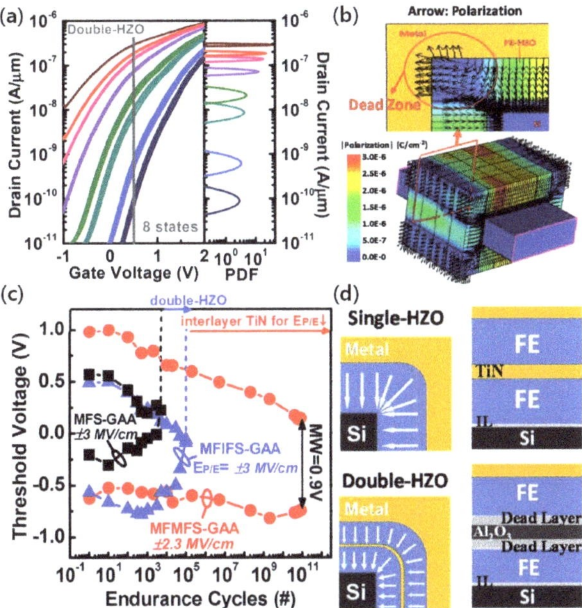

Figure 7. (**a**) I_D-V_G curve with 3-bit operation from double HZO layers; (**b**) decreased polarization in corner region for GAA FeFET with a single HZO layer by TCAD; (**c**) endurance ($>10^{11}$) and MW (=0.9 V) characteristics obtained using double HZO layer; and (**d**) schematic of the effect of polarization field on corner areas when single and double HZO layers are used with permission from Ref. [92], 2022, IEEE.

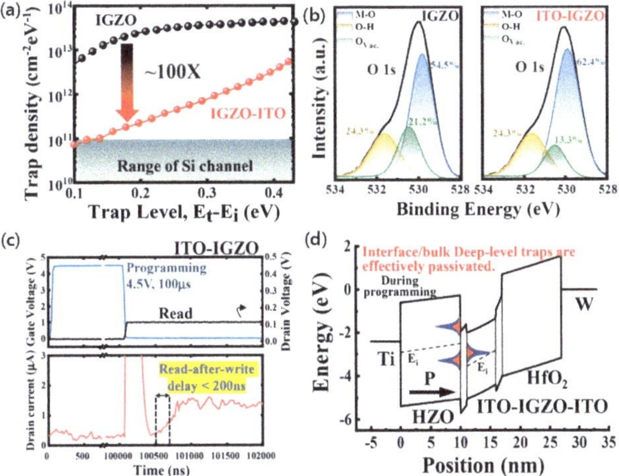

Figure 8. (**a**) Trap density for interface/bulk of deep-level through C-V measurement; (**b**) XPS analysis confirms the effective inhibition of oxygen vacancies in the ITO-IGZO channel device; (**c**) 200 ns delay occurred during read after programming operation; (**d**) The energy band diagram shows that the self-compensation effect is capable of good passivation for defects with permission from Ref. [93], 2022, IEEE.

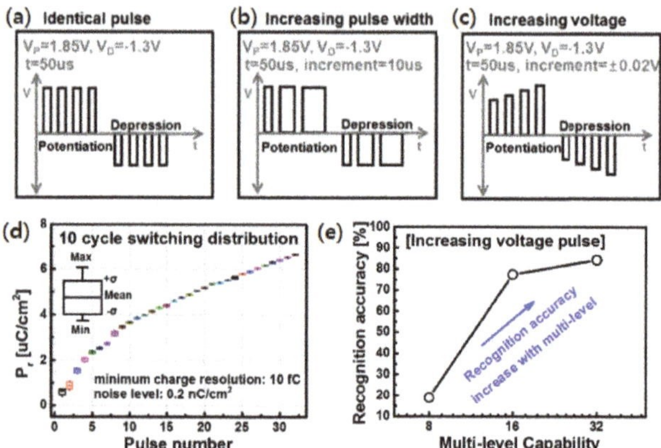

Figure 9. (**a**) Programming identical pulse; (**b**) increasing programming pulse width; (**c**) increasing programming pulse voltage; (**d**) the polarization states of 32 levels did not overlap each other, resulting from the cycle-to-cycle variation of each polarization state; (**e**) as a result of the simulation for three neural networks, the pattern recognition accuracy was 84.34% recorded with permission from Ref. [96], 2017, IEEE.

Chung et al. fabricated FeFETs using a germanium (Ge) channel to study the time response to the HZO gate stack, as shown in Figure 10a [98]. In Figure 10b,c, experiments demonstrate that a single pulse of fewer than 10 ns and a pulse train of 100 ps are suitable for attempting polarization transitions in FeFET. Time response is an indicator of the working speed of FeFET, which means that it is possible to operate in the GHz region when responding in an ultra-high-speed region of less than 10 ns. This group also presents a study that updates the optimal weights for improving the linearity and asymmetry of the channel conductance using the same Ge-FeFET [69]. The weight update applies potentiation and depression pulses to the gate electrode to enter a new conductance value through partial polarization of the ferroelectric. Online networks are required for the effective operation of linear and symmetric conductance. Therefore, if it is not optimized for the voltage or width of the pulse, as the number of pulses increases, the size of the decreasing conductance has nonlinear and asymmetric characteristics, and the network becomes inefficient. Through experiments, the pulse was optimized, and it was proved that the better the linearity, the higher the accuracy, and the lower the accuracy of the nonlinearity pulse. Using these devices, MNIST images were trained with these devices and achieved high accuracy of up to 88%. This means that the results confirm that optimization for the pulse has a great influence on updating the conductance.

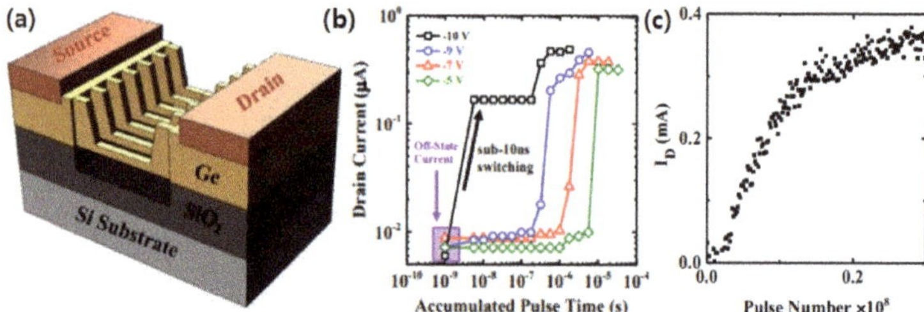

Figure 10. (**a**) 3D structure of nanowire FET with Ge-channel; (**b**) I_D-Accumulated pulse time curve showing polarization switching limits less than 10 ns; and (**c**) accumulated 100 ps pulse train to detect polarization transition with permission from Ref. [98], 2018, IEEE.

Noh et al. conducted a study to combine β-Ga$_2$O$_3$ and HZO, which operate well at high temperatures available in rough environments (desert, space, etc.) [18]. β-Ga$_2$O$_3$ is a wide bandgap of 4.8~4.9 eV and has recently succeeded in melt-grown growth about low-priced large bulk substrates, receiving attention as a next-generation power semiconductor [99,100]. In this letter, FeFET was manufactured using HZO and β-Ga$_2$O$_3$ channels on sapphire substrates. The SS and I$_{on}$/I$_{off}$ characteristics of FeFET were excellent, and MW for use as a synaptic device was also confirmed. Meanwhile, as the number of pulses increases, both the nonlinear formation and the maximum conductance increase, resulting in a trade-off relationship. On-chip learning allowed excellent learning of 94% accuracy in a wide temperature band (20–200 °C) for the MNIST data set. Therefore, ferroelectric materials combined with β-Ga$_2$O$_3$ and HZO can be used to create devices that can overcome high temperatures.

So far, this paper has only considered HZO-based ferroelectric memory devices, but a recent study of P(VDF-TrFE) ferroelectric material-based 2D FeFET is worth reviewing from a structural and application perspective. To realize in-memory computing, Luo et al. implemented the 'logic-in-memory' system, which means digital operation, and the 'neuromorphic computing', which means analog operation, in a single three-terminal device, blurring the boundaries between the two operations [101]. With dual-ferroelectric-coupling effects and 2D transition metal dichalcogenide (TMD) channels, MoS$_2$ with unipolar characteristics was used as the upper electrode, and MoTe$_2$ with ambipolar characteristics was used as the lower electrode. The digital operation was performed using two ferroelectric layers located at the upper and lower ends. These layers induced four conductance states, and the 'OR' and 'AND' were implemented through the upper electrode while the 'XNOR' logic was implemented through the lower electrode, as shown in Figure 11a–c. In addition, one layer was set to a fixed polarization to implement analog operations that perform synaptic behavior like a general FeFET through long-term focus (LTD) measurement and MNIST simulation. This study is considered one of the low-power and high-efficiency in-memory computing hardware technologies that simultaneously conducts logic and synaptic operations when the switching between double-gate and single-gate methods was free.

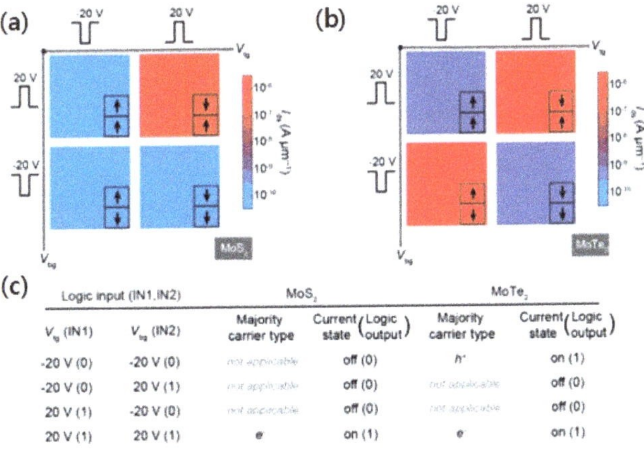

Logic input (IN1, IN2)		MoS$_2$		MoTe$_2$	
V_{tg} (IN1)	V_{bg} (IN2)	Majority carrier type	Current state (Logic output)	Majority carrier type	Current state (Logic output)
-20 V (0)	-20 V (0)	not applicable	off (0)	h*	on (1)
-20 V (0)	20 V (1)	not applicable	off (0)	not applicable	off (0)
20 V (1)	-20 V (0)	not applicable	off (0)	not applicable	off (0)
20 V (1)	20 V (1)	e	on (1)	e	on (1)

Figure 11. Polarization state by applying different voltage pulses to the top and bottom gates of the (**a**) MoS$_2$ FET; and (**b**) MoTe$_2$ FET; (**c**) 'And' and 'XNOR' Boolean logic gate operation results table with permission from Ref. [101], 2018, ACS.

3.2.2. FeRAM

FeRAM, with its 1T-1C structure, has been studied due to its high endurance, low voltage operation, and disturbance-free operation, but it is large footprint and destructive read operation suggest that it is better suited for use as, a memory array than in any neuromorphic application [44,102,103].

FeRAM was developed by Francois et al. as a 1T-1C array with a 16 kbit capacity [102]. The capacitor is less than 300 nm in diameter, and it has been confirmed that it has an operating speed of 30 ns, that data are in a 125 °C environment, that it has endurance characteristics of 10^{11} or more, and that it is compatible with the BEOL process. A 64 kbit array with a larger capacity was demonstrated by Okuno et al. [104]. A high operating speed of 14 ns, an endurance of 10^{11} or higher, and data retention characteristics of 10 years or more were also investigated. Their research attempted to amplify a voltage that changes in accordance with the difference in capacitance obtained through the comparison of the reference capacitor and the ferroelectric capacitor by the sense amplifier. Simulation and experimental results confirmed that the two memory states were distinguished. The dependence on the read voltage margin on the MFM capacitor size was calculated, and it was found that it was possible through a 40 nm process based on a 3D cylinder capacitor. While working on a low-capacity FeRAM array, a high-capacity FeRAM with 8 GB density was demonstrated by Sung et al. [105]. While the existing FeRAM operated by maximizing 2Pr by adjusting the capacitor plate voltage, the demonstrated FeRAM showed potential as a low-power, high-speed memory by using a low fixed capacitor plate voltage that widened the difference in work functions of the upper and lower electrodes. However, the system remained disadvantageous in terms of density, and additional improvements will be needed to improve the 2Pr and speed switching through asymmetric operating voltage, upper/lower electrode development, and HZO composition.

4. Conclusions

We described four types of ferroelectric-based memory devices, focusing on their operational methods, features, and applications. FTJ operates according to the TER whose resistance changes depending on the polarization direction of the ferroelectric. Fe-diodes operate by causing polarization inversion through Schottky barrier modulation at the interface between the metal electrode and the ferroelectric thin film. These two structures are two-terminal devices, so high integration that enables large-scale parallel computing can be implemented. In addition, their analog switching characteristics and uniformity are excellent. FTJs which have a linear I-V and Fe-diode generally have nonlinear I-V characteristics, but FTJs with a nonlinear I-V are also being studied. This is the nonlinear I-V characteristic of the two-terminal device that is advantageous for synaptic devices in neuromorphic applications and low-current devices in large-scale parallel computing with resistance-based crossbar arrays [77,78]. To improve mobility, SS, endurance, and retention characteristics, an IGZO layer or several oxide layers (such as TiO_2 and HfO_x) can be used as insertion layers [58,87,89]. FeFET induces a change in the current of a channel according to polarization switching, and FeRAM uses a ferroelectric material in a capacitor having a 1T-1C structure to perform a destructive operation. FeFET has low leakage current due to the wide bandgap of the HZO and thus shows advantageous high-speed and low-power characteristics, even when the associated device scale is small. Since they operate like MOSFET, researchers have focused on developing memory performance through a structural approach [91–93]. Especially, inducing a channel current change by gate has proved that learning and operation can be separated and have advantages in online learning [97]. FeRAM is being actively researched for use as a memory array, in light of its high endurance characteristics and compatibility with the BEOL process [102,104]. Finally, studies of low fixed plate voltages, have yielded encouraging findings for the development of low-power high-speed memory [105]. In summary, while research has only just begun on the application of memory devices using ferroelectric doped with Zr on Hf, the potential is clearly enormous. If each structure's unique strengths and weaknesses continue to be improved upon, many applications, such as in-memory computing and neuromorphic computing, will be developed.

Author Contributions: Conceptualization, J.Y. and H.B.; resources, H.S., H.L. and S.L.; visualization, H.L. and S.K.; writing—original draft, J.Y.; writing—review and editing, K.H. and H.B.; supervision, K.H. and H.B.; All authors have read and agreed to the published version of the manuscript.

Funding: The article publishing fee was funded by the Basic Science Research Program through the NRF of Korea funded by the Ministry of Education (2022R1F1A1071914).

Data Availability Statement: No new data were created or analyzed in this work. Data sharing is not applicable to this article.

Acknowledgments: This work was supported in part by the Basic Science Research Program through the NRF of Korea funded by the Ministry of Education (2022R1F1A1071914), in part by the Nano·Material Technology Development Program through the NRF of Korea funded by the Ministry of Science, ICT and Future Planning (2009-0082580), in part by the IC Design Education Center (IDEC), in part by the National Research Council of Science & Technology (NST) grant by the Korea government (MSIT) (No. CAP22031-110), in part by "Research Base Construction Fund Support Program" funded by Jeonbuk National University in 2022, and in part by Basic Science Research Program funded by the Ministry of Education (2022R1I1A307258211).

Conflicts of Interest: The authors declare no conflict of interest.

References

1. Song, C.-M.; Kwon, H.-J. Ferroelectrics Based on HfO_2 Film. *Electronics* **2021**, *10*, 2759. [CrossRef]
2. Cappy, A. *Neuro-Inspired Information Processing*; John Wiley & Sons: Hoboken, NJ, USA, 2020.
3. Keshavarzi, A.; Ni, K.; Van Den Hoek, W.; Datta, S.; Raychowdhury, A. FerroElectronics for Edge Intelligence. *IEEE Micro* **2020**, *40*, 33–48. [CrossRef]
4. Wang, Z.; Wu, H.; Burr, G.W.; Hwang, C.S.; Wang, K.L.; Xia, Q.; Yang, J.J. Resistive switching materials for information processing. *Nat. Rev. Mater.* **2020**, *5*, 173–195. [CrossRef]
5. Dai, S.; Zhao, Y.; Wang, Y.; Zhang, J.; Fang, L.; Jin, S.; Shao, Y.; Huang, J. Recent Advances in Transistor-Based Artificial Synapses. *Adv. Funct. Mater.* **2019**, *29*, 1903700. [CrossRef]
6. Sun, B.; Guo, T.; Zhou, G.; Ranjan, S.; Jiao, Y.; Wei, L.; Zhou, Y.N.; Wu, Y.A. Synaptic devices based neuromorphic computing applications in artificial intelligence. *Mater. Today Phys.* **2021**, *18*, 100393. [CrossRef]
7. Han, J.-K.; Yun, S.-Y.; Lee, S.-W.; Yu, J.-M.; Choi, Y.-K. A Review of Artificial Spiking Neuron Devices for Neural Processing and Sensing. *Adv. Funct. Mater.* **2022**, *32*, 2204102. [CrossRef]
8. Banerjee, W. Challenges and Applications of Emerging Nonvolatile Memory Devices. *Electronics* **2020**, *9*, 1029. [CrossRef]
9. Sebastian, A.; Le Gallo, M.; Khaddam-Aljameh, R.; Eleftheriou, E. Memory devices and applications for in-memory computing. *Nat. Nanotechnol.* **2020**, *15*, 529–544. [CrossRef]
10. Jouppi, N.P.; Young, C.; Patil, N.; Patterson, D.; Agrawal, G.; Bajwa, R.; Bates, S.; Bhatia, S.; Boden, N.; Borchers, A.; et al. In-Datacenter Performance Analysis of a Tensor Processing Unit. In Proceedings of the 44th Annual International Symposium on Computer Architecture, New York, NY, USA, 24–28 June 2017.
11. Sze, V.; Chen, Y.-H.; Yang, T.-J.; Emer, J.S. Efficient Processing of Deep Neural Networks: A Tutorial and Survey. *Proc. IEEE* **2017**, *105*, 2295–2329. [CrossRef]
12. Yu, J.-M.; Lee, C.; Kim, D.-J.; Park, H.; Han, J.-K.; Hur, J.; Kim, J.-K.; Kim, M.-S.; Seo, M.; Im, S.G.; et al. All-Solid-State Ion Synaptic Transistor for Wafer-Scale Integration with Electrolyte of a Nanoscale Thickness. *Adv. Funct. Mater.* **2021**, *31*, 2010971. [CrossRef]
13. Wang, W.; Gao, S.; Li, Y.; Yue, W.; Kan, H.; Zhang, C.; Lou, Z.; Wang, L.; Shen, G. Artificial Optoelectronic Synapses Based on TiN_xO_{2-x}/MoS_2 Heterojunction for Neuromorphic Computing and Visual System. *Adv. Funct. Mater.* **2021**, *31*, 2101201. [CrossRef]
14. Ielmini, D.; Wong, H.-S.P. In-memory computing with resistive switching devices. *Nat. Electron.* **2018**, *1*, 333–343. [CrossRef]
15. Raoux, S.; Wełnic, W.; Ielmini, D. Phase Change Materials and Their Application to Nonvolatile Memories. *Chem. Rev.* **2010**, *110*, 240–267. [CrossRef] [PubMed]
16. Kent, A.D.; Worledge, D.C. A new spin on magnetic memories. *Nat. Nanotechnol.* **2015**, *10*, 187–191. [CrossRef] [PubMed]
17. Jan, A.; Rembert, T.; Taper, S.; Symonowicz, J.; Strkalj, N.; Moon, T.; Lee, Y.S.; Bae, H.; Lee, H.J.; Choe, D.-H.; et al. In Operando Optical Tracking of Oxygen Vacancy Migration and Phase Change in few Nanometers Ferroelectric HZO Memories. *Adv. Funct. Mater.* **2023**, 2214970. [CrossRef]
18. Noh, J.; Bae, H.; Li, J.; Luo, Y.; Qu, Y.; Park, T.J.; Si, M.; Chen, X.; Charnas, A.R.; Chung, W.; et al. First Experimental Demonstration of Robust $HZO/\beta\text{-}Ga_2O_3$ Ferroelectric Field-Effect Transistors as Synaptic Devices for Artificial Intelligence Applications in a High-Temperature Environment. *IEEE Trans. Electron Devices* **2021**, *68*, 2515–2521. [CrossRef]
19. Choe, D.-H.; Kim, S.; Moon, T.; Jo, S.; Bae, H.; Nam, S.-G.; Lee, Y.S.; Heo, J. Unexpectedly low barrier of ferroelectric switching in HfO_2 via topological domain walls. *Mater. Today* **2021**, *50*, 8–15. [CrossRef]

20. Si, M.; Luo, Y.; Chung, W.; Bae, H.; Zheng, D.; Li, J.; Qin, J.; Qiu, G.; Yu, S.; Ye, P. A novel scalable energy-efficient synaptic device: Crossbar ferroelectric semiconductor junction. In Proceedings of the 2019 IEEE International Electron Devices Meeting (IEDM), San Francisco, CA, USA, 7–11 December 2019.

21. Kang, S.; Jang, W.-S.; Morozovska, A.N.; Kwon, O.; Jin, Y.; Kim, Y.-H.; Bae, H.; Wang, C.; Yang, S.-H.; Belianinov, A. Highly enhanced ferroelectricity in HfO$_2$-based ferroelectric thin film by light ion bombardment. *Science* **2022**, *376*, 731–738. [CrossRef]

22. Seo, M.; Kang, M.-H.; Jeon, S.-B.; Bae, H.; Hur, J.; Jang, B.C.; Yun, S.; Cho, S.; Kim, W.-K.; Kim, M.-S.; et al. First Demonstration of a Logic-Process Compatible Junctionless Ferroelectric FinFET Synapse for Neuromorphic Applications. *IEEE Electron Device Lett.* **2018**, *39*, 1445–1448. [CrossRef]

23. Martin, L.W.; Rappe, A.M. Thin-film ferroelectric materials and their applications. *Nat. Rev. Mater.* **2016**, *2*, 16087. [CrossRef]

24. Schroeder, U.; Park, M.H.; Mikolajick, T.; Hwang, C.S. The fundamentals and applications of ferroelectric HfO$_2$. *Nat. Rev. Mater.* **2022**, *7*, 653–669. [CrossRef]

25. Mulaosmanovic, H.; Breyer, E.T.; Dünkel, S.; Beyer, S.; Mikolajick, T.; Slesazeck, S. Ferroelectric field-effect transistors based on HfO$_2$: A review. *Nanotechnology* **2021**, *32*, 502002. [CrossRef]

26. Qi, L.; Ruan, S.; Zeng, Y.-J. Review on Recent Developments in 2D Ferroelectrics: Theories and Applications. *Adv. Mater.* **2021**, *33*, e2005098. [CrossRef]

27. Kim, S.J.; Mohan, J.; Summerfelt, S.R.; Kim, J. Ferroelectric Hf$_{0.5}$Zr$_{0.5}$O$_2$ Thin Films: A Review of Recent Advances. *JOM* **2018**, *71*, 246–255. [CrossRef]

28. Kim, J.Y.; Choi, M.-J.; Jang, H.W. Ferroelectric field effect transistors: Progress and perspective. *APL Mater.* **2021**, *9*, 021102. [CrossRef]

29. Park, H.W.; Lee, J.-G.; Hwang, C.S. Review of ferroelectric field-effect transistors for three-dimensional storage applications. *Nano Sel.* **2021**, *2*, 1187–1207. [CrossRef]

30. Park, M.H.; Lee, Y.H.; Mikolajick, T.; Schroeder, U.; Hwang, C.S. Review and perspective on ferroelectric HfO$_2$-based thin films for memory applications. *Mrs Commun.* **2018**, *8*, 795–808. [CrossRef]

31. Böscke, T.S.; Müller, J.; Bräuhaus, D.; Schröder, U.; Böttger, U. Ferroelectricity in hafnium oxide thin films. *Appl. Phys. Lett.* **2011**, *99*, 102903. [CrossRef]

32. Park, M.H.; Lee, Y.H.; Kim, H.J.; Kim, Y.J.; Moon, T.; Kim, K.D.; Muller, J.; Kersch, A.; Schroeder, U.; Mikolajick, T.; et al. Ferroelectricity and Antiferroelectricity of Doped Thin HfO$_2$-Based Films. *Adv. Mater.* **2015**, *27*, 1811–1831. [CrossRef]

33. Schroeder, U.; Yurchuk, E.; Müller, J.; Martin, D.; Schenk, T.; Polakowski, P.; Adelmann, C.; Popovici, M.I.; Kalinin, S.V.; Mikolajick, T. Impact of different dopants on the switching properties of ferroelectric hafniumoxide. *Jpn. J. Appl. Phys.* **2014**, *53*, 08LE02. [CrossRef]

34. Zhou, D.; Müller, J.; Xu, J.; Knebel, S.; Bräuhaus, D.; Schröder, U. Insights into electrical characteristics of silicon doped hafnium oxide ferroelectric thin films. *Appl. Phys. Lett.* **2012**, *100*, 082905. [CrossRef]

35. Park, M.H.; Kim, H.J.; Kim, Y.J.; Lee, W.; Moon, T.; Kim, K.D.; Hwang, C.S. Study on the degradation mechanism of the ferroelectric properties of thin Hf$_{0.5}$Zr$_{0.5}$O$_2$ films on TiN and Ir electrodes. *Appl. Phys. Lett.* **2014**, *105*, 072902. [CrossRef]

36. Kim, H.J.; Park, M.H.; Kim, Y.J.; Lee, Y.H.; Jeon, W.; Gwon, T.; Moon, T.; Kim, K.D.; Hwang, C.S. Grain size engineering for ferroelectric Hf$_{0.5}$Zr$_{0.5}$O$_2$ films by an insertion of Al$_2$O$_3$ interlayer. *Appl. Phys. Lett.* **2014**, *105*, 192903. [CrossRef]

37. Müller, J.; Schröder, U.; Böscke, T.S.; Müller, I.; Böttger, U.; Wilde, L.; Sundqvist, J.; Lemberger, M.; Kücher, P.; Mikolajick, T.; et al. Ferroelectricity in yttrium-doped hafnium oxide. *J. Appl. Phys.* **2011**, *110*, 114113. [CrossRef]

38. Polakowski, P.; Riedel, S.; Weinreich, W.; Rudolf, M.; Sundqvist, J.; Seidel, K.; Muller, J. Ferroelectric deep trench capacitors based on Al:HfO$_2$ for 3D nonvolatile memory applications. In Proceedings of the 2014 IEEE 6th International Memory Workshop (IMW), Taipei, Taiwan, 18–21 May 2014.

39. Mueller, S.; Adelmann, C.; Singh, A.; Van Elshocht, S.; Schroeder, U.; Mikolajick, T. Ferroelectricity in Gd-Doped HfO$_2$ Thin Films. *ECS J. Solid State Sci. Technol.* **2012**, *1*, N123–N126. [CrossRef]

40. Schenk, T.; Schroeder, U.; Pesic, M.; Popovici, M.; Pershin, Y.V.; Mikolajick, T. Electric Field Cycling Behavior of Ferroelectric Hafnium Oxide. *ACS Appl. Mater. Interfaces* **2014**, *6*, 19744–19751. [CrossRef]

41. Müller, J.; Böscke, T.; Müller, S.; Yurchuk, E.; Polakowski, P.; Paul, J.; Martin, D.; Schenk, T.; Khullar, K.; Kersch, A.; et al. Ferroelectric Hafnium Oxide: A CMOS-compatible and highly scalable approach to future ferroelectric memories. In Proceedings of the 2013 IEEE International Electron Devices Meeting, Washington, DC, USA, 9–11 December 2013.

42. Muller, J.; Boscke, T.S.; Schroder, U.; Mueller, S.; Brauhaus, D.; Bottger, U.; Frey, L.; Mikolajick, T. Ferroelectricity in Simple Binary ZrO$_2$ and HfO$_2$. *Nano Lett.* **2012**, *12*, 4318–4323. [CrossRef]

43. Jo, S.; Lee, H.; Choe, D.-H.; Kim, J.-H.; Lee, Y.S.; Kwon, O.; Nam, S.; Park, Y.; Kim, K.; Chae, B.G.; et al. Negative differential capacitance in ultrathin ferroelectric hafnia. *Nat. Electron.* **2023**. [CrossRef]

44. Park, J.Y.; Choe, D.-H.; Lee, D.H.; Yu, G.T.; Yang, K.; Kim, S.H.; Park, G.H.; Nam, S.-G.; Lee, H.J.; Jo, S.; et al. Revival of Ferroelectric Memories Based on Emerging Fluorite-Structured Ferroelectrics. *Adv. Mater.* **2022**, e2204904. [CrossRef]

45. Mikolajick, T.; Slesazeck, S.; Mulaosmanovic, H.; Park, M.H.; Fichtner, S.; Lomenzo, P.D.; Hoffmann, M.; Schroeder, U. Next generation ferroelectric materials for semiconductor process integration and their applications. *J. Appl. Phys.* **2021**, *129*, 100901. [CrossRef]

46. Max, B.; Hoffmann, M.; Slesazeck, S.; Mikolajick, T. Ferroelectric Tunnel Junctions based on Ferroelectric-Dielectric $Hf_{0.5}Zr_{0.5}O_2$ /Al_2O_3 Capacitor Stacks. In Proceedings of the 2018 48th European Solid-State Device Research Conference (ESSDERC), Dresden, Germany, 3–6 September 2018.
47. Mikolajick, T.; Schroeder, U.; Slesazeck, S. The Past, the Present, and the Future of Ferroelectric Memories. *IEEE Trans. Electron Devices* **2020**, *67*, 1434–1443. [CrossRef]
48. Fujii, S.; Kamimuta, Y.; Ino, T.; Nakasaki, Y.; Takaishi, R.; Saitoh, M. First demonstration and performance improvement of ferroelectric HfO_2-based resistive switch with low operation current and intrinsic diode property. In Proceedings of the 2016 IEEE Symposium on VLSI Technology, Honolulu, HI, USA, 14–16 June 2016.
49. Hwang, J.; Goh, Y.; Jeon, S. Effect of Insertion of Dielectric Layer on the Performance of Hafnia Ferroelectric Devices. *IEEE Trans. Electron Devices* **2021**, *68*, 841–845. [CrossRef]
50. Blom, P.W.M.; Wolf, R.M.; Cillessen, J.F.M.; Krijn, M.P.C.M. Ferroelectric Schottky Diode. *Phys. Rev. Lett.* **1994**, *73*, 2107–2110. [CrossRef]
51. Hsiang, K.-Y.; Liao, C.-Y.; Wang, J.-F.; Lou, Z.-F.; Lin, C.-Y.; Chiang, S.-H.; Liu, C.-W.; Hou, T.-H.; Lee, M.-H. Unipolar Parity of Ferroelectric-Antiferroelectric Characterized by Junction Current in Crystalline Phase $Hf_{1-x}Zr_xO_2$ Diodes. *Nanomaterials* **2021**, *11*, 2685. [CrossRef]
52. Luo, Q.; Cheng, Y.; Yang, J.; Cao, R.; Ma, H.; Yang, Y.; Huang, R.; Wei, W.; Zheng, Y.; Gong, T.; et al. A highly CMOS compatible hafnia-based ferroelectric diode. *Nat. Commun.* **2020**, *11*, 1391. [CrossRef]
53. Chen, A. A review of emerging non-volatile memory (NVM) technologies and applications. *Solid-State Electron.* **2016**, *125*, 25–38. [CrossRef]
54. Hoffman, J.; Pan, X.; Reiner, J.W.; Walker, F.J.; Han, J.P.; Ahn, C.H.; Ma, T.P. Ferroelectric Field Effect Transistors for Memory Applications. *Adv. Mater.* **2010**, *22*, 2957–2961. [CrossRef]
55. Mo, F.; Tagawa, Y.; Jin, C.; Ahn, M.; Saraya, T.; Hiramoto, T.; Kobayashi, M. Experimental Demonstration of Ferroelectric HfO_2 FET with Ultrathin-body IGZO for High-Density and Low-Power Memory Application. In Proceedings of the 2019 Symposium on VLSI Technology, Kyoto, Japan, 9–14 June 2019.
56. Mo, F.; Tagawa, Y.; Jin, C.; Ahn, M.; Saraya, T.; Hiramoto, T.; Kobayashi, M. Low-Voltage Operating Ferroelectric FET with Ultrathin IGZO Channel for High-Density Memory Application. *IEEE J. Electron Devices Soc.* **2020**, *8*, 717–723. [CrossRef]
57. Hoffmann, M.; Tan, A.J.; Shanker, N.; Liao, Y.-H.; Wang, L.-C.; Bae, J.-H.; Hu, C.; Salahuddin, S. Fast Read-After-Write and Depolarization Fields in High Endurance n-Type Ferroelectric FETs. *IEEE Electron Device Lett.* **2022**, *43*, 717–720. [CrossRef]
58. Gaddam, V.; Das, D.; Jeon, S. Insertion of HfO_2 Seed/Dielectric Layer to the Ferroelectric HZO Films for Heightened Remanent Polarization in MFM Capacitors. *IEEE Trans. Electron Devices* **2020**, *67*, 745–750. [CrossRef]
59. Okuno, J.; Kunihiro, T.; Konishi, K.; Materano, M.; Ali, T.; Kuehnel, K.; Seidel, K.; Mikolajick, T.; Schroeder, U.; Tsukamoto, M.; et al. 1T1C FeRAM Memory Array Based on Ferroelectric HZO With Capacitor Under Bitline. *IEEE J. Electron Devices Soc.* **2022**, *10*, 29–34. [CrossRef]
60. Breyer, E.T.; Mulaosmanovic, H.; Mikolajick, T.; Slesazeck, S. Reconfigurable NAND/NOR logic gates in 28 nm HKMG and 22 nm FD-SOI FeFET technology. In Proceedings of the 2017 IEEE International Electron Devices Meeting (IEDM), San Francisco, CA, USA, 2–6 December 2017.
61. Breyer, E.T.; Mulaosmanovic, H.; Mikolajick, T.; Slesazeck, S. Perspective on ferroelectric, hafnium oxide based transistors for digital beyond von-Neumann computing. *Appl. Phys. Lett.* **2021**, *118*, 050501. [CrossRef]
62. Berdan, R.; Marukame, T.; Ota, K.; Yamaguchi, M.; Saitoh, M.; Fujii, S.; Deguchi, J.; Nishi, Y. Low-power linear computation using nonlinear ferroelectric tunnel junction memristors. *Nat. Electron.* **2020**, *3*, 259–266. [CrossRef]
63. Jerry, M.; Chen, P.-Y.; Zhang, J.; Sharma, P.; Ni, K.; Yu, S.; Datta, S. Ferroelectric FET Analog Synapse for Acceleration of Deep Neural Network Training. In Proceedings of the 2017 IEEE International Electron Devices Meeting (IEDM), San Francisco, CA, USA, 2–6 December 2017.
64. Wang, P.; Yu, S. Ferroelectric devices and circuits for neuro-inspired computing. *MRS Commun.* **2020**, *10*, 538–548. [CrossRef]
65. Mulaosmanovic, H.; Ocker, J.; Müller, S.; Noack, M.; Müller, J.; Polakowski, P.; Mikolajick, T.; Slesazeck, S. Novel ferroelectric FET based synapse for neuromorphic systems. In Proceedings of the 2017 Symposium on VLSI Technology, Kyoto, Japan, 5–8 June 2017.
66. Han, J.-K.; Yun, S.-Y.; Yu, J.-M.; Jeon, S.-B.; Choi, Y.-K. Artificial Multisensory Neuron with a Single Transistor for Multimodal Perception through Hybrid Visual and Thermal Sensing. *ACS Appl. Mater. Interfaces* **2023**, *15*, 5449–5455. [CrossRef]
67. Lee, D.H.; Park, G.H.; Kim, S.H.; Park, J.Y.; Yang, K.; Slesazeck, S.; Mikolajick, T.; Park, M.H. Neuromorphic devices based on fluorite-structured ferroelectrics. *InfoMat* **2022**, *4*, e12380. [CrossRef]
68. Fan, Z.; Chen, J.; Wang, J. Ferroelectric HfO_2-based materials for next-generation ferroelectric memories. *J. Adv. Dielectr.* **2016**, *6*, 1630003. [CrossRef]
69. Chung, W.; Si, M.; Ye, P.D. First Demonstration of Ge Ferroelectric Nanowire FET as Synaptic Device for Online Learning in Neural Network with High Number of Conductance State and G_{max}/G_{min}. In Proceedings of the 2018 IEEE International Electron Devices Meeting (IEDM), San Francisco, CA, USA, 1–5 December 2018.
70. Si, M.; Ye, P.D. The Critical Role of Charge Balance on the Memory Characteristics of Ferroelectric Field-Effect Transistors. *IEEE Trans. Electron Devices* **2021**, *68*, 5108–5113. [CrossRef]

71. Saha, A.K.; Si, M.; Ni, K.; Datta, S.; Ye, P.D.; Gupta, S. Ferroelectric Thickness Dependent Domain Interactions in FEFETs for Memory and Logic: A phase-Field Model based Analysis. In Proceedings of the 2020 IEEE International Electron Devices Meeting (IEDM), San Francisco, CA, USA, 12–18 December 2020.
72. Lyu, X.; Si, M.; Sun, X.; Capano, M.A.; Wang, H.; Ye, P.D. Ferroelectric and Anti-Ferroelectric Hafnium Zirconium Oxide: Scaling Limit, Switching Speed and Record High Polarization Density. In Proceedings of the 2019 Symposium on VLSI Technology, Kyoto, Japan, 9–14 June 2019.
73. Saitoh, M.; Ichihara, R.; Yamaguchi, M.; Suzuki, K.; Takano, K.; Akari, K.; Takahashi, K.; Kamiya, Y.; Matsuo, K.; Kamimuta, Y.; et al. HfO$_2$-based FeFET and FTJ for Ferroelectric-Memory Centric 3D LSI towards Low-Power and High-Density Storage and AI Applications. In Proceedings of the 2020 IEEE International Electron Devices Meeting (IEDM), San Francisco, CA, USA, 12–18 December 2020.
74. Marinella, M.J.; Agarwal, S.; Hsia, A.; Richter, I.; Jacobs-Gedrim, R.; Niroula, J.; Plimpton, S.J.; Ipek, E.; James, C.D. Multiscale Co-Design Analysis of Energy, Latency, Area, and Accuracy of a ReRAM Analog Neural Training Accelerator. *IEEE J. Emerg. Sel. Top. Circuits Syst.* **2018**, *8*, 86–101. [CrossRef]
75. Yang, S.-T.; Li, X.-Y.; Yu, T.-L.; Wang, J.; Fang, H.; Nie, F.; He, B.; Zhao, L.; Lü, W.-M.; Yan, S.-S.; et al. High-Performance Neuromorphic Computing Based on Ferroelectric Synapses with Excellent Conductance Linearity and Symmetry. *Adv. Funct. Mater.* **2022**, *32*, 2202366. [CrossRef]
76. Yu, S.; Chen, P.-Y. Emerging Memory Technologies: Recent Trends and Prospects. *IEEE Solid-State Circuits Mag.* **2016**, *8*, 43–56. [CrossRef]
77. Yang, R. In-memory computing with ferroelectrics. *Nat. Electron.* **2020**, *3*, 237–238. [CrossRef]
78. Fuller, E.J.; Keene, S.T.; Melianas, A.; Wang, Z.; Agarwal, S.; Li, Y.; Tuchman, Y.; James, C.D.; Marinella, M.J.; Yang, J.J. Parallel programming of an ionic floating-gate memory array for scalable neuromorphic computing. *Science* **2019**, *364*, 570–574. [CrossRef]
79. Yu, J.; Li, Y.; Sun, W.; Zhang, W.; Gao, Z.; Dong, D.; Yu, Z.; Zhao, Y.; Lai, J.; Ding, Q. Energy efficient and robust reservoir computing system using ultrathin (3.5 nm) ferroelectric tunneling junctions for temporal data learning. In Proceedings of the 2021 Symposium on VLSI Technology, Kyoto, Japan, 13–19 June 2021.
80. Chen, L.; Wang, T.-Y.; Dai, Y.-W.; Cha, M.-Y.; Zhu, H.; Sun, Q.-Q.; Ding, S.-J.; Zhou, P.; Chua, L.; Zhang, D.W. Ultra-low power Hf$_{0.5}$Zr$_{0.5}$O$_2$ based ferroelectric tunnel junction synapses for hardware neural network applications. *Nanoscale* **2018**, *10*, 15826–15833. [CrossRef]
81. Ota, K.; Yamaguchi, M.; Berdan, R.; Marukame, T.; Nishi, Y.; Matsuo, K.; Takahashi, K.; Kamiya, Y.; Miyano, S.; Deguchi, J.; et al. Performance Maximization of In-Memory Reinforcement Learning with Variability-Controlled Hf$_{1-x}$Zr$_x$O$_2$ Ferroelectric Tunnel Junctions. In Proceedings of the 2019 IEEE International Electron Devices Meeting (IEDM), San Francisco, CA, USA, 7–11 December 2019.
82. Berdan, R.; Marukame, T.; Kabuyanagi, S.; Ota, K.; Saitoh, M.; Fujii, S.; Deguchi, J.; Nishi, Y. In-memory Reinforcement Learning with Moderately-Stochastic Conductance Switching of Ferroelectric Tunnel Junctions. In Proceedings of the 2019 Symposium on VLSI Technology, Kyoto, Japan, 9–14 June 2019.
83. Luo, Y.-C.; Hur, J.; Yu, S. Ferroelectric Tunnel Junction Based Crossbar Array Design for Neuro-Inspired Computing. *IEEE Trans. Nanotechnol.* **2021**, *20*, 243–247. [CrossRef]
84. Goh, Y.; Hwang, J.; Kim, M.; Lee, Y.; Jung, M.; Jeon, S. Selector-less Ferroelectric Tunnel Junctions by Stress Engineering and an Imprinting Effect for High-Density Cross-Point Synapse Arrays. *ACS Appl. Mater. Interfaces* **2021**, *13*, 59422–59430. [CrossRef]
85. Kim, S.; Lee, K.; Oh, M.-H.; Lee, J.-H.; Park, B.-G.; Kwon, D. Physical Unclonable Functions Using Ferroelectric Tunnel Junctions. *IEEE Electron Device Lett.* **2021**, *42*, 816–819. [CrossRef]
86. Lim, S.; Goh, Y.; Lee, Y.K.; Ko, D.H.; Hwang, J.; Kim, M.; Jeong, Y.; Shin, H.; Jeon, S.; Jung, S.-O. A Highly Integrated Crosspoint Array Using Self-rectifying FTJ for Dual-mode Operations: CAM and PUF. In Proceedings of the ESSCIRC 2022- IEEE 48th European Solid State Circuits Conference (ESSCIRC), Milan, Italy, 19–22 September 2022.
87. Bae, H.; Moon, T.; Nam, S.G.; Lee, K.-H.; Kim, S.; Hong, S.; Choe, D.-H.; Jo, S.; Lee, Y.; Heo, J. Ferroelectric Diodes with Sub-ns and Sub-fJ Switching and Its Programmable Network for Logic-in-Memory Applications. In Proceedings of the 2021 Symposium on VLSI Technology, Kyoto, Japan, 13–19 June 2021.
88. Zhang, Z.; Zhang, F.; Zhang, Y.; Xu, G.; Wu, Z.; Zhang, Q.; Li, Y.; Yin, H.; Luo, J.; Wang, W.; et al. Ultradense One-Memristor Ternary-Content-Addressable Memory Based on Ferroelectric Diodes. *IEEE Electron Device Lett.* **2023**, *44*, 64–67. [CrossRef]
89. Kao, R.-W.; Peng, H.-K.; Chen, K.-Y.; Wu, Y.-H. HfZrO$_x$-Based Switchable Diode for Logic-in-Memory Applications. *IEEE Trans. Electron Devices* **2021**, *68*, 545–549. [CrossRef]
90. Ma, T.P.; Han, J.-P. Why is Nonvolatile Ferroelectric Memory Field-Effect Transistor Still Elusive? *IEEE Electron Device Lett.* **2002**, *23*, 386–388. [CrossRef]
91. Bae, H.; Nam, S.G.; Moon, T.; Lee, Y.; Jo, S.; Choe, D.-H.; Kim, S.; Lee, K.-H.; Heo, J. Sub-ns Polarization Switching in 25 nm FE FinFET toward Post CPU and Spatial-Energetic Mapping of Traps for Enhanced Endurance. In Proceedings of the 2020 IEEE International Electron Devices Meeting (IEDM), San Francisco, CA, USA, 12–18 December 2020.
92. Liao, C.-Y.; Hsiang, K.-Y.; Lou, Z.-F.; Tseng, H.-C.; Lin, C.-Y.; Li, Z.-X.; Hsieh, F.-C.; Wang, C.-C.; Chang, F.-S.; Ray, W.-C.; et al. Endurance > 10^{11} Cycling of 3D GAA Nanosheet Ferroelectric FET with Stacked HfZrO$_2$ to Homogenize Corner Field Toward Mitigate Dead Zone for High-Density eNVM. In Proceedings of the 2022 IEEE Symposium on VLSI Technology and Circuits (VLSI Technology and Circuits), Honolulu, HI, USA, 12–17 June 2022.

93. Chen, C.-K.; Fang, Z.; Hooda, S.; Lal, M.; Chand, U.; Xu, Z.; Pan, J.; Tsai, S.-H.; Zamburg, E.; Thean, A.V.-Y. First Demonstration of Ultra-low D_{it} Top-Gated Ferroelectric Oxide-Semiconductor Memtransistor with Record Performance by Channel Defect Self-Compensation Effect for BEOL-Compatible Non-Volatile Logic Switch. In Proceedings of the 2022 International Electron Devices Meeting (IEDM), San Francisco, CA, USA, 3–7 December 2022.
94. Yu, S. Neuro-Inspired Computing With Emerging Nonvolatile Memorys. *Proc. IEEE* **2018**, *106*, 260–285. [CrossRef]
95. Jeong, D.S.; Hwang, C.S. Nonvolatile Memory Materials for Neuromorphic Intelligent Machines. *Adv. Mater.* **2018**, *30*, e1704729. [CrossRef]
96. Oh, S.; Kim, T.; Kwak, M.; Song, J.; Woo, J.; Jeon, S.; Yoo, I.K.; Hwang, H. HfZrO$_x$-Based Ferroelectric Synapse Device with 32 Levels of Conductance States for Neuromorphic Applications. *IEEE Electron Device Lett.* **2017**, *38*, 732–735. [CrossRef]
97. Nako, E.; Toprasertpong, K.; Nakane, R.; Takenaka, M.; Takagi, S. Experimental demonstration of novel scheme of HZO/Si FeFET reservoir computing with parallel data processing for speech recognition. In Proceedings of the 2022 IEEE Symposium on VLSI Technology and Circuits (VLSI Technology and Circuits), Honolulu, HI, USA, 12–17 June 2022.
98. Chung, W.; Si, M.; Shrestha, P.R.; Campbell, J.P.; Cheung, K.P.; Peide, D.Y. First Direct Experimental Studies of Hf$_{0.5}$Zr$_{0.5}$O$_2$ Ferroelectric Polarization Switching Down to 100-picosecond in Sub-60mV/dec Germanium Ferroelectric Nanowire FETs. In Proceedings of the 2018 IEEE Symposium on VLSI Technology, Honolulu, HI, USA, 18–22 June 2018.
99. Bae, H.; Park, T.J.; Noh, J.; Chung, W.; Si, M.; Ramanathan, S.; Peide, D.Y. First demonstration of robust tri-gate β-Ga$_2$O$_3$ nano-membrane field-effect transistors. *Nanotechnology* **2021**, *33*, 125201. [CrossRef] [PubMed]
100. Noh, J.; Alajlouni, S.; Tadjer, M.J.; Culbertson, J.C.; Bae, H.; Si, M.; Zhou, H.; Bermel, P.A.; Shakouri, A.; Peide, D.Y. High Performance β-Ga$_2$O$_3$ Nano-Membrane Field Effect Transistors on a High Thermal Conductivity Diamond Substrate. *IEEE J. Electron Devices Soc.* **2019**, *7*, 914–918. [CrossRef]
101. Luo, Z.-D.; Zhang, S.; Liu, Y.; Zhang, D.; Gan, X.; Seidel, J.; Liu, Y.; Han, G.; Alexe, M.; Hao, Y. Dual-Ferroelectric-Coupling-Engineered Two-Dimensional Transistors for Multifunctional In-Memory Computing. *ACS Nano* **2022**, *16*, 3362–3372. [CrossRef]
102. Francois, T.; Grenouillet, L.; Coignus, J.; Blaise, P.; Carabasse, C.; Vaxelaire, N.; Magis, T.; Aussenac, F.; Loup, V.; Pellissier, C.; et al. Demonstration of BEOL-compatible ferroelectric Hf$_{0.5}$Zr$_{0.5}$O$_2$ scaled FeRAM co-integrated with 130nm CMOS for embedded NVM applications. In Proceedings of the 2019 IEEE International Electron Devices Meeting (IEDM), San Francisco, CA, USA, 7–11 December 2019.
103. Mikolajick, T.; Park, M.H.; Begon-Lours, L.; Slesazeck, S. From Ferroelectric Material Optimization to Neuromorphic Devices. *Adv. Mater.* **2022**, e2206042. [CrossRef]
104. Okuno, J.; Kunihiro, T.; Konishi, K.; Maemura, H.; Shuto, Y.; Sugaya, F.; Materano, M.; Ali, T.; Kuehnel, K.; Seidel, K. SoC compatible 1T1C FeRAM memory array based on ferroelectric Hf$_{0.5}$Zr$_{0.5}$O$_2$. In Proceedings of the 2020 IEEE Symposium on VLSI Technology, Honolulu, HI, USA, 16–19 June 2020.
105. Sung, M.; Rho, K.; Kim, J.; Cheon, J.; Choi, K.; Kim, D.; Em, H.; Park, G.; Woo, J.; Lee, Y.; et al. Low Voltage and High Speed 1Xnm 1T1C FE-RAM with Ultra-Thin 5nm HZO. In Proceedings of the 2021 IEEE International Electron Devices Meeting (IEDM), San Francisco, CA, USA, 11–16 December 2021.

Disclaimer/Publisher's Note: The statements, opinions and data contained in all publications are solely those of the individual author(s) and contributor(s) and not of MDPI and/or the editor(s). MDPI and/or the editor(s) disclaim responsibility for any injury to people or property resulting from any ideas, methods, instructions or products referred to in the content.

electronics

MDPI

Article

A Novel Structure to Improve the Erase Speed in 3D NAND Flash Memory to Which a Cell-On-Peri (COP) Structure and a Ferroelectric Memory Device Are Applied

Seonjun Choi [1], Jae Kyeong Jeong [1], Myounggon Kang [2] and Yun-heub Song [1,*]

1   Department of Electronics Engineering, Hanyang University, Seoul 04763, Korea;
    msz009@hanyang.ac.kr (S.C.); jkjeong1@hanyang.ac.kr (J.K.J.)
2   Department of Electronics Engineering, Korea National University of Transportation, Chung-ju 27469, Korea;
    mgkang@ut.ac.kr
*   Correspondence: yhsong2008@hanyang.ac.kr; Tel.: +82-2-2220-4136

Abstract: In this paper, a Silicon-Pillar (SP) structure, a new structure to improve the erase speed in the 3D NAND flash structure to which ferroelectric memory is applied, is proposed and verified. In the proposed structure, a hole is supplied to the channel through a pillar in the P+ crystal silicon sub-region located at the bottom of the 3D NAND flash structure to which the COP structure is applied. To verify this, we first confirmed that when the Gate Induced Drain Leakage (GIDL) erasing method used in the 3D NAND structure using the existing Charge Trap Flash (CTF) memory is applied as it is, the operation speed takes more than 10ms, for various reasons. Next, as a result of using the SP structure to solve this problem, even if the conventional erasing method was used until the thickness of the pillar was 20 nm, thanks to the rapidly supplied hole carriers, a fast-erasing rate of 1us was achieved. Additionally, this result is up to 10,000 times faster than the GIDL deletion method. Next, it was confirmed that when the pillar thickness is 10 nm, the erase operation time is greatly delayed by the conventional erasing method, but this can also be solved by appropriately adjusting the operating voltage and time. In conclusion, it was confirmed that, when the proposed SP structure is applied, it is possible to maximize the fast operation performance of the ferroelectric memory while securing the biggest advantage of the 3D NAND flash structure, the degree of integration.

Keywords: ferroelectric; 3D NAND; GIDL; polysilicon

Citation: Choi, S.; Jeong, J.K.; Kang, M.; Song, Y.-h. A Novel Structure to Improve the Erase Speed in 3D NAND Flash Memory to Which a Cell-On-Peri (COP) Structure and a Ferroelectric Memory Device Are Applied. *Electronics* **2022**, *11*, 2038. https://doi.org/10.3390/electronics11132038

Academic Editors: Yi Zhao and ChoongHyun Lee

Received: 10 June 2022
Accepted: 28 June 2022
Published: 29 June 2022

Publisher's Note: MDPI stays neutral with regard to jurisdictional claims in published maps and institutional affiliations.

Copyright: © 2022 by the authors. Licensee MDPI, Basel, Switzerland. This article is an open access article distributed under the terms and conditions of the Creative Commons Attribution (CC BY) license (https:// creativecommons.org/licenses/by/ 4.0/).

1. Introduction

The recent development of NAND flash technology for high density has been extensively utilized in industry. The first vertical channel NAND flash technology was announced by Toshiba in 2007 [1]. The Terabit Cell Array Transistor (TCAT) by Samsung [2] and Stacked Memory Array Transistor (SMArT) by SK Hynix [3] were developed in 2009 based on this structure. Since then, the development of vertical channel structures has grown rapidly. In particular, the layered stack of the vertical channel structure is expected to have a high-end structure of more than 200 layers in 2023 [4–6].

Three-dimensional NAND memory is a high-density, cost-effective technology, but it still has some drawbacks, such as speed, cell size, and power consumption at the system level due to the required peripherals (e.g., charge pump). In particular, the program speed is pointed out as a weak point compared to other memories (ex: Dynamic Random Access Memory (DRAM)) in the NAND flash structure, and attempts to solve this have been studied.

To solve this problem, resistance change memory, such as Phase-Change Random Access Memory (PCRAM), Resistance Random Access Memory (RRAM), and Magnetic Random Access Memory (MRAM) are being actively studied as next-generation memory that can replace charge trap flash (CTF), which is currently used in 3D NAND flash memory. In particular, these resistance change memories have the advantage of relatively fast operating speed and low operating power [6]. However, it is difficult to apply the resistance change memory to a large-scale product, especially when it is applied to the current 3D NAND flash structure, due to the disadvantage that only the one transistor-one resistor structure is possible compared to the CTF device that can be configured with one transistor.

As a result of this problem, most of the resistance-changing memories studied so far do not replace mass storage devices such as Solid-State Disks (SSD) and have limited use in specialized fields such as internal memory. Among these, in the case of ferroelectric memory recently studied using ferroelectric material [7–10], it is only necessary to replace the ONO structure used in the CTF flash with ferroelectric material. Therefore, the NAND flash structure using the ferroelectric memory basically has an advantage in that the existing NAND flash structure and operation method can be used as it is. Due to these strengths, research for applying ferroelectric memory to the 3D NAND structure, the latest structure of NAND flash, has progressed rapidly compared to other memory devices, and there are cases where actual devices have been made to prove the possibility [11].

However, there are several problems in the practical application of ferroelectric memory to 3D NAND flash structures. Among them, the most urgent and major problem is the low operating speed of the Gate Induced Drain Leakage (GIDL) erasing method currently used in 3D NAND flash memory. In the early days of 3D NAND flash development, there was Bulk Erase of the TCAT structure and GIDL deletion of the Bit Cost Scalable (BiCS) structure, but as the structure of 3D NAND flash evolved to the COP structure to improve density, the peri circuit layer was placed underneath. Therefore, although it showed a faster erase speed compared to the GIDL erase method, the bulk erase method, which relies on the p+ sub-region under the string to supply hole carriers to supply the erase voltage to the channel, is deprecated. Therefore, although most of the currently manufactured 3D NAND flash structures use the GIDL erase method, this GIDL erase method has many disadvantages in actual operation because it requires making a hole carrier using the GIDL phenomenon. The biggest problem among them is that the operation speed is extremely slow.

Furthermore, if ferroelectric memory is applied to the 3D NAND flash structure, the GIDL erase method becomes a bigger problem because the operating voltage of the ferroelectric memory is very small compared to the operating voltage of the conventional CTF memory. Because it is the operating voltage that determines the GIDL erase speed, if the erase voltage is decreased to protect the performance of the ferroelectric memory and the ferroelectric film, the operation speed will inevitably be lowered. Therefore, in order to apply ferroelectric memory to the current 3D NAND flash structure, an innovative structure and operation method that can have a fast operation speed even at a low erase voltage is required. If this problem cannot be solved, even if ferroelectric memory is applied to the 3D NAND flash structure, the program speed will only be faster, and the erase speed will be slower than the existing CTF memory. To address this, this paper proposes and validates a Silicon-Pillar (SP) structure capable of supplying high-speed hall carriers by columns inside the 3D NAND flash string.

2. Details of the Proposed Structures

Figure 1 describes the structure proposed in this study by comparing it with the general 3D NAND flash structure. First, it is assumed that a ferroelectric memory is applied to the general 3D NAND flash structure described in (a) and (c). In this structure, the lower P+ crystal sub-region is completely separated from the single memory string and simply plays the role of a substrate for the Peri circuit. Therefore, the supply of hole carriers to increase the channel voltage in the erase operation depends entirely on the GIDL phenomenon that occurs in the N+ doping region (red color) of the GSL and SSL transistor regions. Therefore, in this structure, the erase speed will take about 1.5 ms [12] only for the rise of the channel voltage, so the fast operation speed of the ferroelectric memory will not be reflected at all and will operate very slowly.

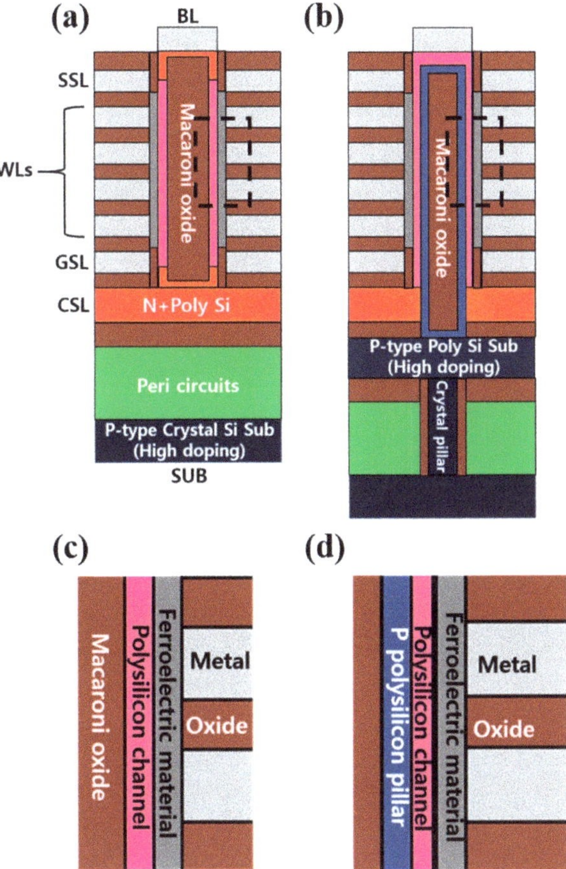

Figure 1. Comparison of the proposed structure with general 3D NAND flash to which ferroelectric memory is applied. The schematic diagrams in (**a**) and (**b**) are a 3D NAND flash structure and a single string of the proposed SP structure. (**c**,**d**) show the details of the proposed structure.

Next, the SP structure described in (b) and (d) is based on the IP structure presented in a previous study [13]. As shown in (b), the proposed SP structure can serve as a passageway for hole carriers rich in crystal pillars connected in the P+ polysilicon sub-region to move upward, unlike the existing 3D NAND flash structure. Next, the P+ polysilicon sub-region on the upper part of the Peri circuits exists to stably supply the holes supplied from the crystal pillar to the upper P polysilicon pillar. Of course, it is ideal that the lower crystal pillar and the upper polysilicon pillar are directly connected one to one. However, in reality, it is almost impossible to align pillars with a diameter of only several tens of nanometers through lithography. Therefore, in order to secure the degree of freedom, an additional sub-region with a large area is absolutely necessary. In the figure, for convenience, the upper polysilicon pillar and the lower crystal silicon pillar are located at the same location, but in the actual structure, they can be freely placed at any position in the P+ polysilicon sub-region.

The supplied hole carriers are quickly supplied to the adjacent intrinsic polysilicon channel on the P polysilicon pillar, and at the same time, the high erase voltage applied from the BL contact and the CSL contact can be transmitted to the channel. Because the expected performance of the proposed SP structure, especially the rising rate of the channel voltage, has already been confirmed in the IP structure published by our previous study, using this structure can improve the very slow erase speed of the existing 3D NAND flash structure. Due to these advantages, the proposed SP structure will become a new structure that can properly utilize the speed of ferroelectric memory in the 3D NAND flash structure.

3. Simulation Results and Discussion

3.1. Simulated Structures, Models, and Parameters

In this paper, Synopsys' Sentaurus TCAD tool [14] was used. The structure editor tool was used to design the proposed structure in Figure 2 above. This structure was used as a Virtual Gate All (GAA) structure during device simulation via the Sentaurus tool's cylindrical command (Rotate 360 degrees). Next, in the composition of each contact, starting with the SUB contact at the bottom, the Common-Source-Line (CSL) contact serves a ground function, and the Ground-Select-Line (GSL) located above it controls a total of 10 word lines (WLs), and the Source-Select-Line (SSL) contact serves as a gate for control Bit-Line(BL). Finally, the uppermost Bit-Line (BL) serves as a drain.

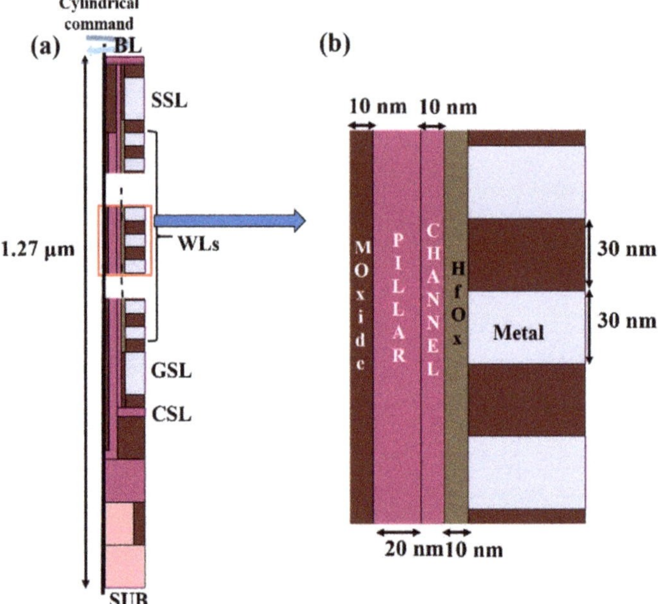

Figure 2. Details of the simulated device. (**a**) Schematic of the 10 stacked memory cells, word lines WL0 to WL9, and select transistors, GSL, SSL, BL, and CSL. (**b**) Expanded view of the simulated structure with dimensions of the constituent parts.

Figure 2b shows an expanded view of the dimensions of the elements used in the simulated device. The Macaroni(M)oxide, pillar, channel, and ferroelectric layer thicknesses were held at 0~20, 30~10, 10, and 10 nm, respectively. Therefore, the total diameter of the simulation device is 100 nm. The word line gate length and the interval between successive word lines were fixed at 30 nm. It is also confirmed that the doping concentration dependence, high field saturation, and trap scattering mobility models were used in the given device simulations. In addition, Shockley–Read–Hall [15], Auger [16], and Hurkx band-to-band recombination models [17] were also used to simulate the operation of the transistor in a 3D NAND flash structure.

Figure 3 shows the doping concentration of the formed structure and additional information on the SSL, GSL, and SUB-regions. First, in the case of the SSL region shown in (b), the length of the SSL gate is set to 100 nm, and this is also the case for the GSL in (c). This is for suppression of leakage current flowing to BL and CSL, respectively, and enhancement of gate control capability. It is known that such a structure or a plurality of gates is used in the actual 3D NAND structure. Second, the doping concentration of the pillar was set to a lower concentration (5×10^{18} cm^{-3}) than that of the lower sub-region. This is in consideration of the difficulties in the process and the change of the threshold voltage due to the increase in the doping concentration. Lastly, the position of the crystal pillar in the sub-region is the same as the upper polysilicon pillar for the convenience of simulation. However, in reality, it will be placed at the optimal position considering the area and performance of the Peri circuits. In addition, the following document [13] was referred to for the selection of the parameters for reflecting the trap characteristics of the polysilicon channel. Finally, among the ferroelectric parameters of the ferroelectric, the Saturation polarization (P_s) was set to 30 uC/cm^2, the Remanent polarization (P_r) was set to 25 uC/cm^2, and the Coercive field (E_c) was set to 2 MV/cm. Additionally, the Time for the polarization was set to 0 s. This setting is too ideal, but in this paper, we want to confirm that the operation of ferroelectric memory can be effectively implemented in existing 3D NAND flash structures, so reliability such as retention and endurance is not considered, and the reliability part is to be confirmed by fabricating actual devices in the next study.

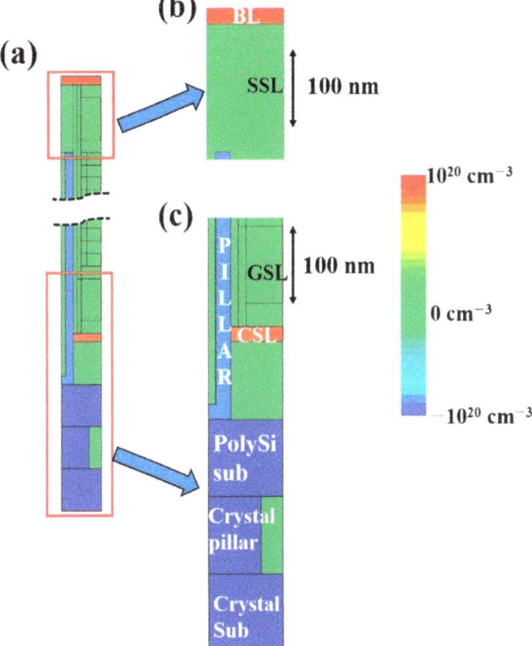

Figure 3. Details of the doping concentration. Schematic of (**a**) the overall doping distribution, (**b**) the SSL region, and (**c**) the GSL and sub-region.

3.2. Analysis of the Erase Speed of Ferroelectric Memory When GIDL Erase Is Used

In this paper, we first analyzed the operation speed, especially the erase speed, when using a ferroelectric memory in a structure using general GIDL erase. In particular, since the ferroelectric memory operates at a lower voltage compared to the existing CTF memory due to its characteristics, it is very important to check the operating speed reflecting this low voltage.

Figure 4 is the result of verifying the erase speed in the GIDL erase structure composed of 10 WLs while having the same geometry as the SP structure described above. First, as shown in T_{PGM} in Figure 4, the program operation that directly applies a voltage to the gate proceeds at a high speed of 80 ns. However, on the contrary, the erase operation that needs to increase the channel voltage due to the channel GIDL phenomenon does not progress at all until 1 ms, and it can be seen that the erasing operation completely proceeds after 10 ms. In particular, the fact that the erase voltage does not increase at all, even when the erase voltage is increased to 8 V, suggests that it is impossible to improve the erase speed by simply increasing the erase voltage.

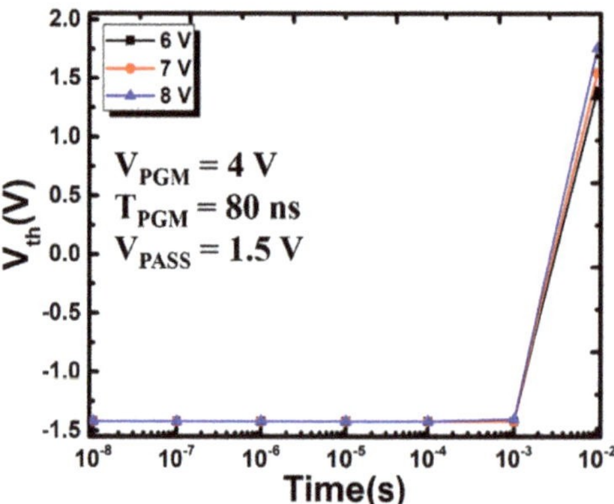

Figure 4. Result of operation speed change according to erase voltage in ferroelectric memory using general GIDL erase structure.

The cause of the slow speed of the erase operation identified in Figure 4 can be explained in Figure 5. First, looking at the change in channel potential through (a), it can be seen that the channel potential rose only 2.2~3 V when the erase operation time progressed to 1 ms. Of course, in this voltage range, E_c (2.5 MV/cm), one of the ferroelectric parameters set for this simulation, is not satisfied, or the erase operation will not proceed because it has a value that is almost meaningless. In fact, a time point at which a channel potential rises to a level at which an erase operation may proceed is 10 ms, and in this case, the channel potential rises to 3.5 to 6.2 V according to a change in an operating voltage (6 to 8 V), and this voltage range greatly exceeds the aforementioned Ec, and thus an erase operation can be performed. Next, looking at the change in the hole carrier density of the channel in the same time zone through (b), it can be seen that it is also very low at about 10^{15} cm^{-3} at the time of 1 ms. With this degree of hole carrier density, it is natural that the erase voltage (6 to 8 V) applied in the BL and CSL contact is not transferred to the channel properly, and this problem may be seen when the normal erase operation is performed only when the hole carrier density exceeds 10^{19} cm^{-3} at a time of 10 ms.

From the above results, it may be seen that the change in the hole carrier density does not show a significant change more than when the voltage exceeds a predetermined value (10^{19} cm^{-3}), but the change in the channel potential gradually increases. Therefore, in order to shorten the operation time, an operation voltage close to 20 V, which is an erase operation voltage of CTF memory, may also be considered. (Of course, in the erase operation, the reduction in power consumption due to the low operating voltage, which is an advantage of ferroelectric memory, should be given up.) However, recklessly increasing the erase operating voltage can be very dangerous. This is because the thickness of the CTF memory layer currently used is about 20 nm, but the thickness of the ferroelectric memory layer is being studied to be about 10 nm, and the breakdown limit of the ferroelectric material itself is known to be only 6 MV/cm.

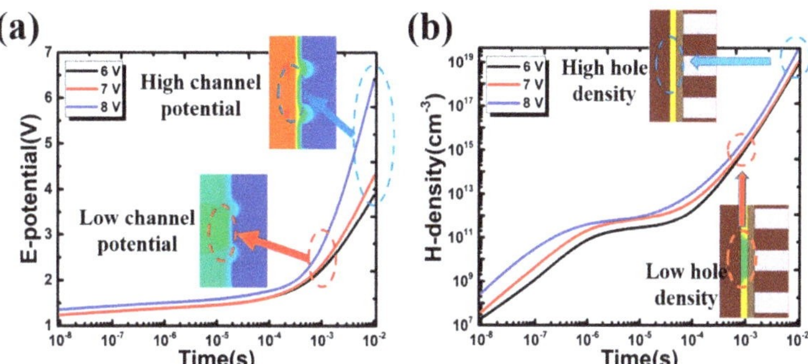

Figure 5. Channel potential change and hole density change by operating voltage (6~8 V) in erase operation. (**a**) shows the channel potential change and (**b**) shows the hole carrier density change in the channel.

As can be seen in Figure 6, the increase in operating voltage sharply increases the E-field and increases stress applied to the ferroelectric thin film, thus increasing the breakdown potential. Until now, various research results have been published on the breakdown of ferroelectric thin films with ferroelectric properties, but assuming 6 MV/cm, which is the breakdown field of a typical ferroelectric thin film [18], the maximum erase operation voltage that can be applied at a thickness of 10 nm is 7.6 V. Of course, assuming a very ideal situation, in order to avoid the possibility of a breakdown of the ferroelectric thin film, it is possible to think of a method of quickly generating hole carriers with a high erase voltage and reducing the erase voltage at the moment when the hole carriers are sufficiently filled in the channel. However, in an actual device, the speed at which the hole carrier is charged and the speed at which the channel voltage is increased may vary for each string by the influence of traps and other defects randomly present on the polysilicon channel or the surface. In addition, it is clear that there will be a difference in the transmission speed of the operation signal in a large-scale string currently used, and even if the difference is only several ns, the breakdown of the ferroelectric layer by high voltage is sufficient. Accordingly, in order to overcome this problem, a structure and an operation method for quickly supplying hole carriers to the channel even with a low operating voltage are required. If this structure and operation method can be secured, the ferroelectric memory can be applied to the 3D NAND structure to catch two rabbits at once: fast operation speed and high integration of the 3D NAND flash structure.

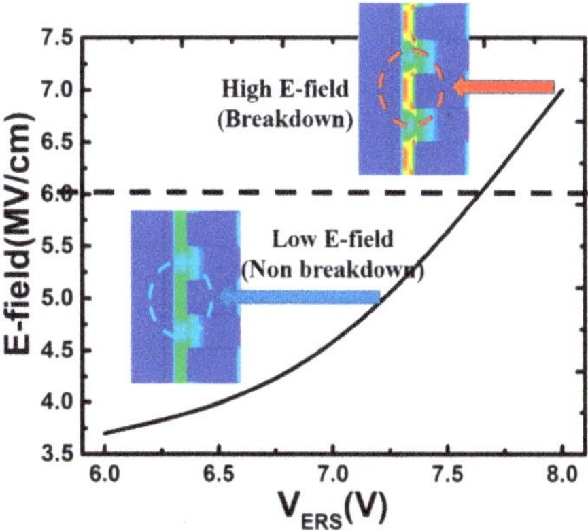

Figure 6. Electric field (E-field) change of the ferroelectric thin film according to the change of the erase operation voltage (6~8 V).

3.3. Analysis of the Erase Speed of Ferroelectric Memory When the Proposed Structure Is Used

In this section, we will check the performance and considerations of the proposed SP structure as a new concept structure mentioned above.

As seen in Figure 7a, the leakage current generated in the SP structure is about 10^{-12} A, which is one order higher than that of the general 3D NAND structure. However, this leakage current can be suppressed by increasing the voltage of the SUB and increasing the potential of the SUB. Instead, what was additionally confirmed was band bending caused by the P-type pillar and the lowering of the operating current due to the disturbance of electrons. This problem can also be solved by increasing the voltage (0 ->2 V) of the SUB contact. Additionally, it can be seen that the voltage rise of the pillar through the SUB contact greatly improves the Subthreshold Swing (SS) of the I-V curve (200 mV/dec -> 86 mV/dec). Unlike the general 3D NAND structure in which the SUB contact cannot exist, the SP structure in which the potential of the entire channel can be kept constant by the pillars shows that a much more stable operation is possible. For this reason, it is shown that the stability in Multi Level Cell (MLC) operation, which is essential for current NAND flash operation, can also be greatly improved. Next, as shown in Figure 7b, the erase performance of the proposed SP structure is very good, and in particular, in terms of erase speed, it can be confirmed that it shows a speed more than 10,000 times faster than that of the GIDL erase method. However, as expected, the erase speed decreases as the thickness of the pillar decreases. In particular, when the thickness is 10 nm, it can be seen that the same speed as the GIDL erase method is shown except for a slight increase in the memory window (0.23 V).

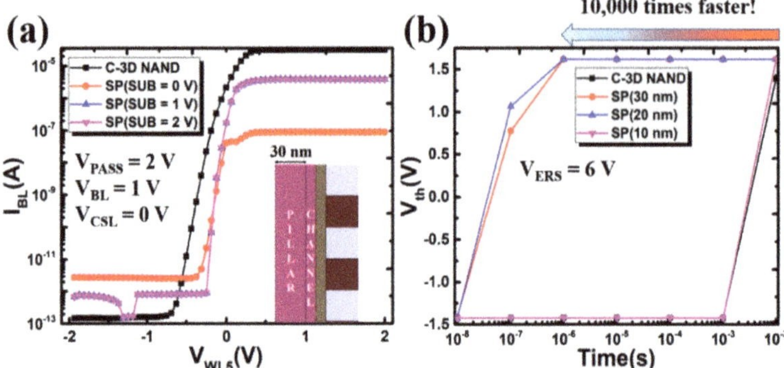

Figure 7. Results of performance analysis of the proposed structure. (**a**) is the result of the I-V operation, and (**b**) is the result of the erase operation.

Figure 8 explains the difference in the erasing speed according to the change in the thickness of the column, as in Figure 7. As expected, it can be seen that the channel potential and hole density rapidly decrease as the thickness of the column decreases. This phenomenon can be said to be because the supply of hole carriers is not smooth due to the decrease in the thickness of the pillar. In particular, when the thickness of the pillar is 10 nm, the channel potential rises by only 0.91 V, and the density of the hole carriers is also close to zero. For this reason, as in the previous results, there is almost no difference between the erase operation speed and the existing GIDL erase operation, and so real erase operation is thought to depend on the GIDL phenomenon, which is operated much later without the supply of hole carriers by the pillar. Therefore, in order to secure a stable erasing operation speed, it is preferable that the thickness of the pillar is 20 nm or more, but it is difficult for the reasons described below. In terms of the actual device, if the pillar thickness is designed to be 30 nm, there is virtually no space for the M-Oxide to be located. This point is practically difficult to use because it gives up the grain boundary suppression function of the polysilicon channel, which is an important function of M-oxide, and it can be said that the structure from 20 nm is realistic. Furthermore, in order to form a more stable polysilicon channel, it is also desirable to increase the thickness of the M-oxide, so it is necessary to enable fast operation even when the thickness of the pillar is 10 nm or less.

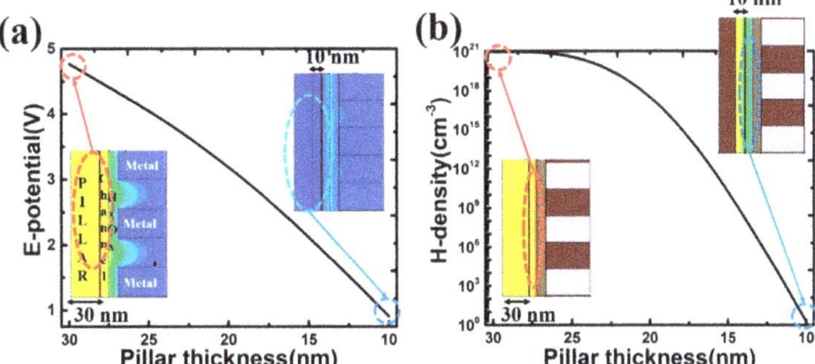

Figure 8. Channel potential change and hole density change by pillar thickness (30~10 nm) in the erase operation. (**a**) shows the channel potential change and hole density, and (**b**) shows hole density change according to the pillar thickness.

3.4. Two-Step Erase Method to Achieve Fast Erase Operation on Structures with a Pillar Thickness of 10 nm or Less

Figure 9 shows the change in hole density in the CSL region, which is the cause of the change in hole density according to the change in the thickness of the pillar shown in Figure 8. When the voltage of the pillar and the voltage of the CSL contact are equal to 6 V, it can be seen that the voltage of the CSL prevents the flow of hole carriers moving through the pillar. However, when the thickness of the pillar is sufficiently thick (30~20 nm), since there is a region not affected by the CSL voltage, hole carriers move through the region to supply the hole carriers to the channel. Conversely, when the thickness of the pillar is thin (10 nm), since the CSL voltage affects the entire pillar, the hole carriers cannot move and are trapped, which prevents the hole carriers from being supplied to the channel.

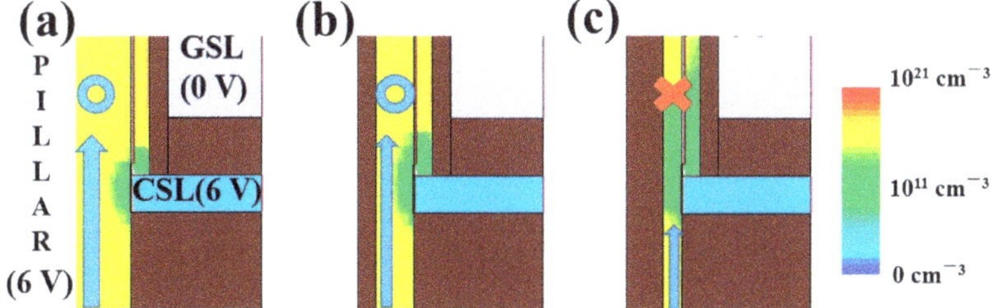

Figure 9. Hole density change by pillar thickness in the erase operation in the CSL region. (**a–c**) show the hole density change according to the pillar thickness change (30 to 10 nm) at 100 ns.

Therefore, in order to solve this problem, it is necessary to use a two-step erase method that does not interfere with the movement of hole carriers by appropriately adjusting the CSL voltage.

In the proposed two-step deletion method, as shown in Figure 10, the voltage of the BL contact and the CSL contact is set to Hole Charging Voltage (V_{hc}) lower than Erase voltage (V_{erase}) during the Hole Charging Time1 (T_{hc1}), and through this, the pillar and channel are charged with hole carriers. The Hole Charging Time2 (T_{hc2}) of the two contacts is properly adjusted so that the hole carrier can be fully transmitted to the channel. With this method, it is possible to sufficiently supply hole carriers to the channel even with a thin pillar. Figure 10c shows the results of comparing the erase performance according to the difference in T_{hc2} when the two-step erase method is used in the SP structure using a pillar with a thickness of 10 nm. Interestingly, when T_{hc2} is 100 ns, the erase speed does not show much difference from the result of the GIDL erase operation but increasing T_{hc2} by 1 us dramatically increases performance.

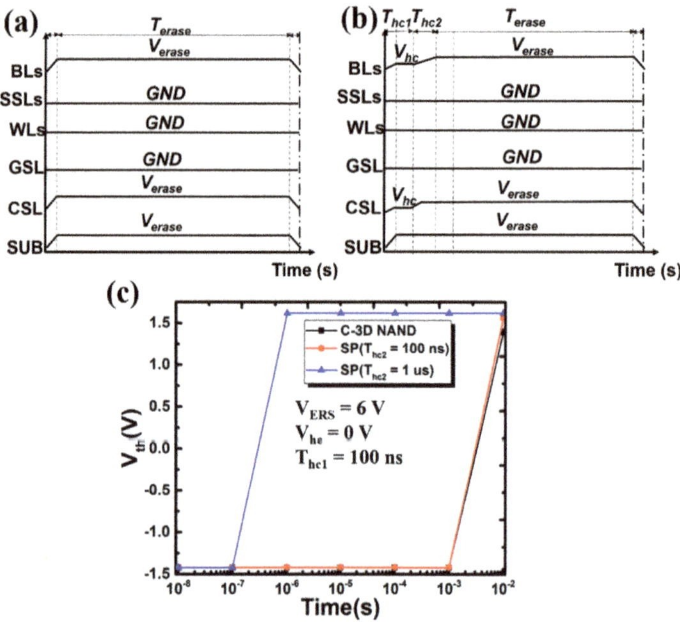

Figure 10. Description and application results of the proposed 2-step operation method. (**a**) describes the existing erase method, (**b**) describes the 2-step method, and (**c**) result of operation speed change according to T_{hc2} in the SP structure (Pillar thickness = 10 nm).

Figure 11 explains how the difference in erase speed of Figure 10c occurs. First, in the T_{hc1} section, the voltage of the CSL and BL contact points is 0 V, and only the SUB voltage is 6 V, so the hole carrier can move quickly through the pillar. The problem is that the T_{hc2} section raises the voltage between the CSL contact and the B contact. If this section is short (100 ns), as shown in (a)-Low channel potential, the hole carriers in the channel are not sufficiently filled due to the rapidly rising CSL voltage. In this state, the supply of hole carriers through the pillar and the change in potential are blocked. Therefore, it can be seen that the CSL contact and the channel region with sufficient hole carriers are far apart, as shown (b)-Low hole density. However, if the time of T_{hc2} is increased to 1us, the distance between the CSL contact point and the point where the hole carriers are sufficient is reduced, as shown (b)-High hole density because the hole carrier supplied through the pillar has enough time to fill the channel. Therefore, finally, the change in the potential of the channel rises from 2.71 V to 4.9 V, and the erase speed also rises sharply accordingly. In conclusion, it was confirmed that even if the thickness of the pillar becomes 10 nm, the proposed structure can show an erase speed of 1 μs if the operation method is properly controlled.

Electronics **2022**, *11*, 2038

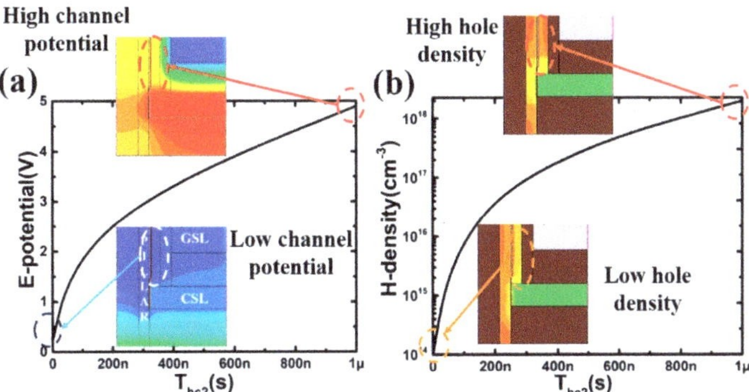

Figure 11. Channel potential change and hole density change according to T_{hc2} change during erase operation. (**a**,**b**) show the channel potential change and the hole density change in the cross-section according to the T_{hc2} change.

4. Conclusions

In this paper, the SP structure, a new structure to improve the erase speed in the 3D NAND flash structure to which ferroelectric memory is applied, was proposed and verified. First, it was confirmed that when the GIDL erase method used in the 3D NAND structure using the existing CTF memory is applied as it is, the operation speed takes more than 10 ms for various reasons. Next, as a result of using the SP structure to solve this problem, even if the existing erase method was used until the thickness of the pillar was 20 nm, a fast erase speed of 1us was achieved thanks to the hole carriers rapidly supplied through the pillar. Next, when the thickness of the pillar is 10 nm, the erase operation time is greatly delayed by the existing erase method, but when the two-step method is used, the hole carrier is sufficiently supplied to the channel, and the erase operation time is maintained at 1 µs. In conclusion, it is confirmed that the application of the proposed SP structure maximizes the fast operating performance of ferroelectric memory and secures the biggest advantage of the 3D NAND flash structure, the degree of integration.

Author Contributions: Methodology, paper writing, investigation, conceptualization, and validation, S.C.; review, J.K.J. and M.K.; funding acquisition, supervision, and editing, Y.-h.S. All authors have read and agreed to the published version of the manuscript.

Funding: This work was supported by a grant from the Samsung Research Funding and Incubation Center of Samsung Electronics under Project Number SRFC-TA1903-04 and was supported by the Technology Innovation Program (20016366) funded By the Ministry of Trade, Industry and Energy (MOTIE, Korea).

Conflicts of Interest: The authors declare no conflict of interest.

References

1. Tanaka, H.; Kido, M.; Yahashi, K.; Oomura, M.; Katsumata, R.; Kito, M.; Fukuzumi, Y.; Sato, M.; Nagata, Y.; Matsuoka, Y.; et al. Bit Cost Scalable Technology with Punch and Plug Process for Ultra High Density Flash Memory. In Proceedings of the 2007 IEEE Symposium on VLSI Technology, Kyoto, Japan, 12–14 June 2007; pp. 14–15.
2. Jang, J.; Kim, H.S.; Cho, W.; Cho, H.; Kim, J.; Shim, S.L.; Jang, Y.; Jeong, H.J.; Son, B.K.; Kim, D.W.; et al. Vertical Cell Array usingTCAT (Terabit Cell Array Transistor) Technology for UltraHigh Density NAND Flash Memory. In Proceedings of the 2009 Symposium on VLSI Technology, Honolulu, HI, USA, 15–17 June 2009; pp. 192–193.
3. Choi, E.S.; Park, S.K. Device Considerations for High Density and Highly Reliable 3D NAND Flash Cell in Near Future. In Proceedings of the 2012 International Electron Devices Meeting, San Francisco, CA, USA, 10–13 December 2012; pp. 211–214.
4. Memory Technology 2020 and Beyond NAND, DRAM, Emerging and Embedded Memory Technology Trends. Available online: https://www.techinsights.com/ko/node/33556 (accessed on 4 June 2022).
5. Johann, A.; Higashitani, M.; Paak, S.S.; Sivaram, S. Past and Future of 3D Flash. In Proceedings of the 2020 IEEE International Electron Devices Meeting (IEDM), Vitually, 12–18 December 2020; pp. 103–106.

6. International Roadmap for Devices and Systemstm 2020 Update More Moore. 2020. Available online: https://irds.ieee.org/images/files/pdf/2020/2020IRDS_MM.pdf (accessed on 4 June 2022).
7. Trentzsch, M.; Flachowsky, S.; Richter, R.; Paul, J.; Reimer, B.; Utess, D.; Jansen, S.; Mulaosmanovic, H.; Müller, S.; Slesazeck, S.; et al. A 28 nm HKMG super low power embedded NVM technology based on ferroelectric FETs. In Proceedings of the 2016 IEEE International Electron Devices Meeting (IEDM), San Francisco, CA, USA, 3–7 December 2016; pp. 11.5.1–11.5.4.
8. Dünkel, S.; Trentzsch, M.; Richter, R.; Moll, P.; Fuchs, C.; Gehring, O.; Majer, M.; Wittek, S.; Müller, B.; Melde, T.; et al. A FeFET based super-low-power ultra-fast embedded NVM technology for 22 nm FDSOI and beyond. In Proceedings of the 2017 IEEE International Electron Devices Meeting (IEDM), San Francisco, CA, USA, 2–6 December 2017; pp. 486–488.
9. Ni, K.; Smith, J.; Ye, H.; Grisafe, B.; Bruce, R.G.; Kummel, A.; Datta, S. A Novel Ferroelectric Superlattice Based Multi-Level Cell Non-Volatile Memory. In Proceedings of the IEEE International Electron Devices Meeting (IEDM), San Francisco, CA, USA, 7–11 December 2019; pp. 669–672.
10. Florent, Y.K.; Pesic, M.; Subirats, A.; Banerjee, K.; Lavizzari, S.; Arreghini, A.; Piazza, L.D.; Potoms, G.; Sebaai, F.; McMitchell, S.R.C.; et al. Vertical Ferroelectric HfO_2 FET based on 3-D NAND Architecture: Towards Dense Low-Power Memory. In Proceedings of the IEEE International Electron Devices Meeting, San Francisco, CA, USA, 1–5 December 2018; pp. 43–46.
11. Kim, M.K.; Kim, I.J.; Lee, J.S. CMOS-compatible ferroelectric NAND flash memory for high-density, low-power, and high-speed three-dimensional memory. *Sci. Adv.* **2021**, *7*, eabe1341. [CrossRef] [PubMed]
12. Caillat, C.; Beaman, K.; Bicksler, A.; Camozzi, E.; Ghilardi, T.; Huang, G.; Liu, H.; Liu, Y.; Mao, D.; Mujumdar, S.; et al. 3DNAND GIDL-Assisted Body Biasing for Erase Enabling CMOS Under Array (CUA) Architecture. In Proceedings of the 2017 IEEE International Memory Workshop (IMW), Monterey, CA, USA, 14–17 May 2017. [CrossRef]
13. Choi, S.; Kim, B.; Jeong, J.K.; Song, Y.H. A Novel Structure for Improving Erase Performance of Vertical Channel NAND Flash With an Indium-Gallium-Zinc-Oxide Channel. *IEEE Trans. Electron Devices* **2019**, *66*, 4739–4744. [CrossRef]
14. *Sentaurus Device User Guide*; Version J-2014.9; Synopsys Inc.: Mountain View, CA, USA, 2014.
15. Fossum, J.G.; Mertens, R.P.; Lee, D.S.; Nijs, J.F. Carrier Recombination and Lifetime in Highly Doped Silicon. *Solid-State Electron.* **1983**, *26*, 569–576. [CrossRef]
16. Huldt, L.; Nilsson, N.G.; Svantesson, K.G. The temperature dependence of bandto-band Auger recombination in silicon. *Applied Physics Lett.* **1979**, *35*, 776–777. [CrossRef]
17. Hurkx, G.A.M.; Klaassen, D.B.M.; Knuvers, M.P.G. A New Recombination Model for Device Simulation Including Tunneling. *IEEE Trans. Electron Devices* **1992**, *39*, 331–338. [CrossRef]
18. Conley, J.F.; Ono, Y.; Solanki, R.; Stecker, G.; Zhuang, W. Electrical properties of HfO_2 deposited via atomic layer deposition using $Hf(NO_3)_4$ and H_2O. *Appl. Phys. Lett.* **2003**, *82*, 3508–3510. [CrossRef]

 *electronics*

MDPI

Communication

Ti/HfO$_2$-Based RRAM with Superior Thermal Stability Based on Self-Limited TiO$_x$

Huikai He [1], Yixin Tan [1,2], Choonghyun Lee [3,*] and Yi Zhao [4]

[1] China Nanhu Academy of Electronics and Information Technology, Jiaxing 314001, China; hehuikai@cnaeit.com (H.H.); yxtan20@mail.ustc.edu.cn (Y.T.)
[2] Institute of Advanced Technology, University of Science and Technology of China, Hefei 230026, China
[3] College of Information Science and Electronic Engineering, Zhejiang University, Hangzhou 310058, China
[4] School of Integrated Circuits, East China Normal University, Shanghai 200241, China; yizhao@ic.ecnu.edu.cn
* Correspondence: chlee01@zju.edu.cn

Abstract: HfO$_2$-based resistive random-access memory (RRAM) with a Ti buffer layer has been extensively studied as an emerging nonvolatile memory (eNVM) candidate because of its excellent resistive switching (RS) properties and CMOS process compatibility. However, a detailed understanding of the nature of Ti thickness-dependent RS and systematic thermal degradation research about the effect of post-metallization annealing (PMA) time on oxygen vacancy distribution and RS performance still needs to be included. Herein, the impact of Ti buffer layer thickness on the RS performance of the Al/Ti/HfO$_2$/TiN devices is first addressed. Consequently, we have proposed a simple strategy to regulate the leakage current, forming voltage, memory window, and uniformity by varying the thickness of the Ti layer. Moreover, it is found that the device with 15 nm Ti shows the minimum cycle-to-cycle variability (CCV) and device-to-device variability (DDV), good retention (10^5 s at 85 °C), and superior endurance (10^4). In addition, thermal degradation of the Al/Ti(15 nm)/HfO$_2$/TiN devices under different PMA times at 400 °C is carried out. It is found that the leakage current increases and the forming voltage and memory window decrease with the increase in PMA time due to the thermally activated oxidation of the Ti. However, when the PMA time increases to 30 min, the Ti can no longer capture oxygen from HfO$_2$ due to the formation of self-limited TiO$_x$. Therefore, the device shows superior thermal stability with a PMA time of 90 min at 400 °C and no degradation of the memory window, uniformity, endurance, or retention. This work demonstrates that the Ti/HfO$_2$-based RRAM shows superior back-end-of-line compatibility with high thermal stability up to 400 °C for over an hour.

Keywords: RRAM; HfO$_2$; Ti buffer layer; thermal stability; self-limited TiO$_x$

Citation: He, H.; Tan, Y.; Lee, C.; Zhao, Y. Ti/HfO$_2$-Based RRAM with Superior Thermal Stability Based on Self-Limited TiO$_x$. *Electronics* **2023**, *12*, 2426. https://doi.org/10.3390/electronics12112426

Academic Editor: Ana-Maria Lepadatu

Received: 18 April 2023
Revised: 23 May 2023
Accepted: 23 May 2023
Published: 26 May 2023

Copyright: © 2023 by the authors. Licensee MDPI, Basel, Switzerland. This article is an open access article distributed under the terms and conditions of the Creative Commons Attribution (CC BY) license (https://creativecommons.org/licenses/by/4.0/).

1. Introduction

Recently, RRAM based on transition metal oxides (TMOs) has attracted increasing attention as one of the most promising candidates for the next generation of eNVM because of its prominent advantages, including its low cost, high integration density, fast switching speed, high endurance, and good CMOS process compatibility [1–9]. In recent decades, RS characteristics have been observed and extensively investigated in various TMOs such as HfO$_2$, TaO$_x$, TiO$_2$, AlO$_x$, NiO$_x$, and ZrO$_x$ [10–15]. Among these TMOs, HfO$_2$ is one of the most representative candidates owing to its high endurance, fast switching speed, and excellent thermal stability [10,16,17].

However, HfO$_2$ deposited by atomic layer deposition (ALD) tends to be a stoichiometric film with few defects or vacancies, resulting in high forming voltage and hard breakdown during the forming operation [10]. This forming process, which involves the application of a voltage typically higher than the operation voltage, induces worse RS performance, larger power consumption, and an additional burden on circuit design. Therefore, to increase the initial oxygen vacancy concentration in the HfO$_2$ to reduce the forming voltage and improve the RS performance, reactive metals such as Ta, Ti, and Hf are generally introduced on the top of the HfO_2 layer as oxygen-scavenging metals [18]. Up to now, excellent RS properties, including superior endurance (10^{10}), good retention characteristics (10 years at room temperature), and low switching energy (0.1 pJ per bit operation), have been demonstrated in Ti/HfO_2-based RRAM [19,20]. Although the importance of the Ti buffer layer is well understood, the detailed interpretation of Ti thickness-dependent RS has not been fully understood, which significantly limits its applications. Recently, other researchers have found that the leakage current increases and the forming voltage decreases as the Ti thickness increases, and the RS properties can be regulated by changing the Ti thickness [21–23]. However, due to the insufficient thickness of the Ti layer, the saturation of the leakage current and the forming voltage with increasing Ti thickness was not observed, resulting in some interesting phenomena being missed. In addition, the effect of the Ti thickness on the RS uniformity has yet to be investigated. Therefore, it is urgent to comprehensively reveal the relationship between Ti thickness and oxygen vacancy concentration in the HfO_2 layer and regulate the RS behavior by changing Ti thickness.

Notably, both the fabrication process of the RRAM device and the post-RRAM process need to be carried out at 400 °C. Furthermore, since HfO_2-based devices operate based on the formation/fracture of conductive filaments originating from the migration of oxygen vacancies, they are vulnerable to structural defects caused by the heat applied during the annealing process. Therefore, to enable the mass production of RRAM, it is critical to meet the thermal budget requirements of the CMOS process. However, most research on HfO_2-based RRAM has focused on improving the RS characteristics of the device. In contrast, there are few reports on the impact of the PMA process on RS performance [24]. Additionally, the cumulative duration of the post-RRAM thermal process, including chemical vapor deposition, rapid thermal annealing, ALD, furnace annealing, and UV curing, usually exceeds one hour. However, up to now, most of the reported thermal stability tests have lasted less than or equal to one hour [25–29]. More importantly, there is no systematic research about the effect of PMA time on oxygen vacancy distribution and RS properties.

In this work, the HfO_2-based RRAM devices with different thicknesses of Ti buffer layer are fabricated. It is found that the leakage current, forming voltage, memory window, and uniformity can be regulated by varying the thickness of the Ti buffer layer. Owing to the low forming voltage, the device with 15 nm Ti presents superior uniformity, good retention, and excellent endurance. In addition, the $Al/Ti(15\ nm)/HfO_2/TiN$ devices with different PMA times at 400 °C are compared and analyzed. It is found that the leakage current (the forming voltage and ON/OFF ratio) increases (decrease) with the increase of the PMA time and remains unchanged when the PMA time increases to 30 min due to the formation of self-limited TiO_x. More importantly, the memory window, uniformity, endurance, and retention show no obvious degradation with a PMA time of 90 min at 400 °C, indicating superior thermal stability of the present devices.

2. Materials and Methods

The fabrication process of the $Al/Ti/HfO_2/TiN$ devices is described in Figure 1a. To fabricate the $Al/Ti/HfO_2/TiN$ devices, 15 nm TiN was first deposited on the heavily doped Si as the bottom electrode by ALD. The precursors used for TiN deposition are ammonia and titanium tetrachloride. Subsequently, pure HfO_2 with a thickness of 8 nm was deposited by ALD using tetrakis(dimethylamino)hafnium (TDMAHf) and H_2O as precursors. The chamber temperature was set at 250 °C. After the HfO_2 deposition, five different thicknesses of the Ti buffer layer (0, 5, 10, 15, and 20 nm) on top of the HfO_2 layer were deposited by the sputtering system. Al capping layer with a thickness of 80 nm to protect Ti from oxidation was then deposited by thermal evaporation using the same sputtering system without breaking the vacuum. The TE layer (Al/Ti) was patterned via the lift-off process. Finally, PMA was performed for 0, 10, 15, 30, 60, and 90 min at 400 °C in the N_2 atmosphere. The schematic of the prepared $Al/Ti/HfO_2/TiN$ devices is shown in Figure 1b.

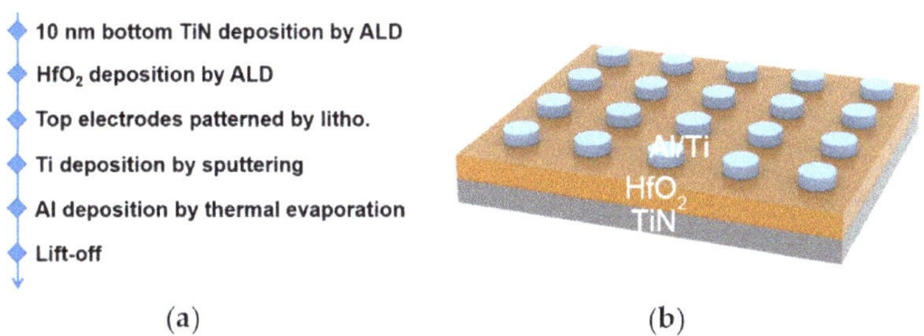

(**a**)　　　　　　　　　　　　　　　　(**b**)

Figure 1. (**a**) Process flow and (**b**) schematic of the $Al/Ti/HfO_2/TiN$ RRAM devices with a Ti buffer layer.

X-ray photoelectron spectroscopy (XPS) characterizations were performed using a Kratos Axis Ultra system to analyze the chemical states of the samples with a monochromated Al anode. The source operated at a voltage of 14 kV and an emission of 8 mA. The vacuum of the analysis chamber was higher than 5×10^{-9} Torr before measurements. The binding energies were calibrated for the sample charging effect by referencing the C 1s peak at 284.8 eV. Before XPS characterization, the TiO_x on the surface was first removed by Ar sputtering in the load lock chamber. To ensure consistency between different samples, special attention needs to be paid to keeping the etching time constant (5 s). The etching rate was roughly 0.7 nm/s.

An Agilent B1500A semiconductor parameter analyzer conducted electrical measurements in the atmospheric ambient. The voltage bias was applied to the top electrode while the bottom electrode was grounded.

3. Results

3.1. Ti Thickness-Dependent RS

In order to reveal the effect of Ti buffer layer thickness on the RS performance, electrical properties, including leakage current (J_g), forming voltage (V_f), and ON/OFF ratio of the $Al/Ti/HfO_2/TiN$ devices with different Ti thicknesses, are displayed in Figure 2. The J_g represents the current measured at the initial resistance with a read voltage of 0.1 V. Notably, over 20 devices were tested for each structure to compare. As shown in Figure 2a, the mean values of J_g of the $Al/Ti/HfO_2/TiN$ devices with five different Ti thicknesses (0, 5, 10, 15, and 20 nm) are 1.27×10^{-11}, 1.18×10^{-10}, 3.52×10^{-10}, 8.48×10^{-10}, and 8.45×10^{-10} A, respectively. Correspondingly, the V_f values are 4.4, 2.4, 1.6, 1.35, and 1.35 V, respectively. It is found that J_g increases and V_f decreases as the Ti thickness increases due to the high oxygen-gettering ability of the Ti. When the Ti thickness increases from 0 to 15 nm, more oxygen vacancies in the HfO_2 are introduced, resulting in a higher J_g and a lower V_f. However, J_g and V_f remain almost unchanged when the Ti thickness increases from 15 to 20 nm. It can probably be explained that the oxidation of Ti is self-limiting and that the oxygen vacancy concentration in HfO_2 reaches a constant value. Yang et al. demonstrated that self-limiting oxidation comes from the residual compressive stress in TiO_2 [30]. When the potential barrier resulting from compressive deformation is higher than the activation energy of oxygen diffusion and reaction, self-limiting oxidation occurs.

After the forming process, bipolar RS behaviors were obtained, where the device could be SET (the switching from high resistance state (HRS) to low resistance state (LRS)) after the positive voltage sweeping and RESET (the switching from LRS to HRS) after the negative voltage sweeping. Figure 2b shows the ON/OFF ratio distribution of the $Al/Ti/HfO_2/TiN$ devices with different Ti thicknesses. Notably, the ON/OFF ratio cannot be obtained in the device without the Ti buffer layer, since no RS was observed after the forming process. It is found that the ON/OFF ratio shows a slight decrease when the Ti thickness increases from 5 to 15 nm, which suggests that the initially created oxygen vacancies also play a vital role in switching cycles. Moreover, the ON/OFF ratio remains almost unchanged when the Ti thickness exceeds 15 nm due to the formation of self-limited TiO_x.

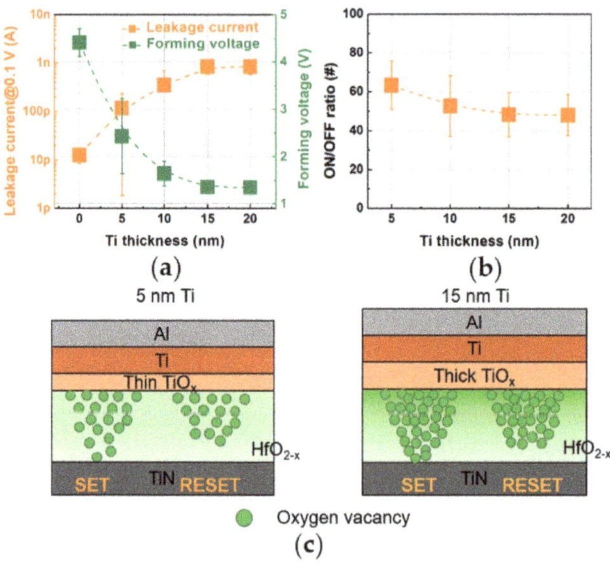

Figure 2. (a) The J_g and V_f of the Al/Ti/HfO$_2$/TiN devices with different Ti thicknesses. (b) ON/OFF ratio distribution of the Al/Ti/HfO$_2$/TiN devices with different Ti thicknesses. (c) Schematics of the role of Ti thickness in the RS performance of the HfO$_2$-based RRAM devices during SET and RESET operations. Green circle represents oxygen vacancies.

Based on the above experimental results, schematic diagrams of the generalized model to understand the role of Ti thickness in the RS performance of the present devices are shown in Figure 2c. When a thin Ti layer is inserted on top of the HfO$_2$, the RS behavior is generally unstable due to the higher V_f originating from pre-existing oxygen vacancies. Note that more oxygen vacancies are introduced in the HfO$_2$ layer, and a lower V_f can be obtained by selecting a thick Ti layer (15 nm). However, the ON/OFF ratio decreases (see Figure 2b), which may result from the shorter oxygen vacancy gap region during switching cycles compared to the sample with a thin Ti layer.

Additionally, to further investigate the effect of Ti buffer layer thickness on the RS uniformity, the CCV and DDV of the Al/Ti/HfO$_2$/TiN devices with different thicknesses of the Ti buffer layer (5, 10, and 15 nm) are displayed in Figure 3. It is worth noting that the data for the devices with 20 nm Ti is not displayed in Figure 3, since the devices with 20 nm Ti exhibited a similar RS behavior to those with 15 nm Ti. Regarding the extraction of SET voltage (V_{SET}), we employ a numerical technique that identifies the voltage at which a sudden change occurs in the current. This abrupt transition signifies the initiation of the SET process. In contrast, the LRS and HRS are determined by measuring the resistance values at 0.1 V after the completion of the SET and RESET processes, respectively. Figure 3a,b present the statistical distribution of V_{SET}, LRS, and HRS obtained by 100 dc sweep cycles. It can be clearly observed that the sample with a thicker Ti buffer layer shows improved control of tail bits in the distribution of V_{SET}, HRS, and LRS, indicating that the device with a 15 nm Ti buffer layer exhibits a minimum CCV. In addition, we statistically analyze 200 *I-V* curves collected in 10 devices for each device with different Ti thicknesses (5, 10, and 15 nm) and quantify the DDV of the V_{SET}, HRS, and LRS by calculating the coefficient of variation (C_V) as the standard deviation (σ) divided by the mean value (μ). As displayed in Figure 3c,d, both V_{SET}, HRS, and LRS follow a Gaussian distribution:

$$y = y_0 + A \times \exp\left(-\frac{(x-\mu)^2}{2\sigma^2}\right), \tag{1}$$

where y_0 and A are constants, μ is the expectation, representing the mean value of the voltage, and σ is the standard deviation, representing the concentration of the voltage distribution. C_V is the ratio of σ and μ, reflecting the concentration of the distribution. Furthermore, the V_{SET}, HRS, and LRS of the device with 15 nm Ti show the narrowest distribution, and the DDV of the V_{SET}, HRS, and LRS is 12.85%, 17.07%, and 17.16%, respectively. The above results demonstrate that the device with a 15 nm Ti shows the best RS uniformity, which may originate from the initial high oxygen vacancy concentration, reduced forming voltage, and TiO$_x$ formation [26,31].

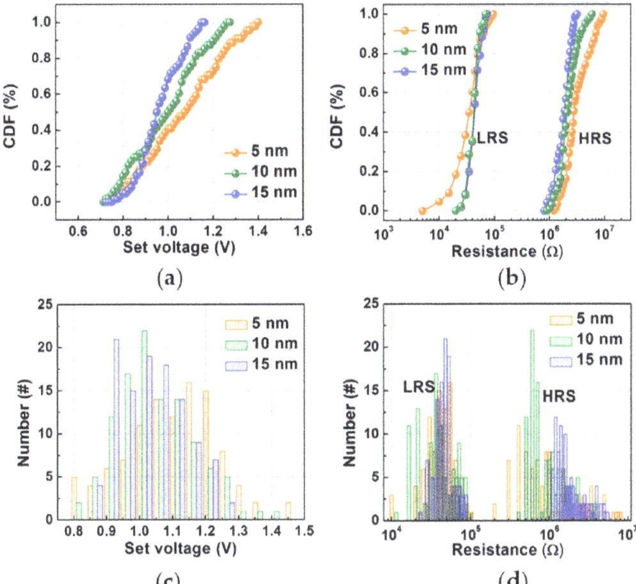

Figure 3. The statistical distribution of (**a**) V_{SET} and (**b**) LRS/HRS obtained by 100 dc sweep cycles. Cumulative distribution of (**c**) V_{SET} and (**d**) LRS/HRS collected from 200 DC cycles of 10 randomly selected devices.

Since the HfO$_2$-based RRAM device with a 15 nm Ti buffer layer shows the best uniformity, the RS behaviors of the present devices are systematically investigated, as displayed in Figure 4. After the application of a forming voltage (~1.3 V), the present device exhibited stable bipolar RS behavior when the voltage was swept between -1.5 V and 1.5 V, as displayed in Figure 4a. While a continuous voltage swept from 0 V to 1.5 V$\rightarrow-1.5$ V$\rightarrow$0 V, an obvious hysteresis loop was obtained, where SET occurred at about 1 V and RESET happened at about -1 V. It is worth noting that a compliance current of 50 µA was applied during the forming and SET processes to prevent irreversible breakdown, while no compliance current was applied during the RESET process. This hysteresis behavior is reproducible in the consecutive 100 voltage sweeps, demonstrating stable RS characteristics in the Al/Ti/HfO$_2$/TiN devices. The retention characteristics were measured at 85 °C, as shown in Figure 4b. The ON/OFF ratio is larger than 50, and the current values of the HRS and LRS show no degradation within 10^5 s, indicating superior retention characteristics.

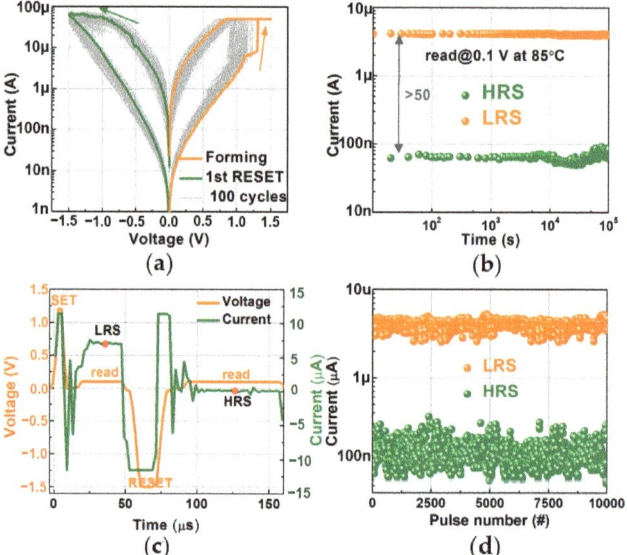

Figure 4. (**a**) The forming process, the first RESET process, and subsequent 100 cycles of *I-V* curve of the Al/Ti(15 nm)/HfO$_2$/TiN device. The arrow represents the sweeping direction. (**b**) Current values of the HRS and the LRS with a read voltage of 0.1 V at 85 °C. (**c**) The measured current when one sequence of SET, read, RESET, and read pulses is applied. (**d**) Endurance of the Al/Ti(15 nm)/HfO$_2$/TiN device.

In addition to stable bipolar RS behavior and good retention characteristics, the present device exhibits superior endurance. When one sequence of SET (1.2 V, 1 μs), read (0.1 V, 30 ms), RESET (−1.5 V, 10 μs), and read (0.1 V, 50 ms) pulses was applied, it was observed that the current value after the SET pulse is several tens of times that of the current value after the RESET pulse (see Figure 4c), demonstrating the presence of the RS behavior. To obtain the endurance for one cycle, one can record one current data point in each read pulse. By repeatedly applying the above pulse sequence, an endurance plot showcasing the current levels of HRS and LRS for 10^4 cycles was achieved (see Figure 4d), indicating the superior endurance of the present device.

3.2. Systematic Thermal Degradation Research about the Effect of PMA Time on Oxygen Vacancy Distribution and RS Performance

The electrical properties of the Al/Ti(15 nm)/HfO$_2$/TiN devices with different PMA times were compared and analyzed to investigate the thermal budget effects on the RS characteristics. Figure 5a presents the distribution of J_g and V_f of the Al/Ti(15 nm)/HfO$_2$/TiN devices with increasing PMA time from 0 to 90 min. The mean values of V_f in the present devices with increasing PMA times (0, 10, 15, 30, 60, and 90 min) are 1.35, 1.05, 1.00, 0.96, 0.96, and 0.96 V, respectively. Correspondingly, the J_g values are 1.1×10^{-9}, 2.44×10^{-8}, 3.36×10^{-8}, 4.00×10^{-8}, 3.77×10^{-8}, and 4.09×10^{-8} A, respectively. Notably, for the devices with PMA for over 30 min, the V_f (0.96 V) is close to the V_{SET} (0.9–1.0 V), indicating that the devices show forming-free characteristics. It is obvious that J_g increases and V_f decreases with the increase of the PMA time from 0 to 30 min due to thermally activated oxidation of the Ti buffer layer. However, J_g and V_f remain almost unchanged when the PMA time is over 30 min, which indicates that self-limiting oxidation has occurred.

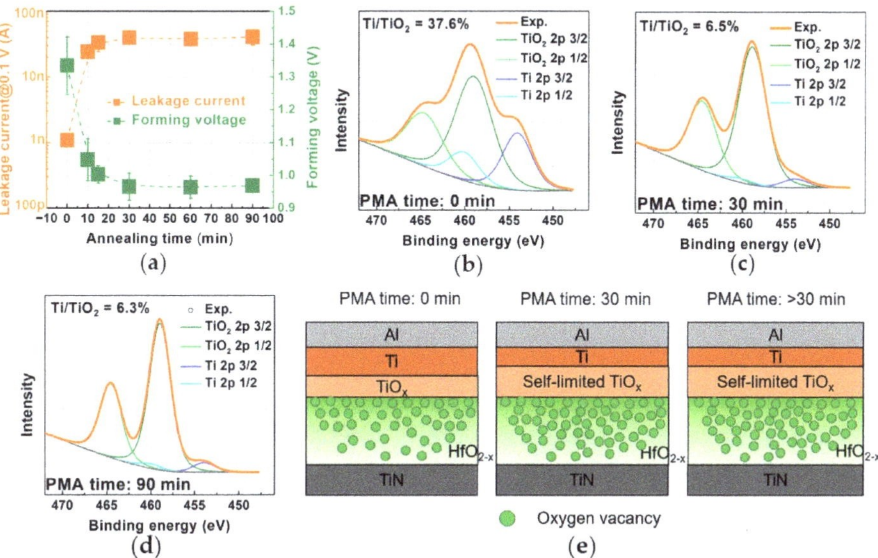

Figure 5. (**a**) The distribution of J_g and V_f of the Al/Ti(15 nm)/HfO$_2$/TiN devices with increasing PMA time from 0 to 90 min. Ti 2p spectra of three samples with the structure of Ti/HfO$_2$/TiN using PMA times of (**b**) 0 min, (**c**) 30 min, and (**d**) 90 min, respectively. All spectra are normalized and fitted with Ti 2p and TiO$_2$ 2p peaks. (**e**) Schematic diagrams of the model to understand the role of the self-limited TiO$_x$ in thermal stability of the Al/Ti/HfO$_2$/TiN device with increasing PMA time.

Furthermore, to demonstrate the occurrence of self-limiting oxidation, XPS characterizations were employed to reveal the atomic composition changes of Ti on the surface of three samples with the structure of Ti/HfO$_2$/TiN using PMA times of 0, 30, and 90 min. Notably, the TiO$_x$ on the surface is first removed by Ar sputtering in the load lock chamber of the XPS system. Figure 5b–d show the normalized Ti 2p core-level spectra of three samples with increased PMA time. Quantitative peak analysis was performed by fitting spin-orbit splitting components with Doniach Sunjic (Ti) and Gaussian-Lorentzian (TiO$_2$) line shapes to the spectrum. Each Ti 2p orbital doublet peak was fitted with an area ratio of 1:2, and the spin orbit splitting of Ti and TiO$_2$ was 6.1 eV and 5.60 eV, respectively, consistent with the values in the literature [32]. To highlight the changes in the Ti oxidation among these three samples, the ratio between the Ti and the TiO$_2$ intensity was calculated and further expressed as Ti/TiO$_2$. When the PMA time increases from 0 to 30 min, the Ti/TiO$_2$ ratio decreases from 37.6% to 6.5%, demonstrating enhanced Ti oxidation at the Ti/HfO$_2$ interface because of the thermally activated oxidation of the Ti. More importantly, the Ti/TiO$_2$ ratio is nearly unchanged when the PMA time further increases to 90 min, suggesting that the oxidation of Ti has stopped due to the formation of self-limited TiO$_x$.

In order to better understand the role of the self-limited TiO$_x$ in the thermal stability of the Al/Ti/HfO$_2$/TiN device, schematic diagrams of the model with increasing PMA time are depicted in Figure 5e. For the device without PMA, due to the high oxygen-gettering ability of the Ti, the Ti captures oxygen from the HfO$_2$ layer and is spontaneously oxidized to TiO$_x$. Meanwhile, a large number of oxygen vacancies are introduced in the HfO$_2$ layer. When the PMA time increases to 30 min, the TiO$_x$ layer becomes thicker because of the thermally activated oxidation of the Ti. Therefore, more oxygen vacancies are introduced in the HfO$_2$ layer, resulting in higher J_g and lower V_f. When the PMA time further increases, the TiO$_x$ layer is thick enough that the remote Ti cannot capture oxygen from HfO$_2$, leading to the occurrence of self-limiting oxidation. Once the self-limited TiO$_x$ is formed, the J_g and V_f become saturated (see Figure 5a).

Furthermore, the *I-V* curves of the Al/Ti(15 nm)/HfO$_2$/TiN devices with increasing PMA time from 0 to 90 min are compared in Figure 6a. To confirm the fair comparison of the devices with different PMA times, a consistent compliance current of 100 µA was applied during the SET process. It is observed that all the devices show similar bipolar RS behavior. However, the HRS and LRS degraded clearly with the increase in PMA time. To further verify the relationship between PMA time and RS performance, the distribution of HRS/LRS and ON/OFF ratio with increasing PMA time from 0 to 90 min is summarized in Figure 6b. When the PMA time increases from 0 to 30 min, the HRS (LRS) degrades from 2.75 MΩ to 519 KΩ (from 57 KΩ to 16 KΩ), and the ON/OFF ratio exhibits a slight decrease from 48.2 to 34.8. This trend can be explained by the formation of permanent vacancies in the film due to the thermally activated oxidation of the Ti buffer layer. Notably, the HRS/LRS and ON/OFF ratios remain nearly constant when the PMA time is over 30 min, similar to the trend of the J_g and V_f caused by the formation of self-limited TiO$_x$.

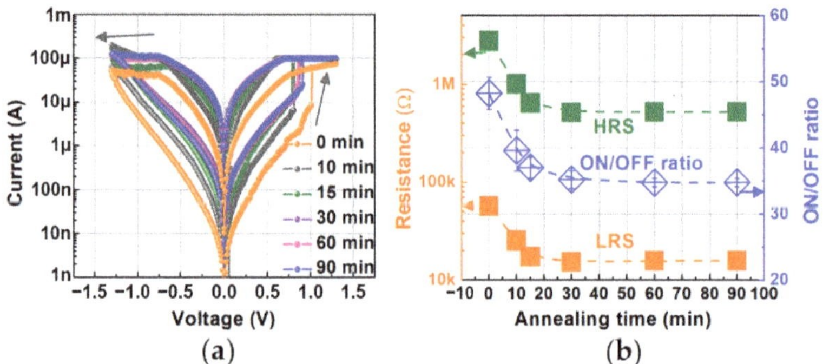

Figure 6. (a) $I-V$ curves of the Al/Ti(15 nm)/HfO$_2$/TiN devices with increasing PMA time from 0 to 90 min. The PMA temperature is fixed at 400 °C. The arrow represents the sweeping direction. (b) The distribution of HRS/LRS and ON/OFF ratio of the Al/Ti(15 nm)/HfO$_2$/TiN devices with increasing PMA time from 0 to 90 min. The PMA temperature is fixed at 400 °C.

Additionally, to further verify the thermal stability of the Al/Ti(15 nm)/HfO$_2$/TiN devices, the switching properties of the devices with PMA times of 0 and 90 min are compared in Figure 7. Although the device with a PMA time of 90 min shows lower V_{SET} and HRS/LRS, both devices have similar distributions of V_{SET} and HRS/LRS, indicating that the uniformity of the Al/Ti(15 nm)/HfO$_2$/TiN device shows no degradation after PMA (see Figure 7a,b). In addition, after PMA for 90 min, the device still exhibits superior retention characteristics (10^4 s) and good endurance (10^3), as shown in Figure 7c,d. These results demonstrate that the Ti/HfO$_2$-based device shows high thermal stability up to 400 °C for over an hour.

Table 1 summarizes the performance comparison with previous works using PMA. Most of the reported HfO$_2$-based or TaO$_x$-based RRAM with PMA [25–28] generally used Pt as the bottom electrode or the capping layer, leading to poor compatibility with the CMOS process. In addition, the duration of most reported thermal stability tests was less than or equal to one hour [25–29]. In our work, the devices show forming-free characteristics with good CMOS compatibility. More importantly, the duration of PMA was more than 1 h in this work, and the effect of PMA time on oxygen vacancy distribution and RS performance was systematically studied.

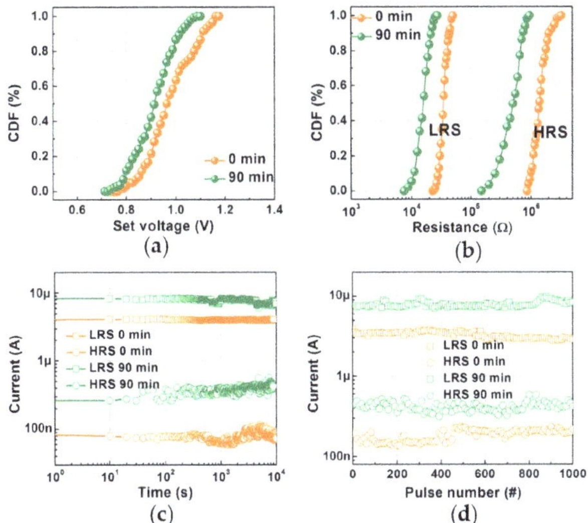

Figure 7. The statistical distribution of (**a**) V_{SET} and (**b**) LRS/HRS obtained by 100 dc sweep cycles of the Al/Ti(15 nm)/HfO$_2$/TiN devices with PMA time of 0 and 90 min. The temperature is fixed at 400 °C. (**c**) Endurance and (**d**) retention characteristics of the Al/Ti(15 nm)/HfO$_2$/TiN devices with PMA times of 0 and 90 min. The temperature is fixed at 400 °C.

Table 1. Performance comparison with previous works using PMA.

Structure	Forming Voltage	Temp.	Time	Ref.
Ta/TaO$_x$/Pt	Forming free	400 °C	30 s	[25]
Pt/Ti/HfO$_2$/Pt	Forming free	500 °C	10 min	[26]
Al/Ta/HfO$_2$/Pt	8.79 V	400 °C	60 min	[27]
Pt/Ti/HfO$_2$/Pt	Forming free	300 °C	60 s	[28]
TiN/Ti/HfO$_2$/TiN	3.8 V	400 °C	60 min	[29]
TiN/Ti/TiN/ HfO$_2$/TiN	1.2 V	400 °C	180 min	[31]
Al/Ti/HfO$_2$/TiN	Forming free	400 °C	90 min	This work

4. Conclusions

In this work, the effect of the Ti buffer layer thickness on the RS performance has been systematically investigated. It was found that the device with a thick Ti shows improved uniformity and RS properties compared with the device with a thin Ti. The Ti buffer layer plays a vital role in the engineering of the interfacial oxidation reaction, which acts as an oxygen-scavenging layer to enhance RS uniformity. In addition, systematic thermal degradation research about the effect of PMA time on oxygen vacancy distribution and RS performance was employed based on the Al/Ti(15 nm)/HfO$_2$/TiN devices. It was found that the leakage current increases and the forming voltage and memory window decrease with the increase in PMA time due to the thermally activated oxidation of the Ti buffer layer. However, when the PMA time is over 30 min, the forming voltage and memory window remain unchanged because of self-limiting oxidation. Therefore, the device shows superior thermal stability with a PMA time of 90 min at 400 °C and no degradation of the memory window, CCV, endurance, or retention.

Author Contributions: Conceptualization, H.H.; methodology, H.H.; software, H.H.; validation, H.H.; formal analysis, H.H. and Y.T.; investigation, H.H.; resources, H.H.; data curation, H.H.; writing—original draft preparation, H.H.; writing—review and editing, H.H. and Y.T.; visualization, H.H.; supervision, C.L. and Y.Z.; project administration, H.H. All authors have read and agreed to the published version of the manuscript.

Funding: This research was supported in part by the Zhejiang Province Natural Science Foundation of China under Grant LZ19F040001, the National Key Research and Development Program of China under Grant 2020AAA0109001, the Key Research and Development Program of Zhejiang Province under Grant 2021C01039, and the "Ling Yan" Program for Tackling Key Problems of Zhejiang Province, "Research on Sensing and Computing-in-Memory Integrated Chip for Image Applications" under Grant 2022C01098.

Data Availability Statement: Not applicable.

Conflicts of Interest: The authors declare no conflict of interest.

References

1. CHUA, L.O. Memristor-the missing circuit element. *IEEE Trans. Circuit Theory* **1971**, *18*, 507–519. [CrossRef]
2. Strukov, D.B.; Snider, G.S.; Stewart, D.R.; Williams, R.S. The missing memristor found. *Nature* **2009**, *459*, 1154. [CrossRef]
3. Yang, R.; Huang, H.M.; Guo, X. Memristive synapses and neurons for bio-inspired computing. *Adv. Electron. Mater.* **2019**, *5*, 1900287. [CrossRef]
4. Pan, F.; Gao, S.; Chen, C.; Song, C.; Zeng, F. Recent Progress in Resistive Random Access Memories: Materials, Switching Mechanisms, and Performance. *Mater. Sci. Eng. R Rep.* **2014**, *83*, 1–59. [CrossRef]
5. Wong, H.S.P.; Lee, H.Y.; Yu, S.; Chen, Y.S.; Wu, Y.; Chen, P.S.; Lee, B.; Chen, F.T.; Tsai, M.J. Metal-Oxide RRAM. *Proc. IEEE* **2012**, *100*, 1951–1970. [CrossRef]
6. Ielmini, D. Resistive Switching Memories Based on Metal Oxides: Mechanisms, Reliability and Scaling. *Semicond. Sci. Technol.* **2016**, *31*, 063002. [CrossRef]
7. Lanza, M.; Waser, R.; Ielmini, D.; Yang, J.J.; Goux, L.; Suñe, J.; Kenyon, A.J.; Mehonic, A.; Spiga, S.; Rana, V.; et al. Standards for the Characterization of Endurance in Resistive Switching Devices. *ACS Nano* **2021**, *15*, 17214–17231. [CrossRef] [PubMed]
8. Sebastian, A.; Gallo, M.L.; Khaddam-Aljameh, R.; Eleftheriou, E. Memory devices and applications for in-memory computing. *Nat. Nanotechnol.* **2020**, *15*, 529–544. [CrossRef]
9. Lanza, M.; Wong, H.-S.P.; Pop, E.; Ielmini, D.; Strukov, D.; Regan, B.C.; Larcher, L.; Villena, M.A.; Yang, J.J.; Goux, L.; et al. Recommended methods to study resistive switching devices. *Adv. Electron. Mater.* **2019**, *5*, 1800143. [CrossRef]
10. Lee, H.Y.; Chen, P.S.; Wu, T.Y.; Chen, Y.S.; Wang, C.C.; Tzeng, P.J.; Lin, C.H.; Chen, F.; Lien, C.H.; Tsai, M.J. Low Power and High Speed Bipolar Switching with a Thin Reactive Ti Buffer Layer in Robust HfO$_2$ Based RRAM. In Proceedings of the 2008 IEEE International Electron Devices Meeting (IEDM), San Francisco, CA, USA, 14–17 December 2008; pp. 297–300.
11. Kwon, D.H.; Kim, K.M.; Jang, J.H.; Jeon, J.M.; Lee, M.H.; Kim, G.H.; Li, X.S.; Park, G.S.; Lee, B.; Han, S.; et al. Atomic Structure of Conducting Nanofilaments in TiO$_2$ Resistive Switching Memory. *Nat. Nanotechnol.* **2010**, *5*, 148–153. [CrossRef]
12. Torrezan, A.C.; Strachan, J.P.; Medeiros-Ribeiro, G.; Williams, R.S. Sub-Nanosecond Switching of a Tantalum Oxide Memristor. *Nanotechnology* **2011**, *22*, 485203. [CrossRef] [PubMed]
13. Wu, Y.; Lee, B.; Wong, H.S.P. Al$_2$O$_3$-Based RRAM Using Atomic Layer Deposition (ALD) with 1-μA RESET Current. *IEEE Electron. Device Lett.* **2010**, *31*, 1449–1451. [CrossRef]
14. Kim, D.C.; Seo, S.; Ahn, S.E.; Suh, D.S.; Lee, M.J.; Park, B.H.; Yoo, I.K.; Baek, I.G.; Kim, H.J.; Yim, E.K.; et al. Electrical Observations of Filamentary Conductions for the Resistive Memory Switching in NiO Films. *Appl. Phys. Lett.* **2006**, *88*, 202102. [CrossRef]
15. Lin, C.Y.; Wu, C.Y.; Wu, C.Y.; Lee, T.C.; Yang, F.L.; Hu, C.; Tseng, T.Y. Effect of Top Electrode Material on Resistive Switching Properties of ZrO$_2$ Film Memory Devices. *IEEE Electron. Device Lett.* **2007**, *28*, 366–368. [CrossRef]
16. Yang, J.J.; Zhang, M.X.; Strachan, J.P.; Feng, F.M.; Pickett, M.D.; Kelley, R.D.; Ribeiro, G.M.; Williams, R.S. High switching endurance in TaO$_x$ memristive devices. *Appl. Phys. Lett.* **2010**, *97*, 232102. [CrossRef]
17. Zhuo, V.Y.Q.; Jiang, Y.; Zhao, R.; Shi, L.P.; Yang, Y.; Chong, T.C.; Robertson, J. Improved Switching Uniformity and Low-Voltage Operation in -Based RRAM Using Ge Reactive Layer. *IEEE Electron. Device Lett.* **2013**, *34*, 1130. [CrossRef]
18. Chen, Y.Y.; Goux, L.; Clima, S.; Govoreanu, B.; Degraeve, R.; Kar, G.S.; Fantini, A.; Groeseneken, G.; Wouters, D.J.; Jurczak, M. Endurance/retention trade-off on HfO$_2$/metal cap 1T1R bipolar RRAM. *IEEE Trans. Electron. Devices* **2013**, *60*, 1114–1121. [CrossRef]
19. Chen, Y.Y.; Govoreanu, B.; Goux, L.; Degraeve, R.; Fantini, A.; Kar, G.S.; Wouters, D.J.; Groeseneken, G.; Kittl, J.A.; Jurczak, M.; et al. Balancing SET/RESET pulse for >10^{10} endurance in HfO$_2$/Hf 1T1R bipolar RRAM. *IEEE Trans. Electron. Devices* **2012**, *59*, 3243–3249. [CrossRef]
20. Govoreanu, B.; Kar, G.S.; Chen, Y.; Paraschiv, V.; Kubicek, S.; Fantini, A.; Radu, I.P.; Goux, L.; Clima, S.; Degraeve, R.; et al. 10 × 10 nm^2 Hf/HfO$_x$ crossbar resistive RAM with excellent performance, reliability and low-energy operation. In Proceedings of the 2011 International Electron Devices Meeting (IEDM), Washington, DC, USA, 5–7 December 2011; pp. 31.6.1–31.6.4.

21. Rahaman, S.Z.; Lin, Y.D.; Lee, H.Y.; Chen, Y.S.; Chen, P.S.; Chen, W.S.; Hsu, C.H.; Tsai, K.H.; Tsai, M.J.; Wang, P.H. The Role of Ti Buffer Layer Thickness on the Resistive Switching Properties of Hafnium Oxide-Based Resistive Switching Memories. *Langmuir* **2017**, *33*, 4654–4665. [CrossRef]

22. Fang, Z.; Wang, X.P.; Sohn, J.; Weng, B.B.; Zhang, Z.P.; Chen, Z.X.; Tang, Y.Z.; Lo, G.-Q.; Provine, J.; Wong, S.S.; et al. The Role of Ti Capping Layer in HfOx-Based RRAM Devices. *IEEE Electron. Device Lett.* **2014**, *35*, 912–914. [CrossRef]

23. Young-Fisher, K.G.; Bersuker, G.; Butcher, B.; Padovani, A.; Larcher, L.; Veksler, D.; Gilmer, D.C. Leakage Current–Forming Voltage Relation and Oxygen Gettering in HfOx RRAM Devices. *IEEE Electron. Device Lett.* **2013**, *34*, 750–752. [CrossRef]

24. Tsai, T.; Chang, H.; Jiang, F.; Tseng, T. Impact of Post-Oxide Deposition Annealing on Resistive Switching in HfO$_2$-Based Oxide RRAM and Conductive-Bridge RAM Devices. *IEEE Electron. Device Lett.* **2015**, *36*, 1146–1148. [CrossRef]

25. Jiang, Y.; Tan, C.C.; Li, M.H.; Fang, Z.; Weng, B.B.; He, W.; Zhuo, V.Y.Q. Forming-Free TaO$_x$ Based RRAM Device with Low Operating Voltage and High On/Off Characteristics. *ECS J. Solid State Sci. Technol.* **2015**, *4*, 137–140. [CrossRef]

26. Wu, L.; Liu, H.; Lin, J.; Wang, S. Self-compliance and high performance Pt/HfO$_x$/Ti RRAM achieved through annealing. *Nanomaterials* **2020**, *10*, 457. [CrossRef]

27. Park, J.; Park, E.; Kim, S.G.; Jin, D.G.; Yu, H.Y. Analysis of the thermal degradation effect on a HfO$_2$-based memristor synapse caused by oxygen affinity of a top electrode metal and on a neuromorphic system. *ACS Appl. Electron. Mater.* **2021**, *3*, 5584. [CrossRef]

28. Chen, H.; Li, L.; Wang, J.; Zhao, G.; Li, Y.; Lan, J.; Tay, B.K.; Zhong, G.; Li, J.; Huang, M. Performance Optimization of Atomic Layer Deposited HfO$_x$ Memristor by Annealing With Back-End-of-Line Compatibility. *IEEE Electron. Device Lett.* **2022**, *43*, 1141–1144. [CrossRef]

29. Chuang, K.C.; Lin, K.Y.; Luo, J.D.; Li, W.S.; Li, Y.S.; Chu, C.Y.; Cheng, H.C. Effects of Post-Metal Annealing on the Electrical Characteristics of HfO$_x$-Based Resistive Switching Memory Devices. *Jpn. J. Appl. Phys.* **2017**, *56*, 06GF10. [CrossRef]

30. Yang, L.; Wang, C.Z.; Lin, S.; Cao, Y.; Liu, X. Early stage of oxidation on titanium surface by reactive molecular dynamics simulation computers. *Comput. Mater. Contin.* **2018**, *55*, 177–188.

31. Sun, J.; Tan, J.B.; Chen, T. HfO$_x$-Based RRAM Device With Sandwich-Like Electrode for Thermal Budget Requirement. *IEEE Trans. Electron. Devices* **2020**, *67*, 4193–4200. [CrossRef]

32. Yang, J.J.; Strachan, J.P.; Xia, Q.; Ohlberg, D.A.A.; Kuekes, P.; D.Kelley, J.R.; Stickle, W.F.; Stewart, D.R.; Medeiros-Ribeiro, G.; Williams, R.S. In-operando and non-destructive analysis of the resistive switching in the Ti/HfO2/TiN-based system by hard X-ray photoelectron spectroscopy. *Adv. Mater.* **2010**, *22*, 4036.

Disclaimer/Publisher's Note: The statements, opinions and data contained in all publications are solely those of the individual author(s) and contributor(s) and not of MDPI and/or the editor(s). MDPI and/or the editor(s) disclaim responsibility for any injury to people or property resulting from any ideas, methods, instructions or products referred to in the content.

Communication

HfO$_x$/Ge RRAM with High ON/OFF Ratio and Good Endurance

Na Wei [1], Xiang Ding [2], Shifan Gao [2], Wenhao Wu [1,*] and Yi Zhao [1,3]

[1] China Nanhu Academy of Electronics and Information Technology, Jiaxing 314001, China
[2] College of Information Science and Electronic Engineering, Zhejiang University, Hangzhou 310058, China
[3] School of Integrated Circuits, East China Normal University, Shanghai 200241, China
* Correspondence: wuwenhao@cnaeit.com

Abstract: A trade-off between the memory window and the endurance exists for transition-metal-oxide RRAM. In this work, we demonstrated that HfO$_x$/Ge-based metal-insulator-semiconductor RRAM devices possess both a larger memory window and longer endurance compared with metal-insulator-metal (MIM) RRAM devices. Under DC cycling, HfO$_x$/Ge devices exhibit a 100× larger memory window compared to HfO$_x$ MIM devices, and a DC sweep of up to 20,000 cycles was achieved with the devices. The devices also realize low static power down to 1 nW as FPGA's pull-up/pull-down resistors. Thus, HfO$_x$/Ge devices act as a promising candidates for various applications such as FPGA or compute-in-memory, in which both a high ON/OFF ratio and decent endurance are required.

Keywords: Germanium; HfO$_x$ RRAM; MIGe; endurance; memory window

1. Introduction

Resistive random access memory (RRAM) has been intensively investigated for its diversified applications, including embedded memory, storage class memory, FPGA, and in-memory computation of neural networks [1–3]. Metal Oxide (i.e., HfO$_x$, TaO$_x$, NiO, and TiO$_2$) is commonly applied as the switching layer sandwiched between two metal electrodes to build a metal-insulator-metal (MIM) structure. Among various switching layer materials, hafnium oxide (HfO$_x$) stands out owing to its technical maturity, fab-friendliness, and decent device performance [4–6]. The choice of metal electrodes affects the behavior of HfO$_x$-based RRAM a lot. Normally, when noble metal (i.e., Pt and Ru) is constructed as the top and bottom electrodes [7–9], the devices have unipolar switching because the conductive filament (CF) is annihilated by thermal diffusion. However, the switching behavior of unipolar RRAM is unstable, and the current during the annihilation of CF is too large for the applications of embedded memory, FPGA, and in-memory computation. As for bipolar RRAM, the materials of the top electrode are usually Ti, W, Al, and TiN [4,10–15]. Table 1 shows the benchmark of bipolar HfO$_x$-based RRAM with different electrodes. It is reported that the endurance is $10^6 \sim 10^7$ in reference [11,14], but the memory window (i.e., the ratio between high and low resistance states) is ~10. Reference [12] improved endurance to 10^{10} by utilizing oxygen plasma treatment within Ti/HfO$_2$/TiN structured RRAM, while the memory window is 10^2 and the operation voltage is larger than 3 V. Moreover, reference [10] realized a larger memory window up to 10^5 as the thickness of HfO$_2$ is 24.7 nm. It is observed that endurance can be improved by increasing the extra available oxygen ions in HfO$_2$, and the memory window can be enlarged using a thicker HfO$_2$ film. Nevertheless, both a large memory window and long endurance achieved in one RRAM device are difficult when the HfO$_2$ film is less than 10 nm. In other words, there is generally a trade-off between the memory window and the endurance of MIM RRAM [16].

Citation: Wei, N.; Ding, X.; Gao, S.; Wu, W.; Zhao, Y. HfO$_x$/Ge RRAM with High ON/OFF Ratio and Good Endurance. *Electronics* **2022**, *11*, 3820. https://doi.org/10.3390/electronics11223820

Academic Editor: Andrea Boni

Received: 24 October 2022
Accepted: 15 November 2022
Published: 20 November 2022

Publisher's Note: MDPI stays neutral with regard to jurisdictional claims in published maps and institutional affiliations.

Copyright: © 2022 by the authors. Licensee MDPI, Basel, Switzerland. This article is an open access article distributed under the terms and conditions of the Creative Commons Attribution (CC BY) license (https://creativecommons.org/licenses/by/4.0/).

Table 1. Benchmark of bipolar HfO_x-based RRAM.

Ref.	Structure	Window	Retention	Endurance	$V_F/V_{SET}/V_{RESET}$	I_{CC}
[10]	$TiN/HfO_2/Pt$	10^5	10^4 s	NS [1]	FF [2]$/-4.3/6$ V	NS
[11]	$W/Zr/HfO_2/TiN$	>10	NS	>10^6	$2/0.5/-1.25$ V	50 μA
[12]	$Ti/HfO_2/TiN$	10^2	10^4 s	10^{10}	$NS/3/-3.5$ V	1 mA
[13]	$Al/HfO_x/Al$	10^4	NS	NS	$1/1.8/0.8$ V	1 μA–1 mA
[14]	$Ti/HfO_2/TiN$	>10	10^4 s	>10^7	$FF/0.5/-0.5$ V	NS
[4]	$TiN(Ti)/HfO_x/W$	~10	NS	>10^4	$2.5/0.5/-1$ V	500 μA
[15]	$Ti/HfO_x/Pt$	40	>10^5	>10^3	$2.5/0.5/-0.7$ V	1 mA

[1] Not Specified. [2] Forming Free.

As of today, MIM-based HfO_x RRAM is gradually entering the market at 40 nm technology nodes and beyond in embedded or standalone memory applications [17,18]. For advanced applications such as FPGA's pull-up/pull-down resistors [19], or compute-in-memory (CIM) [20], both a low leakage in the high resistance state (HRS) and a large memory window are required to achieve good energy efficiency. Since MIM RRAMs with the HfO_x switching layer often exhibit modest memory windows, an alternative stack with a higher ON/OFF ratio is highly desired. In this work, we experimentally studied the metal-insulator-semiconductor (MIS) RRAM devices with the HfO_x switching layer and Ge or Si bottom electrodes in order to achieve a higher memory window with good endurance. In particular, a memory window over 10^5 was achieved with $Pd/HfO_x/p$-Ge RRAM devices, and the conductance mechanism was analyzed in detail. Furthermore, stable DC cycles with a large memory window were demonstrated for $Pd/HfO_x/p$-Ge devices up to 20,000 DC sweep cycles. Compared with $TiN/HfO_x/Pt$ devices and $Pd/HfO_x/p^+$-Si devices, $Pd/HfO_x/p$-Ge devices possess better DC endurance under the same test condition. These results suggest HfO_x/Ge RRAM device is a promising candidate for FPGA and CIM applications.

2. Materials and Methods

Two types of devices were fabricated, including MIS and MIM devices. For MIS devices, $Pd/HfO_x/p$-Ge devices and $Pd/HfO_x/p^+$-Si devices were fabricated. The process flow is depicted in Figure 1a. After the wafer cleaning, 7 nm HfO_x was deposited on Ge or Si substrates by Atomic Layer Deposition (ALD). The ozone post oxidation (OPO) is not processed here in contrast with the process of previous HfO_x/Ge RRAM [21,22]. It is not only because the interface is vital for previous RRAM stack used in MOSFET, but the stack variable here needs to be the same as the one in MIM devices. Subsequently, the lithography process and Pd deposition were carried out, and a lift-off process was utilized to form the top electrode. Al was finally deposited using thermal evaporation as a contact metal to Ge or Si substrates. To distinguish the Ge-based and Si-based devices, hereinafter MIGe is used to refer $Pd/HfO_x/p$-Ge devices, and MISi is used to refer $Pd/HfO_x/p^+$-Si devices. For MIM devices, a typical $TiN/HfO_x/Pt$ device was fabricated by the following process flow. First, back electrode Pt was sputtered on a SiO_2 substrate. Next, 7 nm HfO_x as the switching layer and 15 nm TiN as the top electrode were sequentially deposited by ALD. Afterward, the contact metal W was deposited by sputtering and patterned by lithography. Lastly, a $W/TiN/HfO_x$ stack was etched layer by layer using Inductive Coupled Plasma (ICP) tool to expose the bottom electrode.

The DC I-V characterizations were carried out by Agilent B1500A semiconductor parameter analyzer. SET and RESET operations were achieved by sweeping from a non-zero voltage V_{start} to a larger voltage V_{end}, where V_{end} was large enough to trigger the switching event. During the SET operation, current compliance (CC) was exerted by B1500A on MIGe/MISi devices (CC = 50 μA) and MIM devices (CC = 1 mA). The READ operations were achieved by applying single-point voltage on the device and measuring the currents for resistance calculation.

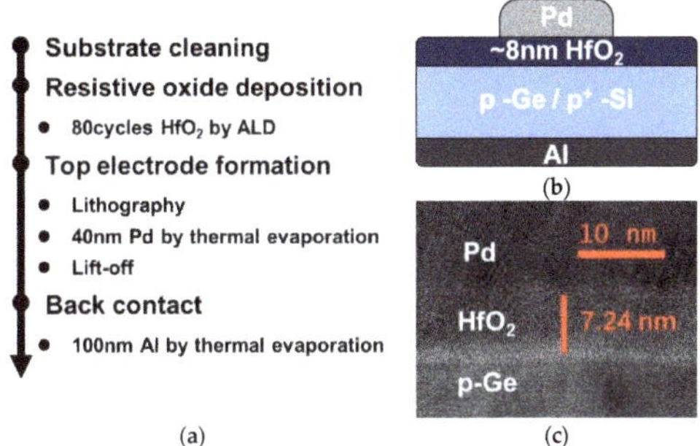

Figure 1. MIGe RRAM devices' (**a**) process flow, (**b**) structure schematic and (**c**) TEM graph.

3. Results and Discussion

Typical DC I-V characteristics were measured for $Pd/HfO_x/p$-Ge and $TiN/HfO_x/Pt$ devices, as shown in Figure 2. For MIM devices, stable low resistance states (LRS) can be achieved by 1 mA CC, and the resistance of high resistance states (HRS) is roughly 10^4 times the virgin-state resistance. The memory window of MIM devices is around 10^3, while a larger memory window of over 10^5 is realized in MIGe devices. It is worth noting that the HRS current of MIGe devices is approximately equal to the current of virgin states, implying a complete annihilation of the conductive filament (CF). Furthermore, the low operation current of MIGe (CC = 50 µA) leads to the advantage of low operating power. The RESET current of MIM devices is around several mA while that of MIGe devices is around 100 µA, in line with the SET current compliance. Thus, the write power of MIGe devices is estimated to be dozens of times smaller than that of MIM devices. For MIM devices, if a lower CC is applied to reduce the operation power, the memory window will also shrink significantly.

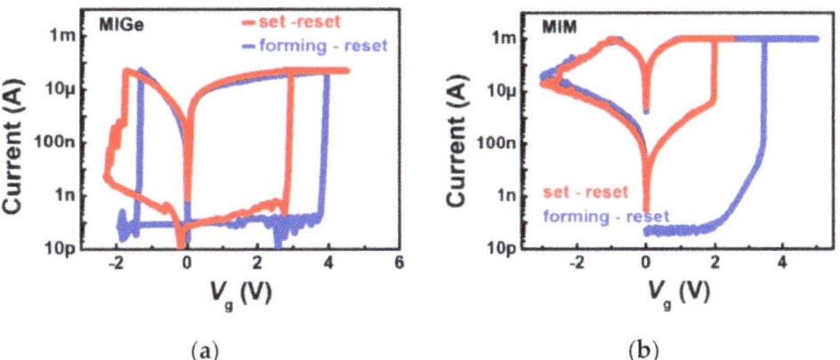

Figure 2. DC I-V_g curves of (**a**) $Pd/HfO_x/p$-Ge devices and (**b**) $TiN/HfO_x/Pt$ devices.

It is worth pointing out that the effective voltage drops on the oxide stack (V_{ox}) of MIGe devices are less than the voltage applied on the gate/top electrode (V_g). There are two situations: (a) The MIGe device works like a MOS when it is at HRS due to the negligible CF; (b) When the MIGe device works at LRS, the CF is conductive, so the device is not equivalent to a MOS. In the case of HRS, the voltage on the gate contributes to a series of oxide capacitance (C_{ox}) and substrate capacitance (C_s). The total capacitance (C_{tot}) is approximate to oxide capacitance at negative bias; thus, the voltage drop on the oxide stack is equal to V_g. While the device is positively biased, the surface potential, which is equal to the voltage dropped on the substrate (V_s), is extracted using a quasi-static technique [23,24]. The surface potential is expressed as Equation (1):

$$\varphi_s\left(V_g\right) = \int_{V_{acc}}^{V_g}\left[1 - \frac{C\left(V_g\right)}{C_{ox}}\right]dV_g + \Delta \tag{1}$$

where, Δ is correction factor and expressed as:

$$\Delta = \int_{V_{acc}}^{V_{FB}} \left[1 - \frac{C(V_g)}{C_{ox}}\right] dV_g \tag{2}$$

To obtain V_{FB}, C_{FB} needs to be calculated firstly using Equation (3):

$$C_{FB} = \frac{1}{\frac{1}{C_{ox}} + \frac{1}{C_s}}, \quad C_s = \frac{\varepsilon_s \varepsilon_0}{D_{deby}} = \frac{\varepsilon_s \varepsilon_0}{\sqrt{kT\varepsilon_s\varepsilon_0/q^2 NA}} \tag{3}$$

NA is 5×10^{16} /cm^3, and ε_s is the relative permittivity of Germanium. Derived from Equation (1), the voltage across the RRAM stack and substrate varies with the gate voltage, as plotted in Figure 3. It is indicated that the practical voltage across the RRAM stack is the same as what is in MIM RRAM. Moreover, the substrate doping concentration does not affect the switching behavior except for the operation voltage due to the partial voltage on the substrate. Hence, the scalability will not be affected by the substrate doping concentration. As for the LRS state, the voltage across the RRAM stack and the substrate will be discussed further in this paper.

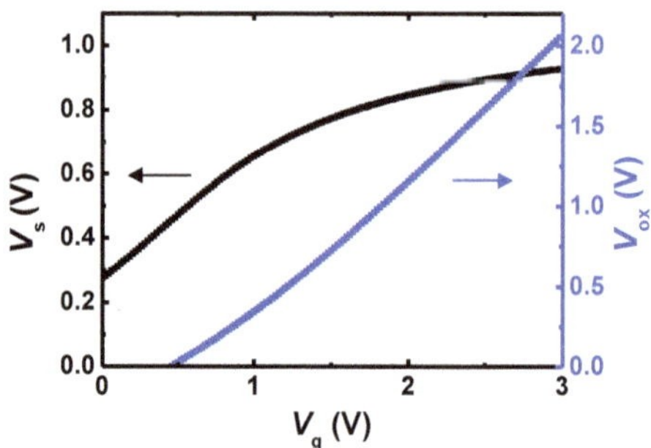

Figure 3. The voltage drops across the RRAM stack (V_{ox}, the curve in blue) and substrate (V_s, the curve in black) when the device is positively biased.

To further understand the conduction mechanisms of MIGe devices, double logarithmic J-E curves in LRS were plotted. As shown in Figure 4a, the fitting results suggest that two different conductive mechanisms exist for MIGe devices. When the applied voltage on the top electrode (TE) is less than V_1, the conduction behavior is ohmic because $\ln(J)$ is proportional to $\ln(E)$, and the slope is around 1. When the voltage on the TE increases above V_1, the slope of the curve reduces to smaller than 1. The current density vs. electric field data fits well with the Schottky emission equation in which $\ln(J) \propto E^{1/2}$ [25]. Based on the Schottky emission equation, the barrier height Φ_B is calculated to be 0.35 eV, while the electron effective mass in HfO$_x$ is approximated to be about 0.11 m$_0$ [26]. The conductive current through CF is contributed by electrons, which are the minority carriers for MIGe devices. A comparatively low current through CF, especially at the moment of CF forming, could help limit the overgrowth of the CF region [27]. For MIM devices, the J-E curve is symmetric, as shown in Figure 4b, and the slopes are both around 1. Plenty of free electrons are available from the two metal electrodes leading to a high operating current.

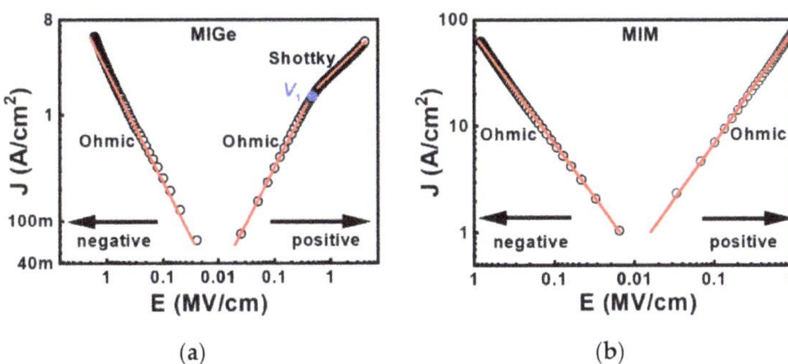

Figure 4. *J-E* curves of (**a**) Pd/HfO$_x$/p-Ge devices and (**b**) TiN/HfO$_x$/Pt devices in low resistance state.

Figure 5 exhibits the energy band diagram for the illustration of electronic transport mechanisms when the MIGe device works at HRS. As shown in Figure 5d, the voltage drop, V_g, is divided between the oxide switching layer and Ge substrate. When negatively biased, the resistance of the Ge substrate is negligible, and the resistance of CF (R_{CF}) can be extracted from I-V_g. Then the voltage dropped on the Ge substrate (V_s) could be derived as $V_g - I^*R_{CF}$, where I^*R_{CF} is approximately the voltage on the switching oxide (V_{ox}). Therefore, V_g is almost equal to V_{ox} when the device is negatively biased ($V_g < 0$). On the other hand, when $V_g > V_1$, V_g is mostly distributed to V_s. As depicted in Figure 5b, the conductive filament consists of abundant oxygen vacancies (V_O), which can gather to form a quasi-continuous defect energy band in the bandgap of HfO$_x$ [28]. In this case, the filament consisting of V_O can be treated as a metallic conducting path between the TE and Ge substrate at negative biases (Figure 5a). However, when the V_g is positive and larger than V_1, the electrons from the Ge substrate need to overcome an energy barrier of 0.35 eV to reach the defect levels of the CF (Figure 5c). From the experimental data, it is estimated that the energy level of CF is close to the conduction band minimum (CBM) of Ge and roughly 1.75 eV below the CBM of HfO$_2$. This energy level is consistent with previous studies, which identified V_O levels to be ranging from 1.2 to 2.1 eV below the CBM of HfO$_2$ [29–31].

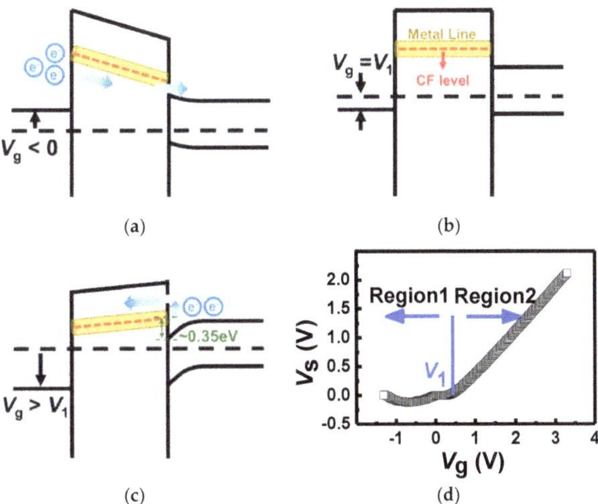

Figure 5. Energy band diagram of current transmission at (**a**) region 1, (**b**) region 2, (**c**) $V_g = V_1$. (**d**) V_s-V_g curve. (V_s: the voltage dropped on Ge substrate; V_g: the voltage applied on the gate.).

Furthermore, the DC endurance was characterized for Pd/HfO$_x$/p-Ge, Pd/HfO$_x$/p$^+$-Si, and TiN/HfO$_x$/Pt devices. The sequence of SET-READ-RESET-READ cycles was used, and devices with the same dimension (40 µm × 40 µm) were characterized. The cycling results of MIGe and MIM devices were plotted in Figure 6, which suggests different endurance failure phenomena of MIGe and MIM devices. For MIM devices, it can be observed that a sudden hard breakdown happened during the RESET, as shown in Figure 6d. Its endurance failure behavior is similar to those that originated from the depletion of O^{2-} which induced the RESET difficulty and increased vacancy concentrations [19,32]. In contrast, the MIGe devices maintained a stable endurance window of 100× or more but failed when the HRS and LRS converged into an intermediate state, as shown in Figure 6a,b. Although the distribution of HRS appears to be wide on account of the randomness of oxygen ion movement, pulse programming is an effective approach to improve the uniformity of HRS [10]. Pulse characterizations were implemented further for MIGe and MIM devices (data not shown here). The results of pulse endurance are in correspondence with the results of DC endurance. The MIGe device is still functional after 10^5 fully successful switching cycles, whereas the MIM device broke down after 10^4 pulse switching cycles. The effective operation cycles of the MIGe device (>10^5) are sufficient for CIM [33] since the SET/RESET operations will not be executed frequently.

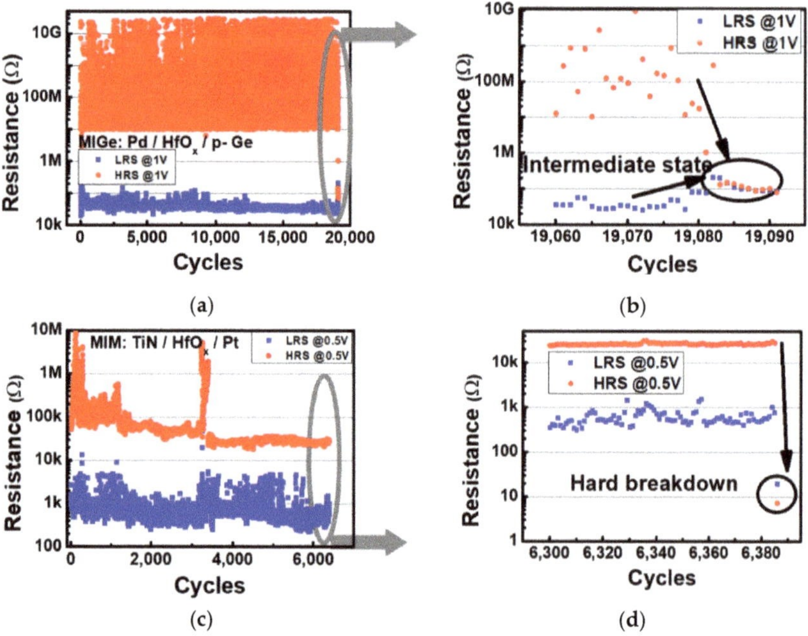

Figure 6. DC endurance cycles of (**a**) Pd/HfO$_x$/p-Ge devices (10 MΩ for minimum of HRS and 500 kΩ for maximum of LRS in test code) and (**c**) TiN/HfO$_x$/Pt devices (50 kΩ for minimum of HRS and 5 kΩ for maximum of LRS in test code); DC endurance failure of (**b**) Pd/HfO$_x$/p-Ge devices and (**d**) TiN/HfO$_x$/Pt devices.

In addition to the larger and more stable memory window achieved by MIGe devices, it is also demonstrated that the devices have superior endurance over the other two types of devices. To further confirm the superior endurance performance of MIGe devices compared to MISi and MIM, DC endurance tests were further carried out for multiple devices of each sample. Figure 7a summarizes the mean DC endurance of MIGe, MISi, and MIM devices. The mean DC endurance of MIGe devices is around 10^4, while that of MIM devices is ten times smaller. Moreover, the static power of FPGA's pull-up/pull-down resistors when implemented with RRAM was calculated [2]. For FPGA's configuration memory, the 1T2R structure was widely investigated. In these two resistors, usually one is ON, and the other is OFF. Therefore, the static power mainly depends on the OFF state resistor. Figure 7b compares the static powers of MIGe, MISi, and MIM devices at a 1.8 V supply. The results suggest that MIGe RRAM is a promising candidate for FPGA's pull-up/pull-down resistors.

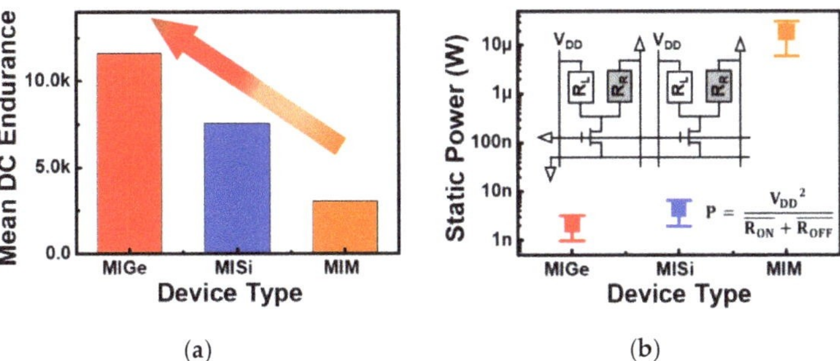

Figure 7. MIGe, MISi and MIM devices' (**a**) mean DC endurance and (**b**) static power (V_{dd} = 1.8 V) of FPGA's pull-up/pull-down resistors.

4. Conclusions

In conclusion, a large memory window of over 10^5 was demonstrated for Pd/HfO$_x$/p-Ge RRAM devices. Furthermore, the devices have been proven to have better endurance and a more stable memory window than MISi and MIM RRAM devices. Moreover, a low static power down to 1 nW was observed in MIGe devices as FPGA's pull-up/pull-down resistors. Therefore, HfO$_x$/p-Ge-based RRAM is a promising candidate for FPGA applications.

Author Contributions: Conceptualization, N.W.; methodology, N.W.; software, N.W.; validation, N.W. and X.D.; formal analysis, N.W., X.D., S.G., W.W. and Y.Z.; investigation, N.W.; resources, W.W. and Y.Z.; data curation, N.W.; writing—original draft preparation, N.W.; writing—review and editing, N.W., X.D., S.G., W.W. and Y.Z.; visualization, N.W.; supervision, W.W. and Y.Z.; project administration, W.W. All authors have read and agreed to the published version of the manuscript.

Funding: This research was supported in part by the Zhejiang Province Natural Science Foundation of China under Grant LZ19F040001, National Key Research and Development Program of China under Grant 2020AAA0109001, Key Research and Development Program of Zhejiang Province under Grant 2021C01039, and the "Ling Yan" Program for Tackling Key Problems of Zhejiang Province, "Research on Sensing and Computing-in-Memory Integrated Chip for Image Applications" under Grant 2022C01098.

Conflicts of Interest: The authors declare no conflict of interest.

References

1. Zahoor, F.; Zulkifli, T.Z.A.; Khanday, F.A. Resistive random access memory (RRAM): An overview of materials, switching mechanism, performance, multilevel cell (MLC) storage, modeling, and applications. *Nanoscale Res. Lett.* **2020**, *15*, 1–26. [CrossRef]
2. Liauw, Y.Y.; Zhang, Z.; Kim, W.; El Gamal, A.; Wong, S.S. Nonvolatile 3D-FPGA with monolithically stacked RRAM-based configuration memory. In Proceedings of the 2012 IEEE International Solid-State Circuits Conference, San Francisco, CA, USA, 19–23 February 2012; pp. 406–408. [CrossRef]
3. Prezioso, M.; Merrikh-Bayat, F.; Hoskins, B.D.; Adam, G.C.; Likharev, K.K.; Strukov, D.B. Training and operation of an integrated neuromorphic network based on metal-oxide memristors. *Nature* **2015**, *521*, 61–64. [CrossRef]
4. Fang, Y.; Yu, Z.; Wang, Z.; Zhang, T.; Yang, Y.; Cai, Y.; Huang, R. Improvement of HfO$_x$-based RRAM device variation by inserting ALD TiN buffer layer. *IEEE Electron. Device Lett.* **2018**, *39*, 819–822. [CrossRef]
5. Lv, H.; Xu, X.; Liu, H.; Liu, R.; Liu, Q.; Banerjee, W.; Sun, H.; Long, S.; Li, L.; Liu, M. Evolution of conductive filament and its impact on reliability issues in oxide-electrolyte based resistive random access memory. *Sci. Rep.* **2015**, *5*, 9964. [CrossRef]

6. Chen, Y.Y.; Roelofs, R.; Redolfi, A.; Degraeve, R.; Crotti, D.; Fantini, A.; Clima, S.; Govoreanu, B.; Komura, M.; Goux, L.; et al. Tailoring switching and endurance/retention reliability characteristics of HfO_2/Hf RRAM with Ti, Al, Si dopants. In Proceedings of the 2014 Symposium on VLSI Technology (VLSI-Technology): Digest of Technical Papers, Honolulu, HI, USA, 9–12 June 2014; pp. 1–2. [CrossRef]
7. Kim, Y.M.; Lee, J.S. Reproducible resistance switching characteristics of hafnium oxide-based nonvolatile memory devices. *J. Appl. Phys.* **2008**, *104*, 114115. [CrossRef]
8. Kang, J.; Park, I.S. Asymmetric current behavior on unipolar resistive switching in Pt/HfO_2/Pt resistor with symmetric electrodes. *IEEE Trans. Electron. Devices* **2016**, *63*, 2380–2383. [CrossRef]
9. Park, I.S.; Jung, Y.C.; Ahn, J. Anode dependence of set voltage in resistive switching of metal/HfO_2/metal resistors. *Appl. Phys. Lett.* **2014**, *105*, 223512. [CrossRef]
10. Zhao, L.; Chen, H.-Y.; Wu, S.-C.; Jiang, Z.; Yu, S.; Hou, T.-H.; PhilipáWong, H.-S.; Nishi, Y. Multi-Level Control of Conductive Nano-Filament Evolution in HfO_2 ReRAM by Pulse-Train Operations. *Nanoscale* **2014**, *6*, 5698–5702. [CrossRef] [PubMed]
11. Chen, W.; Lu, W.; Long, B.; Li1, Y.; Gilmer, D.; Bersuker, G.; Bhunia, S.; Jha, R. Switching Characteristics of W/Zr/HfO_2/TiN ReRAM Devices for Multi-Level Cell Non-Volatile Memory Applications. *Semicond. Sci. Technol.* **2015**, *30*, 075002. [CrossRef]
12. Chand, U.; Huang, C.-Y.; Jieng, J.-H.; Jang, W.-Y.; Lin, C.-H.; Tseng, T.-Y. Suppression of endurance degradation by utilizing oxygen plasma treatment in HfO_2 resistive switching memory. *Appl. Phys. Lett.* **2015**, *106*, 153502. [CrossRef]
13. Chen, Y.-C.; Chang, Y.-F.; Wu, X.; Zhoua, F.; Guo, M.; Lin, C.-Y.; Hsieh, C.-C.; Fowler, B.; Chang, T.-C.; Lee, J.C. Dynamic conductance characteristics in HfO_x-based resistive random access memory. *RSC Adv.* **2017**, *7*, 12984–12989. [CrossRef]
14. Su, Y.-T.; Liu, H.-W.; Chen, P.-H.; Chang, T.-C.; Tsai, T.-M.; Chu, T.-J.; Pan, C.-H.; Wu, C.-H.; Yang, C.-C.; Wang, M.-C.; et al. A method to reduce forming voltage without degrading device performance in hafnium oxide-based 1T1R resistive random access memory. *IEEE J. Electron. Devices Soc.* **2018**, *6*, 341–345. [CrossRef]
15. Wu, L.; Liu, H.; Lin, J.; Wang, S. Self-compliance and high performance Pt/HfO_x/Ti RRAM achieved through annealing. *Nanomaterials* **2020**, *10*, 457. [CrossRef] [PubMed]
16. Nail, C.; Molas, G.; Blaise, P.; Piccolboni, G.; Sklenard, B.; Cagli, C.; Perniola, L. Understanding RRAM endurance, retention and window margin trade-off using experimental results and simulations. In Proceedings of the 2016 IEEE International Electron Devices Meeting (IEDM), San Francisco, CA, USA, 3–7 December 2016; pp. 4–5. [CrossRef]
17. Golonzka, O.; Arslan, U.; Bai, P.; Bohr, M.; Baykan, O.; Chang, Y.; Fischer, K. Non-volatile RRAM embedded into 22FFL FinFET technology. In Proceedings of the 2019 Symposium on VLSI Technology, Kyoto, Japan, 9–14 June 2019; pp. T230–T231. [CrossRef]
18. Fackenthal, R.; Kitagawa, M.; Otsuka, W.; Prall, K.; Mills, D.; Tsutsui, K.; Hush, G. 19.7 A 16Gb ReRAM with 200MB/s write and 1GB/s read in 27 nm technology. In Proceedings of the 2014 IEEE International Solid-State Circuits Conference Digest of Technical Papers (ISSCC), San Francisco, CA, USA, 9–13 February 2014; pp. 338–339. [CrossRef]
19. Zambelli, C.; Castellari, M.; Olivo, P.; Bertozzi, D. Correlating power efficiency and lifetime to programming strategies in rram-based fpgas. In Proceedings of the 2018 New Generation of CAS (NGCAS), Valletta, Malta, 20–23 November 2018; pp. 21–24. [CrossRef]
20. Yu, S.; Shim, W.; Peng, X.; Luo, Y. RRAM for compute-in-memory: From inference to training. *IEEE Trans. Circuits Syst. I Regul. Pap.* **2021**, *68*, 2753–2765. [CrossRef]
21. Wei, N.; Chen, B.; Zheng, Z.; Cai, Z.; Zhang, R.; Cheng, R.; Lee, S.-W.; Zhao, Y. Ge-based Non-Volatile Logic-Memory Hybrid Devices for NAND Memory Application. In Proceedings of the 2018 IEEE International Electron Devices Meeting (IEDM), San Francisco, CA, USA, 1–5 December 2018; pp. 173–176. [CrossRef]
22. Wei, N.; Zhang, Y.; Chen, B.; Zhao, Y. Ge-based non-volatile memories. *Jpn. J. Appl. Phys.* **2020**, *59*, SM0802. [CrossRef]
23. Kuhn, M. A Quasi-Static Technique for MOS C-V And Surface State Measurements. *Solid State Electron.* **1970**, *13*, 873–885. [CrossRef]
24. Berglund, C.N. Surface States at Steam-Grown Silicon-Silicon Dioxide Interfaces. *IEEE Trans. Electron. Devices* **1966**, *13*, 701–705. [CrossRef]
25. Lim, E.W.; Ismail, R. Conduction mechanism of valence change resistive switching memory: A survey. *Electronics* **2015**, *4*, 586–613. [CrossRef]
26. Monaghan, S.; Hurley, P.K.; Cherkaoui, K.; Negara, M.A.; Schenk, A. Determination of electron effective mass and electron affinity in HfO_2 using MOS and MOSFET structures. *Solid-State Electron.* **2009**, *53*, 438–444. [CrossRef]
27. Degraeve, R.; Fantini, A.; Roussel, P.; Goux, L.; Costantino, A.; Chen, C.Y.; Jurczak, M. Quantitative endurance failure model for filamentary RRAM. In Proceedings of the 2015 Symposium on VLSI Technology (VLSI Technology), Kyoto, Japan, 16–18 June 2015; pp. T188–T189. [CrossRef]
28. Zhao, L.; Ryu, S.W.; Hazeghi, A.; Duncan, D.; Magyari-Köpe, B.; Nishi, Y. Dopant selection rules for extrinsic tunability of HfO_x RRAM characteristics: A systematic study. In Proceedings of the 2013 Symposium on VLSI Technology, Kyoto, Japan, 11–13 June 2013; pp. T106–T107.
29. Takeuchi, H.; Ha, D.; King, T.J. Observation of bulk HfO_2 defects by spectroscopic ellipsometry. *J. Vac. Sci. Technol. A Vac. Surf. Film.* **2004**, *22*, 1337–1341. [CrossRef]
30. Xiong, K.; Robertson, J.; Gibson, M.C.; Clark, S.J. Defect energy levels in HfO_2 high-dielectric-constant gate oxide. *Appl. Phys. Lett.* **2005**, *87*, 183505. [CrossRef]

31. Wu, X.; Mei, S.; Bosman, M.; Raghavan, N.; Zhang, X.; Cha, D.; Pey, K.L. Evolution of Filament Formation in $Ni/HfO_2/SiO_x/Si$-Based RRAM Devices. *Adv. Electron. Mater.* **2015**, *1*, 1500130. [CrossRef]
32. Chen, P.Y.; Yu, S. Compact modeling of RRAM devices and its applications in 1T1R and 1S1R array design. *IEEE Trans. Electron. Devices* **2015**, *62*, 4022–4028. [CrossRef]
33. Yin, S.; Sun, X.; Yu, S.; Seo, J. High-Throughput in-Memory Computing for Binary Deep Neural Networks with Monolithically Integrated RRAM and 90 nm CMOS. *IEEE Trans. Electron. Devices* **2020**, *67*, 4185–4192. [CrossRef]

 *electronics*

Review

Understanding and Controlling Band Alignment at the Metal/Germanium Interface for Future Electric Devices

Tomonori Nishimura

Department of Materials Engineering, School of Engineering, The University of Tokyo, Tokyo 113-8656, Japan; nishimura@material.t.u-tokyo.ac.jp; Tel.: +81-3-5841-7184

Abstract: Germanium (Ge) is a promising semiconductor as an alternative channel material to enhance performance in scaled silicon (Si) field-effect transistor (FET) devices. The gate stack of Ge FETs has been much improved based on extensive research thus far, demonstrating that the performance of Ge FETs is much superior to that of Si FETs in terms of the on-state current. However, to suppress the performance degradation due to parasitic contact resistance at the metal/Ge interface in advanced nodes, the reduction of the Schottky barrier height (SBH) at the metal/Ge interface is indispensable, yet the SBH at the common metal/Ge interface is difficult to control by the work function of metal due to strong Fermi level pinning (FLP) close to the valence band edge of Ge. However, the strong FLP could be alleviated by an ultrathin interface layer or a low free-electron-density metal, which makes it possible to lower the SBH for the conduction band edge of Ge to less than 0.3 eV. The FLP alleviation is reasonably understandable by weakening the intrinsic metal-induced gap states at the metal/Ge interface and might be a key solution for designing scaled Ge n-FETs.

Keywords: germanium; metal/semiconductor interface; Schottky barrier height; Fermi level pinning

Citation: Nishimura, T. Understanding and Controlling Band Alignment at the Metal/Germanium Interface for Future Electric Devices. *Electronics* **2022**, *11*, 2419. https://doi.org/10.3390/electronics11152419

Academic Editors: Yi Zhao and ChoongHyun Lee

Received: 22 June 2022
Accepted: 1 August 2022
Published: 3 August 2022

Publisher's Note: MDPI stays neutral with regard to jurisdictional claims in published maps and institutional affiliations.

Copyright: © 2022 by the author. Licensee MDPI, Basel, Switzerland. This article is an open access article distributed under the terms and conditions of the Creative Commons Attribution (CC BY) license (https://creativecommons.org/licenses/by/4.0/).

1. Introduction

Silicon (Si) complementary metal-oxide-semiconductor (CMOS) technologies have been continuously advancing with new materials and technologies, such as high-k gate dielectrics, metal gates and strain, to overcome the physical scaling limit. Furthermore, the channel material has been replaced recently in a p-channel field effect transistor (p-FET) by silicon–germanium (SiGe), which has a smaller effective hole mass [1]. From the viewpoint of channel materials in Si CMOS technologies, since germanium (Ge) has smaller effective masses for both the electrons and hole than Si and SiGe [2] and belongs to the group XIV elements, which has a high process compatibility with Si, meaning it is an attractive candidate for an alternative channel material for further advanced nodes. Although Ge oxide has some disadvantages, such as thermodynamic instability and water solubility [3–7], many studies have been performed to realize high-performance Ge FETs expected from the potential of bulk Ge properties. At first, Chui [8] and Shang [9] demonstrated a high-performance Ge p-FET, which is much superior to that of Si, with the nitride passivation of the gate stack in 2002. Thereafter, by advancing the understanding of the Ge gate stack design, both the Ge n- and p-FET performance have been greatly improved [10–23]. The progress of the effective mobility improvement in inverted FET channels over the last 20 years is summarized in Figure 1. With regard to the equivalent oxide thickness (EOT) scaling in the Ge gate stack, feasibility down to approximately 0.5 nm has been demonstrated [24,25]. Additionally, Ge FETs with nanowire or gate-all-around channel structures that assume further scaling have been reported [26,27]. In the future, for designing practical Ge FETs, it is also important to consider the disadvantage of a higher dielectric constant of Ge in terms of drain-induced barrier lowering and the advantage of easy planarization by thermal annealing in various atmospheres [28–30] in terms of structural dispersion.

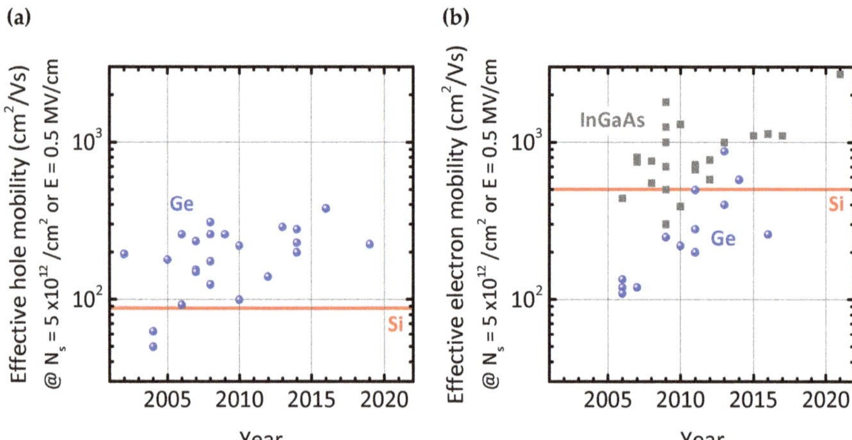

Figure 1. Progress of effective carrier mobility in inversion channels [at inversion carrier density $N_s = 5 \times 10^{12}/cm^2$ or effective electric field $E = 0.5$ MV/cm] of (**a**) p-FET and (**b**) n-FET. The line of "Si" corresponds to the universal mobility of Si FETs fabricated on (100) surface. In contrast to Ge p-FET, the performance of Ge n-FET was worse than Si n-FET at initial stage, but has been improved to be 2 times higher than Si universality. On the other hand, InGaAs ($In_{0.53}Ga_{0.47}As$) exhibits much higher electron mobility than Ge. However, there are other challenges such as small capacitance of inversion channel due to lower DOS, process compatibility with Si, control of stoichiometry, etc.

On the other hand, a typical factor that degrades performance in practically scaled FET devices is an increase in parasitic resistance. The parasitic resistance is composed of interconnecting and contact metal resistance, semiconductor resistance around the source/drain and contact resistance just at the metal/semiconductor interface; the contact resistance becomes more dominant by scaling down [31]. It is preferable to reduce the contact resistance as little as possible and below 10^{-9} Ωcm^2 is required for advanced nodes [32].

The contact resistivity at the metal/semiconductor interface is described as [33]:

$$\rho \propto \exp\left(C \frac{\Phi_b}{\sqrt{N}} \right), \tag{1}$$

where Φ_b and N are the Schottky barrier height at the interface and the density of the electrically activated impurities in the semiconductor, respectively. Therefore, to reduce the contact resistivity, a decrease in the Schottky barrier height and an increase in the activated impurity density are required based on the precise control of the Schottky barrier height and impurity activation and diffusion. Both are tough challenges for Ge CMOS, but this paper addresses the former.

2. Fermi Level Pinning at the Metal/Ge Interface

The Schottky barrier height is one of the band alignment parameters at the metal/semiconductor interface and the barrier for the conduction band (Φ_{bn}) at the ideal interface with no charge transfer is described as:

$$\Phi_{bn} = \Phi_m - \chi, \tag{2}$$

where Φ_m and χ are vacuum work functions of the metal and electron affinity of the semiconductor, respectively. This case is also called the Schottky limit and Φ_{bn} decreases (increased) as much as the decrease (increase) of Φ_m (Figure 2a).

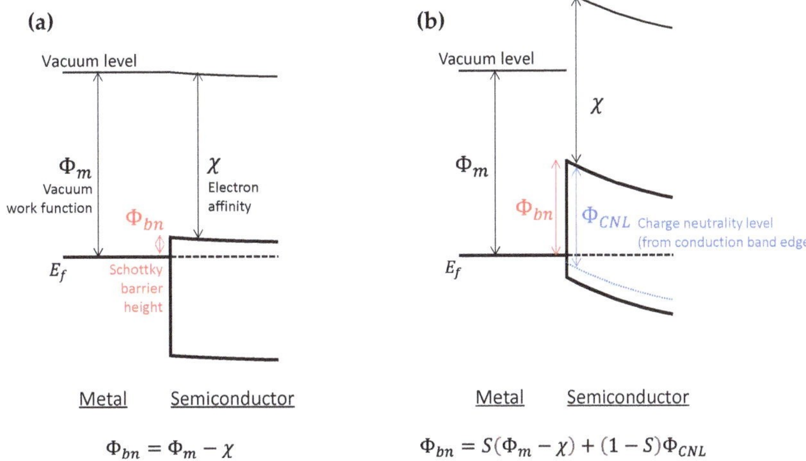

Figure 2. (**a**) Schematic band alignment at metal/semiconductor interface with no charge transfer (no Fermi level pinning (FLP)). This is called Schottky limit and Schottky barrier height for conduction band of semiconductor Φ_{bn} is equal to subtraction of electron affinity of semiconductor χ from vacuum work function of metal Φ_m. Therefore, Schottky barrier height at the interface is controllable by vacuum work function of metal. (**b**) Schematic band alignment at metal/semiconductor interface with FLP. Due to charge transfer at the interface, Fermi level at the interface relative to band edge of semiconductor is shifted from Schottky limit in direction to charge neutrality level, where the charge neutral level is defined as Φ_{CNL} from conduction band edge of semiconductor. In the case that S parameter is 0, Φ_{bn} is equal to be Φ_{CNL} irrespective of vacuum work function of metal and is not modulated by vacuum work function of metal. This is called Bardeen limit.

However, in most cases of actual metal/semiconductor interfaces, Φ_{bn} is not as sensitive to Φ_m. It is known as Fermi level pinning (FLP) and is generally described as a charge transfer between the metal and semiconductor through the interface states (D_{it}) in the band gap of the semiconductor [34] (Figure 2b). Assuming a constant D_{it} in the energy gap of the semiconductor, the band alignment is described as:

$$\Phi_{bn} = S(\Phi_m - \chi) + (1 - S)\Phi_{CNL}, \quad S = \frac{1}{1 + \frac{q\delta D_{it}}{\varepsilon}}, \tag{3}$$

where q, δ and ε are the elementary charge, dipole length and dielectric constant at the interface, respectively. Φ_{CNL} is the charge neutrality level of the semiconductor at the interface from the conduction band edge of the semiconductor. According to the equation, S is a parameter of sensitivity to Φ_m and takes a value between 0 and 1. S and Φ_{CNL} can be regarded as the FLP strength and FLP energy level. When S is 0, the Schottky barrier height becomes completely constant irrespective of Φ_m. This case corresponds to the Bardeen limit.

Next, we discuss the relationship between the vacuum work function of the metal and Schottky barrier height at the metal/Ge interface. The systematic trend was first reported by Thanailakis in 1973 [35]. After that, analyses focusing on FLP were reported by Marshall [36], Dimoulas [37] and Nishimura [38,39]. The relationships between the Schottky barrier height at metal/n-Ge(100) and the vacuum work function of metal are summarized in Figure 3a. The vacuum work function of metals is found in references [40,41]. In three recent reports, the Fermi level at the metal/Ge(100) interface is close to the valence band edge of Ge and is little dependent on the vacuum work function of the metal and a large Schottky barrier comparable to the band gap of Ge is formed on n-Ge. The FLP strength of the S parameter at the metal/Ge(100) interface is evaluated from the slope in Figure 2a to be 0.02–0.05, which is much less than that at the typical metal/Si interface (~0.3) [34]. The FLP energy level of Φ_{CNL} at the interface is estimated from the intersection of the fitted data line with the Schottky limit line in Figure 2a to be 0.49–0.58 eV. Both the S parameter and Φ_{CNL} are almost the same among the results in recent studies [36–39].

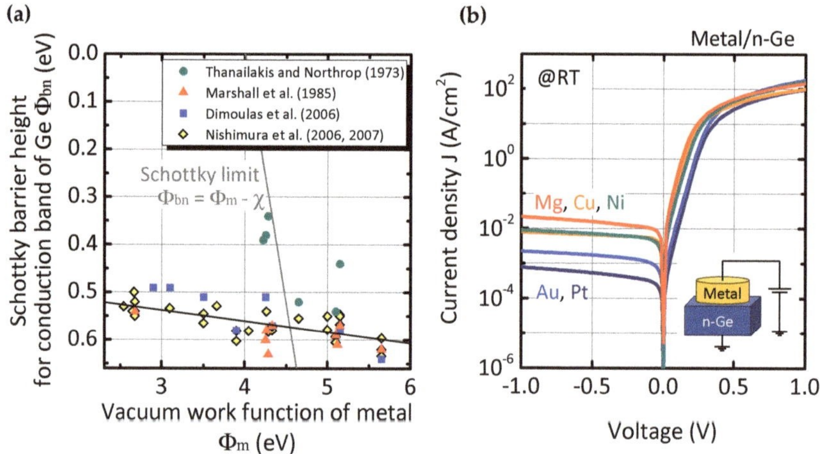

Figure 3. (a) Schottky barrier height for conduction band of Ge Φ_{bn} at metal/Ge(100) interface as a function of vacuum work function of metal Φ_m. The barrier Φ_{bn} values in the graph are estimated from I-V or C-V characteristics of metal/n-Ge diodes. Except for the result reported from Thanailakis, Schottky barrier height for conduction band of Ge at metal/Ge interface is almost constant, insensitive to vacuum work function of metal, and comparable to band gap of Ge (0.66 eV). The S parameter estimated from the slope in the plots is 0.02–0.05, which corresponds to interface states of $\sim 10^{14}$ states/cm^2/eV. To realize Ge device with electron channel, it is indispensable to overcome the strong FLP. Reproduced with permission from ref. [35], A. Thanailakis and E.C. Northrop; published by Elsevier, 1973, from ref. [36] E.D. Marshall et al.; published by Springer Nature, 2011, from ref. [37], A. Dimoulas et al.; published by AIP publishing, 2006, from ref. [38], T. Nishimura et al., published by Japan Society of Applied Physics, 2006, from ref. [39], T. Nishimura et al.; published by AIP publishing, 2007. (b) Typical I-V characteristics of metal/n-Ge(100) diodes. Schottky barrier height can be evaluated from the relationship of $J_s = AT^2 \exp(-\Phi_b/kT)$, where J_s, A and T is saturation current density, Richardson constant and measured temperature, respectively. The saturation current density J_s is estimated from linear extrapolation of current density at $V = 0$ in the graph. Due to the strong FLP, rectified character appears irrespective of metals on n-Ge. On the other hand, typical metal/p-Ge diodes exhibit ohmic characteristics (not shown).

From the correlation between the S parameter and D_{it} shown in Equation (3), the FLP strength can be quantitatively considered as the interface state density for the sake of convenience, irrespective of the mechanism of FLP. For example, assuming a δ of 0.5 nm and ε of 10 ε_0, the S parameter of 0.02 at the metal/Ge(100) interface is equivalent to a D_{it} of 5×10^{14} states/cm^3/eV. It implies that the interface might be electrically very poor, considering that the atomic density on the Ge(100) surface is almost 6×10^{14}/cm^2.

The band alignment shown in Figure 2a was determined from the electric characteristics (such as I-V or C-V characteristics) of metal/n-Ge(100) junctions (typical I-V characteristics are shown in Figure 3b). On the other hand, it is difficult to determine from metal/p-Ge(100) junctions. The Schottky barrier height for the valence band edge of Ge at metal/p-Ge(100) is very low and ohmic contact is easily formed on the junction at room temperature. This is a large advantage for p-type Ge devices in terms of choice for the contact metal, but a critical disadvantage for n-type Ge devices. To reduce the contact resistance at the metal/n-Ge interface towards the realization of practical n-type

Ge devices, the strong FLP close to the valence band edge must be overcome based on an understanding of its FLP mechanism.

3. Dominant FLP Mechanism at the Metal/Ge Interface

First, typical FLP mechanisms discussed for various metal/semiconductor interfaces are reviewed, before a discussion of the dominant FLP mechanism at the metal/Ge interface. FLP mechanisms can be categorized into intrinsic or extrinsic mechanisms. Here, "intrinsic FLP" means that the FLP is caused by an intrinsic charge transfer due to the metal/semiconductor interface formation itself, while "extrinsic FLP" means that the FLP is caused by a charge transfer through the actual interface states, such as defects introduced by the interface formation process. Metal-induced gap states (MIGS) [42–46] and chemical bond models [47,48] are typical intrinsic FLP mechanisms. In the MIGS model, a charge transfer is caused by a wave function tailing from the metal into the finite semiconductor band gap. In the chemical bond model, it is through local chemical bonding between metal and semiconductor atoms, considering an energy gain in the charge transfer against the band gap of the semiconductor. On the other hand, the unified defect model (UDM) [49,50] and disorder-induced gap states (DIGS) [51,52] are typical extrinsic FLP mechanisms. The UDM has been discussed regarding the metal/III-V semiconductor interfaces and is based on the formation of a universal defect energy level irrespective of the adsorbed element on the semiconductor. In the DIGS model, the interface state density related to the FLP strength is characterized by the strain and distortion of bonds, defects and dangling bond densities. The FLP energy level is determined by the sp^3 hybrid orbital energy, which is universally located at approximately 5 eV from the vacuum level in diamond- and zinc-blend-type semiconductors.

However, it is difficult to identify a dominant FLP mechanism at the metal/Ge interface from the FLP strength, as follows. Since 1970, it has been implied that pure covalent bonding semiconductors such as Ge and Si exhibit strong FLP [53]. The FLP strength of various semiconductors has been systematically investigated and the correlation of FLP strength with the ionic bond character in semiconductors has been discussed from the viewpoint of surface states, which is based on the relationship between the ionic bonding character and electronic structure modulation. In the previously mentioned intrinsic MIGS and chemical bond model, FLP strength is primarily characterized by the electronic gap of the semiconductor because the charge transfers in MIGS and the chemical bond model are dependent on the decay of wave function tailing in the semiconductor and the energy gain of the charge transfer to the valence band or conduction band of the semiconductor. Considering the narrow band gap of Ge (0.66 eV at room temperature), it seems reasonable that strong FLP occurs at the metal/Ge interface. In the UDM and DIGS, simply saying, the FLP strength is determined by the actual interface states density. For example, the surface cleaning method before metal deposition and the reduction of metal diffusion into the semiconductor inside by cooling during metal deposition may be effective in reducing the interface state density and weakening the FLP at the metal/6H-SiC [54] and metal/GaAs [55] interfaces, while the alleviation of FLP at the metal/Ge interface by appropriate surface treatment has not been reported. However, the narrow band gap of Ge indirectly implies that the bond structure in the Ge crystals is easily broken and so a high density of interface states may possibly be formed, which implies that the extrinsic FLP mechanism causes the strong FLP of Ge.

It is also difficult to identify the dominant FLP mechanism from the viewpoint of the FLP energy level. The experimentally determined effective charge neutrality level is close to the valence band edge of Ge. In intrinsic mechanisms, the MIGS describes the FLP energy level as an energy level that corresponds to the charge neutrality level of the bulk semiconductor, where the dominance of the conduction band character and valence band character is changed in the band gap. (It is a branch point in a one-dimensional system.) The charge neutrality level Φ_{CNL} of Ge is calculated to be 0.48 [44] and 0.63 [56]. On the other hand, in extrinsic mechanisms, the DIGS model describes the FLP energy of

Ge. The energy in the DIGS model is the sp^3 hybrid orbital energy, which is the same as the charge neutrality level of the bulk semiconductor [51]. This result is also consistent with the fact that vacancies in Ge work as acceptors [57] and that defective Ge exhibits p-type characteristics [58,59].

Therefore, another approach is needed to understand the dominant FLP mechanism. Here, we present an interesting result of comparing Ge and Si with SiGe [60]. A key property of SiGe is the complete solid solution of Si and Ge. Therefore, SiGe includes natural bond distortion and disordering even in perfect crystals. Moreover, SiGe used in the experiment is not a single crystal substrate, but epitaxially grown on a Si substrate, which suggests that SiGe includes extrinsic defects, dislocations and strain. These structural disorders are detected, for example, as the broadening of the diffraction peak in the XRD measurement. On the other hand, the band gap of SiGe monotonically increases with the decreasing Ge ratio. Therefore, the band gap becomes simply wider in the order of Ge, SiGe and Si. Therefore, the FLP mechanism, which dominantly characterizes the metal/Ge interface, could be speculated by which metal/Ge and metal/SiGe interfaces exhibit a stronger FLP. To analyze the band alignment at the metal/SiGe interface, the Schottky barrier height is estimated from the rectified I-V characteristics of metal/n-SiGe(100) diodes, as shown in Figure 4a. The saturation current density at the metal/SiGe junction is obviously metal dependent compared with that at the metal/Ge junction shown in Figure 2b. Figure 4b shows the relationship between the band gap, crystallinity and FLP strength of Ge, SiGe and Si. The S parameter for the metal/SiGe(100) interface is estimated to be 0.07. This value is less than that for the metal/Si (100) of 0.16, which is consistent with reports from Archer [61], but obviously more than that for the metal/Ge(100) of 0.02. The D_{it} of 1.4×10^{13}/cm^2/eV, which corresponds to the S parameter for SiGe, is much lower than that for Ge, even though SiGe has various structural disorders. These experimental facts imply that the FLP of Ge may be dominantly characterized by an intrinsic mechanism. Furthermore, the band alignment at the metal/Ge interface does not seem sensitive to the surface orientation of Ge [39], which suggests that local structural characteristics such as bonding at the interface do not strongly characterize the band alignment. In other words, the FLP seems to be characterized by bulk Ge characteristics. This means that MIGS seems to be the most reasonable FLP mechanism to describe the FLP at the metal/Ge interface. If the strong FLP was extrinsic, the passivation and/or generation of extrinsic interface states should be carefully considered. However, it appears to be intrinsic so that some sort of breakthrough is needed.

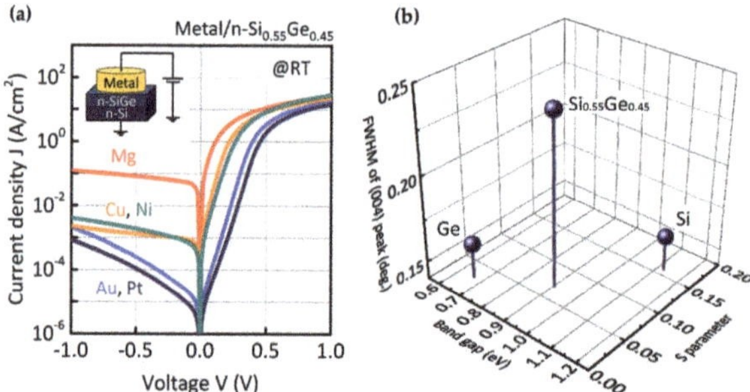

Figure 4. (**a**) Typical I-V characteristics of metal/n-Si$_{0.55}$Ge$_{0.45}$(100) diodes measured at room temperature. The off-state current density on high work function metal diodes is increased at negative bias region, which might be caused by leakage through extrinsic defects in SiGe. However, saturation current density is obviously more dependent on metals compared with that of Ge diodes depicted in Figure 2b. (**b**) The correlation among FLP strength (S parameter), band gap and crystallinity (FWHM of (400) diffraction peak in XRD) of semiconductors. The S parameter is simply increased with band gap of semiconductor independent of its crystallinity, which suggests intrinsic FLP mechanism dominantly characterizes the FLP at metal/Ge interface.

4. FLP Alleviation by Ultrathin Interface Layer

In this chapter, focusing on the fact that such a strong FLP does not occur at a typical metal/insulator/Ge interface, an approach to alleviate the strong FLP at the metal/Ge interface is discussed. As mentioned in the introduction, various Ge gate stacks have been analyzed by C-V characteristics to improve the performance of Ge FETs [8–23]. In these C-V characteristics, the accumulation, depletion and inversion of the Ge layer are distinctly observed, which indicates that the surface potential of Ge is modulatable; in other words, the insulator/Ge interface is not strongly pinned as with the metal/Ge interface. On the other hand, the flat band voltage of the Ge Metal/Insulator/Semiconductor (MIS) capacitor obviously depends on the vacuum work function of the metal [62], which also suggests that a strong FLP does not occur at the metal/insulator interface. Therefore, our interest arising from the band alignment of the Ge MIS interfaces was how the FLP-free Ge MIS interfaces behave after extremely thinning the insulator. In the best case, we hoped that enough current could flow in the on state at the FLP-free MIS junction.

We selected GeO_2, which is an oxide of the substrate, as an ultrathin interface insulator to simplify the system. Additionally, GeO_2 could form an electrically better interface with Ge by the suppression of GeO desorption compared with other insulators [63,64]. Al was chosen as an electrode metal with a low vacuum work function to form low and high Schottky barrier heights for n- and p-Ge at the FLP weakened interface, respectively. Figure 5a shows typical I-V characteristics of Al/Ge(100) diodes with and without an ultrathin GeO_2 layer. Surprisingly, the Al/n-Ge diode characteristics change from Schottky to ohmic by inserting a GeO_2 layer, whereas the Al/p-Ge diode characteristics change from ohmic to Schottky [65]. This is definitely understandable by a shift of the Fermi level at the interface from the valence band edge of Ge to the near conduction band edge. The ultrathin GeO_2 interface layer between the Al and Ge substrates is also confirmed by the cross-sectional TEM image shown in Figure 5b.

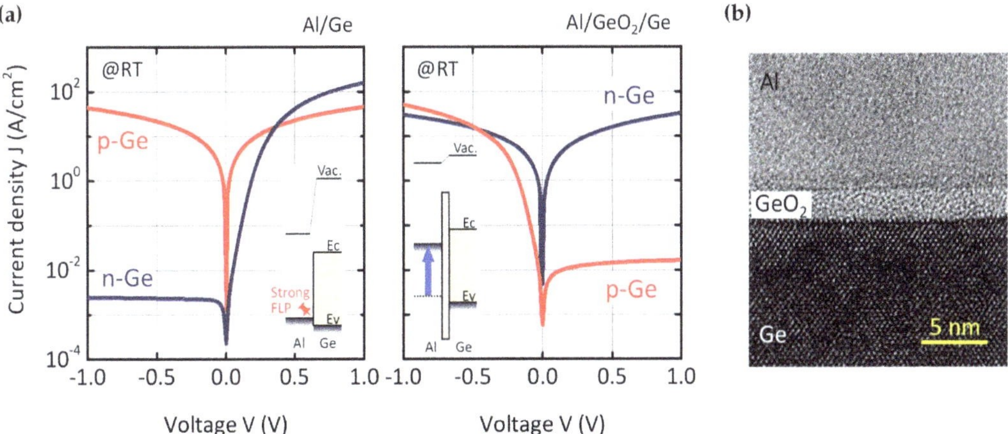

Figure 5. (a) Characteristics I-V of Al/Ge(100) diodes and those of Al/ultrathin GeO_2/Ge(100) ones. Due to the strong FLP close to valence band edge of Ge, as depicted in schematic, Al/Ge diodes exhibit Schottky characteristic and ohmic one for n- and p-Ge, respectively. On the contrary, Al/GeO_2/Ge diodes show ohmic characteristic and Schottky one for n- and p-Ge, respectively, which indicates that the Fermi level at the interface is shifted towards to conduction band edge of Ge. (b) Cross sectional TEM image of Al/GeO_2/Ge contact. It is observed that 2-nm-thick amorphous GeO_2 layer is formed at the interface.

Furthermore, the correlation between the Schottky barrier height at metal/ultrathin GeO_2/Ge exhibits an obvious vacuum work function dependence, as shown in Figure 6.

The vacuum work function dependence of the Schottky barrier height indicates that the ultrathin insulator surely has a role in alleviating the FLP and that the Schottky–ohmic reversal of Al/ultrathin GeO$_2$/Ge diodes shown in Figure 4a is not caused only by the pinning level shift. As summarized in Figure 7a,b, Schottky barrier height modulation and FLP alleviation have been observed by Al$_2$O$_3$ [66,67], TiO$_2$ [68–70], MgO [71,72], GeON [73], TiON [74], GeN [75] and SiN [76], in addition to GeO$_2$ [65,68]. These experimental results imply that the FLP alleviation by inserting an ultrathin insulator into the metal/Ge interface is attributed to the insulating property of the interface layer or interface passivation by the oxygen or nitrogen included in the interface layer.

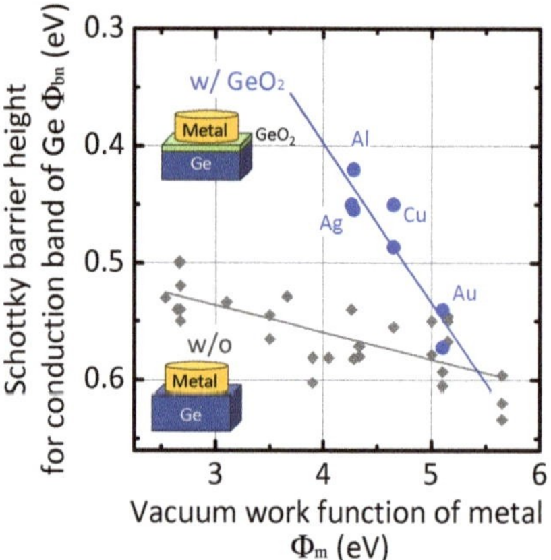

Figure 6. The correlation between band alignment at metal/1-nm-thick GeO$_2$/Ge(100) and vacuum work function of metal. That at metal/Ge(100) [38,39] are also shown. Schottky barrier height were estimated from saturation current density in I-V characteristics measured at room temperature and Richardson constant of Ge. The S parameter for metal/GeO$_2$/Ge interface is estimated to be 0.14 and it is obviously more than that of direct metal/Ge one, which indicates that the FLP at metal/Ge interface is weakened by ultrathin GeO$_2$ interlayer. Adapted with permission from ref. [38], T. Nishimura et al., published by Japan Society of Applied Physics, 2006, from ref. [39], T. Nishimura et al.; published by AIP publishing, 2007.

Next, from the viewpoint of the role of insulator in FLP alleviation, the dominant FLP mechanism at metal/Ge is considered again. The FLP alleviation by ultrathin insulators is explainable in various models, as follows. In the intrinsic MIGS, the wave function tailing to the semiconductor decays due to the wide energy gap and physical distance of the interfacial layer. In the intrinsic chemical bond model, the energy gain in the charge transfer with metal will be reduced by replacing the semiconductor with a wide gap insulator. In extrinsic models, nonideal semiconductor structures, such as defects or dangling bonds, resulting in interface states are passivated by oxygen or nitrogen. However, it is noted that the role of the interface layer in each model is different. It is expected that intrinsic MIGS is gradually weakened with an increasing insulator thickness as a result of the thickness effect, whereas the extrinsic FLP mechanisms are immediately suppressed by forming an insulator/Ge interface by the interface passivation effect. Hence, from an impact of interlayer thickness on the FLP strength, the appropriate mechanism to describe the FLP at the metal/Ge interface could be speculated from another aspect.

interface corresponds to a part of the vacuum work function of the metal. Lang calculated the vacuum work function of metal with a simple Jellium model [80,81]. In that model, the vacuum work function is composed of a bulk term and surface term; the former term includes the interaction energy and kinetic energy in the bulk and the latter corresponds to the surface dipole due to wave function tailing from the metal to the vacuum. The model is significantly simple, but the result well describes the vacuum work function of simple metals such as alkaline and alkali-earth metals. Here, it is a notable point that the surface term is reduced with a decreasing free-electron density in the metal, which indicates that the metal/vacuum interface dipole caused by wave function tailing is reduced. Therefore, MIGS at the metal/semiconductor, which is described as a dipole caused by wave function tailing, could be weakened by the decreasing free-electron density in the metal based on the similarity of the physical description of the "surface term of the vacuum work function" and "MIGS".

The free-electron density in single-element metals commonly used to analyze FLP at the metal/semiconductor interface is approximately 10^{22}–10^{23}/cm^3 [82], whereas, for example, in compound metals such as silicide, it is one digit less, approximately 10^{21}/cm^3 [83]. The free-electron density in germanide, which is a compound of metal and Ge, is expected to be almost the same as that in silicide and the free carrier densities in Y-germanide and Gd-germanide were evaluated to be 7×10^{19} and 9×10^{19}/cm^3, respectively, by Hall effect measurements. Considering the previously mentioned analogy between MIGS and the surface term of the vacuum work function, the FLP at the germanide/Ge interface is expected to be weaker than that at the single-element metal/Ge interface.

Germanide/n-Ge(100) diodes show rectified I-V characteristics (Figure 9a), but the saturation current densities seem to have a strong metal dependence compared with those of single-element metals/n-Ge diodes (shown in Figure 2b), although the swept range of the vacuum work function of germanide might be much narrower than that of single-element metals due to compound formation with Ge. The relationship between the vacuum work function of the metal and the Schottky barrier height at the metal/Ge interface is shown in Figure 9b. Here, the work function of crystallized germanide is assumed from the metal–Ge composition ratio in its crystal structure, the group electronegativity [84] and the relationship between the electronegativity and vacuum work function [85]. The FLP at the germanide/Ge interface is weaker, as expected, and the S parameter for the germanide/Ge(100) interface is estimated to be 0.17, which is much larger than that for the single-element metal/Ge(100) interface. Additionally, deviation from the strong FLP trend at the single-element metal/Ge interface has also been reported at the YbGe$_x$/Ge [86], epitaxial Mn$_5$Ge$_3$/Ge(111) [87], epitaxial HfGe$_2$/Ge(100) [88] and epitaxial Fe$_3$Si/Ge(111) [89] interfaces. These results seem reasonably understandable by the weakening of MIGS, even though interface epitaxiality may sufficiently affect the band alignment in detail. Furthermore, in the case of the metal/Si interface, FLP at the silicide/Si interface is also much weaker than that at the single-element metal/Si interface [90,91]. In addition to interfacial layer insertion, a low free-electron-density metal is also effective in alleviating the FLP at the metal/Si interface, which suggests that MIGS might also describe the FLP at the metal/Si interface well.

At the metal/semiconductor interface with strong MIGS, the band alignment is dominantly determined by the charge neutrality level, which is characterized by the bulk properties of the semiconductor. However, by weakening MIGS at the interface, the assumption is that other FLP mechanisms or interface structures possibly affect band alignment. Actually, the Schottky barrier height at the germanide/Ge interface has a noticeable surface orientation dependence [92], as shown in Figure 10a. It was also demonstrated that ohmic contact can be formed for n-Ge at room temperature without the heavy doping of impurities by employing an appropriate metal and surface orientation of Ge, although a direct metal/Ge interface is formed (Figure 10b). However, here, care is taken that not only Ge surface orientation, but also other properties, such as germanide orientation, might be different for each Ge orientation because the germanide/Ge interface was formed by thermal reaction of the elemental metal/Ge interface.

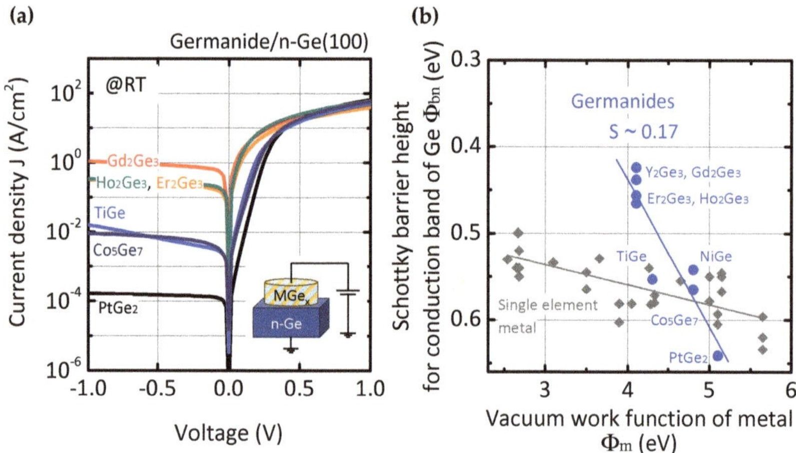

Figure 9. (**a**) Typical I-V characteristics of germanide/n-Ge(100) diodes measured at room temperature. All germanides were formed by thermal reaction of deposited single-element metal with Ge at 500 °C. The off-state current density of germanide/n-Ge diode is much dependent on metal compared with that of single-element metal/n-Ge diode shown in Figure 2b. Adapted with permission from ref. [92], T. Nishimura et al., published by IOP publishing, 2008. (**b**) The relationship between the band alignment at germanide/Ge(100) interface and vacuum work function of metal. The vacuum work function of germanide is estimated from the group electronegativity [$X_{A_m B_n} = {}^{m+n}\sqrt{X_A^m X_B^n}$] [84] and the correlation between electronegativity and vacuum work function [$\Phi_m = 2.27 X_{Pauling} + 0.34$] [85]. It is obvious that the FLP at germanide/Ge interface is much weaker than that at simple element metal/Ge one.

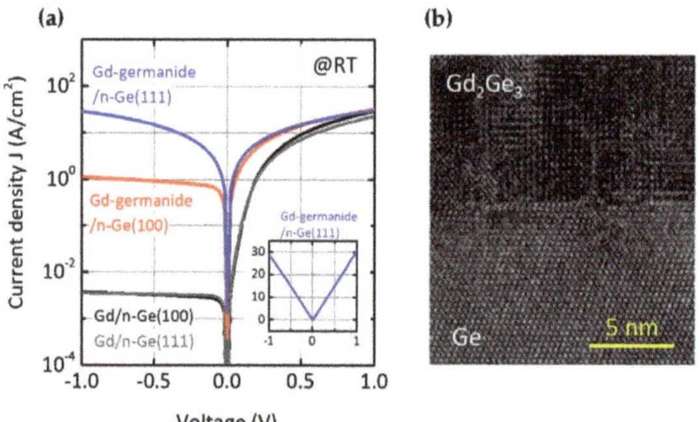

Figure 10. (**a**) I-V characteristics of Gd and Gd-germanide/n-Ge(100) and (111) diodes. The I-V characteristics of Gd/Ge diodes are not dependent on Ge surface orientation, while those of Gd-germanide/Ge ones are strongly dependent. In addition, Gd-germanide/n-Ge(111) diode surprisingly exhibits ohmic character, as also shown in the inset. The interface structure dependence in I-V characteristics of germanide/Ge diodes is possibly understandable by appearance of other mechanisms which are masked by MIGS at typical single-element metal/Ge interface. Adapted with permission from ref. [92], T. Nishimura et al., published by IOP publishing, 2008. (**b**) Cross sectional TEM image of Gd-germanide/Ge(111) interface formed by thermal reaction of Gd with Ge. It is confirmed that direct germanide/Ge interface is certainly formed and the Schottky barrier height lowering is never caused by unintentional interface layer growth at the interface.

The difference between single-element metals and compound metals such as silicide and germanide on FLP at epitaxial metal/Ge and /Si interfaces has been discussed based on first principle calculations [93]. According to the calculations, the Fermi level at the single-element metal/semiconductor interface is determined by MIGS around the dangling bond states of the semiconductor, while that at silicide/Si or germanide/Ge is determined by that of the metal, which provides the difference in FLP strength. The calculated Si and Ge surface orientation dependence on the band alignment is not completely consistent with the experimental results [92], but the difference in the Schottky barrier height at the A- and B-type epitaxial $NiSi_2$/Si interfaces [94] is successfully reproduced. This suggests a key to further understanding the impact of the microscopic interface structure on band alignment from the viewpoint of local atomic configuration.

On the other hand, as previously implied, the interface fabrication process is different between the germanide/Ge and single-element metal/Ge interfaces. The germanide/Ge interface is generally formed by a thermal reaction, while the single-element metal/Ge interface is by deposition. To purely focus on the impact of the low-electron-density metal on the FLP, an example of band alignment at a low free carrier density metal/Ge interface formed by deposition is discussed. Bismuth (Bi) is a famous semimetal and its carrier density is approximately 10^{17}/cm^3, which is much less than that of common single-element metals [95]. Bi can be thermally evaporated in a vacuum chamber in the same manner as other single-element metals. There is no choice to pick up various semimetals with different vacuum work functions, such as single-element metals, but the vacuum work function of Bi is fortunately 4.22 eV [40]. It is energetically far from the FLP energy level of Ge of approximately 4.6 eV from the vacuum level, so the alleviation of the strong FLP is detectable as an obvious modulation of the Schottky barrier height. Additionally, Bi cannot form a compound (germanide) with Ge [96], which means that the possibility of band alignment modulation by Bi-germanide formation can be ruled out. As shown in Figure 11, Bi/n-Ge diodes still show rectified I-V characteristics, but the off-state current density is very high compared with the single-element metal cases. Moreover, the I-V characteristics of Bi/p-Ge(100) diodes are definitely rectified. In the relationship between the vacuum work function of metal and the Schottky barrier height at the metal/Ge interface, Bi obviously deviates from the trend of strong FLP observed at the single-element metal/Ge interface and is close to the Schottky limit [97], which suggests that MIGS at the Bi/Ge interface is efficiently suppressed. The FLP-free interface is also obtained at the Bi/Si interface [97] and the FLP-free Bi/2D semiconductor interface has also been reported [98]. Furthermore, the Schottky barrier height at the Bi/Ge interface is obviously changed within 0.05 eV, depending on the surface orientation of Ge. Although the details of the surface orientation dependence have yet to be clarified, considering the polycrystallinity of Bi on the Ge substrate, the impact of Ge surface orientation on the band alignment might directly appear in this system.

As denoted above, the low-electron-density metal effect on the FLP supports that MIGS is the most reasonable mechanism to describe the strong FLP. Certain metals (e.g., WSi_x [99], a-TiNGe [100], TaN [101,102], Sn [103], graphene [104] and CNT [105]) show deviation from the strong FLP trend at the single-element metal/Ge interface, as summarized in Figure 12, but these results also seem understandable as MIGS reduction from the metal side, similar to germanide and Bi. Further clarification about the correlation between the interface structure and band alignment at low-electron-density metal/Ge and /Si are current big challenges to build a guideline to precisely control the band alignment at direct metal/Ge and /Si interfaces.

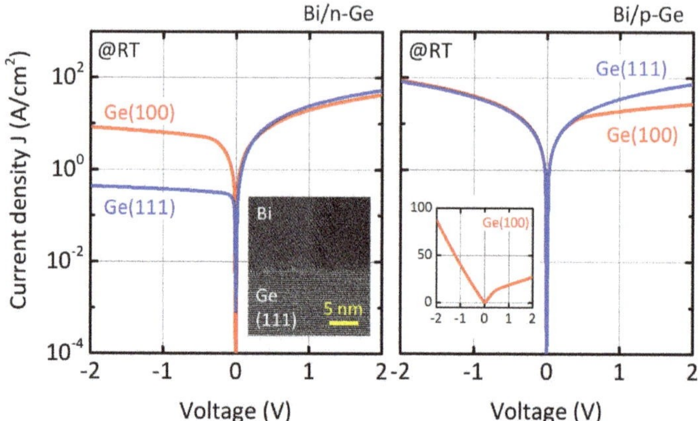

Figure 11. I-V characteristics of Bi/n-Ge and /p-Ge diodes with various Ge surface orientation. The off-state current density of Bi/n-Ge diodes are much higher than that of single-element metal/Ge ones, shown in Figure 2b, and Bi/p-Ge(100) diode shows rectified I-V characteristics, as also shown in the inset. The Bi/Ge interfaces were fabricated by deposition process, which is same to fabricate the single-element metal/Ge interfaces. As shown in the cross sectional TEM image in inset, direct Bi/Ge interface formation is confirmed. However, Schottky barrier height for conduction band of Ge at Bi/Ge(100) and /Ge(111) are estimated from saturation current density to be 0.37 and 0.44 eV, respectively. This band alignment is out of trend of strong FLP observed at single-element metal/Ge interface. The Ge surface orientation dependence on the Schottky barrier height at Bi/Ge interfaces might purely be caused by difference of Ge surface orientation.

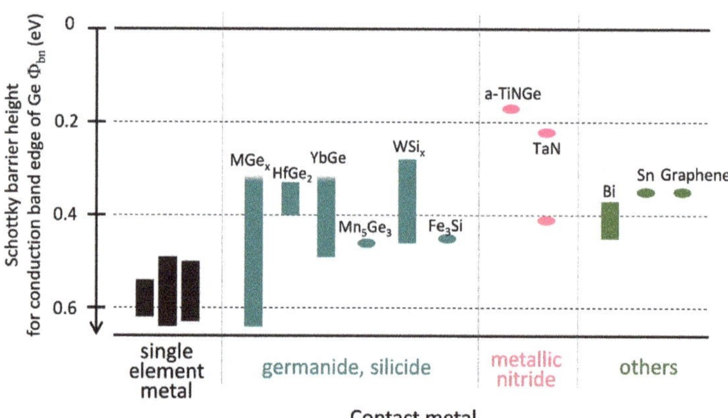

Figure 12. Summary of reported Schottky barrier height at low free-electron-density metal/n-Ge interface. Schottky barrier height for Ge is very controllable by appropriate metal and interface structure even at direct metal/Ge interface. In some case of metallic nitrides, Schottky barrier height for conduction band of Ge is much less than half of Ge band gap (0.33 eV). Considering the process compatibility of nitride metal with Si technology, those metals may also be possible candidates for practical application. Data from refs. [36–39,86–89,92,97,99–104].

6. Reduction of Contact Resistivity

Finally, the reduction of contact resistivity at the metal/Ge interface is discussed. The contact resistivity at the MIGS-effective metal/Si interface is fully described based on the field emission and thermionic-field emission current [106]. Therefore, here, the contact resistivity at the metal/Ge interface is discussed based on the band alignment (Schottky barrier height), although it has been reported that MIGS might affect the electronic structure of the semiconductors at the metal/2D semiconductor interface [107]. As mentioned in the previous sections, the ultrathin insulator layer at the metal/Ge interface enhances the controllability of the Schottky barrier height, but adds series resistance due to the tunnel barrier. Hence, for the ohmic contact formation to lightly doped semiconductors with a lower contact resistance, the best thickness can be determined based on the balance between FLP alleviation and insulator resistance. Since the tunnel resistance is decreased by reducing the potential barrier of the insulator, it is better for the reduction of contact resistivity to select an insulator with a small band offset between the insulator and semiconductor. For contact with the conduction band of Ge, TiO_2 [69], ZnO [108] and WSi_x [99] have been proposed with the demonstration of FLP alleviation.

For contact in aggressively scaled n-type Ge devices, tunnel resistance caused by both the interface insulator layer and semiconductor depletion layer must be considered because of heavily doping of impurities in Ge. In the case of contact for the conduction band of Si with doping over $10^{20}/cm^3$, it is expected that even TiO_2 might lose the advantage of a small band offset due to an increasing tunnel distance in TiO_2 [109]. As a solution, a more conductive interface layer with a low band offset and heavy doping has been proposed. For example, Al-doped ZnO and indium tin oxide (ITO) might be possible candidates for contact with the conduction band of Ge. Low contact resistivity for lightly doped n-Ge has been obtained experimentally [110], calculating that the resistivity would be reduced to approximately 10^{-9} Ωcm^2 [111].

Furthermore, a low-electron-density metal does not have tunnel resistance due to its metallic property. The contact resistivity at the NiGe/n-Ge interface [112,113] still exhibits high resistivity. However, the speculation is that other metals, such as rare-earth germanide, which forms a low Schottky barrier height, could significantly reduce the resistivity. Although the bulk resistivity of rare-earth germanide is slightly high ($\sim 10^{-4}$ Ωcm), rare-earth germanide can be thinned down to 5 nm while maintaining a low Schottky barrier height of 0.3 eV [114]. This result suggests that the total contact resistivity, which includes the bulk resistivity of germanide, is possibly reduced below 10^{-9} Ωcm^2 when the tunneling effective mass of an electron of 0.12 m_0 [115] and the high doping density up to $\sim 10^{20}/cm^3$ are assumed. The contact resistivities at various metal/n-Ge interfaces, including other interface layers [116,117], are summarized in Figure 13.

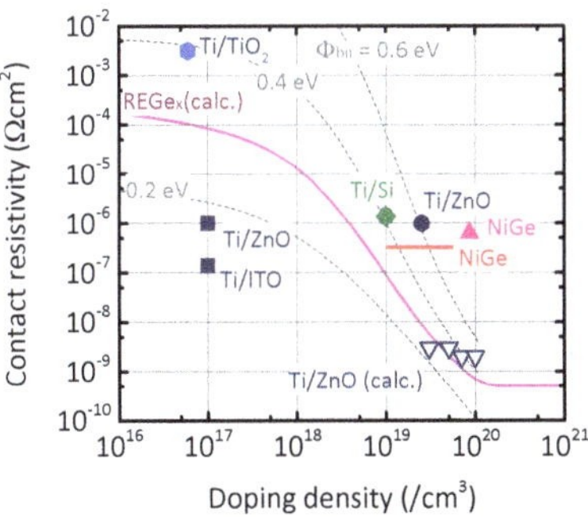

Figure 13. Contact resistivity at various metal/n-Ge interfaces. The calculated contact resistivities with various Schottky barrier heights are denoted in broken lines, where tunneling effective mass of electron of 0.12 m_0 and relative dielectric constant of 16 is assumed. The doped ZnO and ITO looks possible to achieve $\sim 10^{-9}$ Ωcm^2 by increasing doping density. The calculated total resistivity at rare earth metal germanide ($REGe_x$)/n-Ge interface is also shown with solid line, assuming that Schottky barrier height is 0.3 eV and bulk resistivity of 5-nm-thick $REGe_x$ ($\sim 5 \times 10^{-10}$ Ωcm^2). Data from refs. [108,110–113,116,117].

7. Conclusions

To realize scaled Ge FET devices, the reduction of contact resistivity is indispensable. However, it was found that the band alignment at common metal/Ge interfaces is far from the Schottky limit and a high Schottky barrier height for the conduction band edge of Ge is formed irrespective of the metal work function. This is described as a strong FLP close to the valence band edge of Ge. Therefore, this is a major hurdle to achieve high-performance Ge n-channel devices, including FETs.

The origin of strong FLP is expected to be so-called MIGS, based on the comparison of FLP strength between the metal/Ge interface and metal/SiGe interface focusing on intrinsic and extrinsic structural disorder in SiGe. This MIGS dominant FLP at the metal/Ge interface is also reasonably supported by the ultrathin insulator effect and low free-electron-density metal effect, as later denoted.

To alleviate the strong FLP at the metal/Ge interface, two kinds of approaches are demonstrated. One is the insertion of an ultrathin interface layer at the metal/Ge interface. This approach is motivated by research on the FLP free-metal/insulator/Ge gate stacks. The inserted interface layer has a role in reducing the tailing of the wave function into the band gap of the semiconductor by decaying in the layer. As experimental facts, various interlayers exhibit FLP alleviation and the FLP alleviation is enhanced with an increasing interface layer thickness. The other approach is applying a low free-electron metal as a contact metal. The analogy between the surface term of the vacuum work function of metal in a simple Jellium model and MIGS at the metal/semiconductor interface implies that the strength of MIGS is also characterized by the metal and MIGS can be reduced by decreasing the free-electron density in metal, which is a viewpoint different from before. Both the FLP alleviation at the germanides/Ge interfaces and the deviation from the strong FLP trend at some special metal/Ge interfaces are reasonably understandable based on the character of MIGS.

By further understanding the metal/Ge interface, the band alignment at the metal/Ge interface becomes more controllable. For this situation, assuming the well-controlled activation and diffusion of impurities in Ge, contact resistivity of less than 10^{-9} Ωcm^2 seems achievable. Therefore, to realize practical Ge n-channel devices in future nodes, further refined guidelines for interface design to reduce contact resistivity based on a further understanding of band alignment at the MIGS-weakened metal/Ge interface are desired.

Funding: This work was partially supported by JST CREST (JPMJCR0843), Harmonic Ito Foundation and JSPS KAKENHI Grant Numbers 22760244, 25420320, 17K06374 and 22K04212.

Acknowledgments: The author is grateful to Akira Toriumi and Kosuke Nagashio for fruitful discussions.

Conflicts of Interest: The author declares no conflict of interest.

References

1. Yeap, G.; Lin, S.S.; Chen, Y.M.; Shang, H.L.; Wang, P.W.; Lin, H.C.; Peng, Y.C.; Sheu, J.Y.; Wang, M.; Chen, X.; et al. 5nm CMOS Production Technology Platform Featuring Full-Fledged EUV, and High Mobility Channel FinFETs with Densest 0.021μm2 SRAM Cells for Mobile SoC and High Performance Computing Applications. In Proceedings of the 2019 IEEE International Electron Devices Meeting (IEDM), San Francisco, CA, USA, 7–11 December 2019; pp. 36.7.1–36.7.4.
2. Fischetti, M.V.; Laux, S.E. Band Structure, Deformation Potentials, and Carrier Mobility in Strained Si, Ge, and SiGe Alloys. *J. Appl. Phys.* **1996**, *80*, 2234–2252. [CrossRef]
3. Nagashio, K.; Lee, C.H.; Nishimura, T.; Toriumi, A. Thermodynamics and Kinetics for Suppression of GeO Desorption by High Pressure Oxidation of Ge. *MRS Proc.* **2009**, *1155*, C06-02. [CrossRef]
4. Kita, K.; Suzuki, S.; Nomura, H.; Takahashi, T.; Nishimura, T.; Toriumi, A. Direct Evidence of GeO Volatilization from GeO_2/Ge and Impact of Its Suppression on GeO_2/Ge Metal–Insulator–Semiconductor Characteristics. *Jpn. J. Appl. Phys.* **2008**, *47*, 2349–2353. [CrossRef]
5. Wang, S.K.; Kita, K.; Lee, C.H.; Tabata, T.; Nishimura, T.; Nagashio, K.; Toriumi, A. Desorption Kinetics of GeO from GeO_2/Ge Structure. *J. Appl. Phys.* **2010**, *108*, 054104. [CrossRef]
6. Wang, S.K.; Kita, K.; Nishimura, T.; Nagashio, K.; Toriumi, A. Kinetic Effects of O-Vacancy Generated by GeO_2/Ge Interfacial Reaction. *Jpn. J. Appl. Phys.* **2011**, *50*, 10PE04. [CrossRef]

7. Murthy, M.K.; Hill, H. Studies in Germanium Oxide Systems: III, Solubility of Germania in Water. *J. App. Ceram. Soc.* **1965**, *48*, 109–110. [CrossRef]

8. Chui, C.O.; Kim, H.; Chi, D.; Triplett, B.B.; McIntyre, P.C.; Saraswat, K.C. A Sub-400C Germanium MOSFET Technology with High-k Dielectric and Metal Gate. In Proceedings of the Digest. International Electron Devices Meeting, San Francisco, CA, USA, 8–11 December 2002; pp. 437–440.

9. Shang, H.; Okorn-Schmidt, H.; Chan, K.K.; Copel, M.; Ott, J.A.; Kozlowski, P.M.; Steen, S.E.; Cordes, S.A.; Wong, H.-S.P.; Jones, E.C.; et al. High Mobility P-Channel Germanium MOSFETs with a Thin Ge Oxynitride Gate Dielectric. In Proceedings of the Digest. International Electron Devices Meeting, San Francisco, CA, USA, 8–11 December 2002; pp. 441–444.

10. Whang, S.J.; Lee, S.J.; Gao, F.; Wu, N.; Zhu, C.X.; Pan, J.S.; Tang, L.J.; Kwong, D.L. Germanium p- & n-MOSFETs Fabricated with Novel Surface Passivation (Plasma-PH$_3$/ and Thin AlN) and TaN/HfO$_2$/ Gate Stack. In Proceedings of the IEDM Technical Digest. IEEE International Electron Devices Meeting, San Francisco, CA, USA, 13–15 December 2004; pp. 307–310.

11. Kamata, Y.; Kamimuta, Y.; Ino, T.; Iijima, R.; Koyama, M.; Nishiyama, A. Dramatic Improvement of Ge P-MOSFET Characteristics Realized by Amorphous Zr-Silicate/Ge Gate Stack with Excellent Structural Stability through Process Temperatures. In Proceedings of the IEEE International Electron Devices Meeting, IEDM Technical Digest, Washington, DC, USA, 5–7 December 2005; pp. 429–432.

12. Zimmerman, P.; Nicholas, G.; De Jaeger, B.; Kaczer, B.; Stesmans, A.; Ragnarsson, L.-A.; Brunco, D.P.; Leys, F.E.; Caymax, M.; Winderickx, G.; et al. High Performance Ge PMOS Devices Using a Si-Compatible Process Flow. In Proceedings of the 2006 International Electron Devices Meeting, San Francisco, CA, USA, 11–13 December 2006; pp. 26.1.1–26.1.4.

13. Ritenour, A.; Hennessy, J.; Antoniadis, D.A. Investigation of Carrier Transport in Germanium MOSFETs with WN/Al$_2$O$_3$/AlN Gate Stacks. *IEEE Electron Device Lett.* **2007**, *28*, 746–749. [CrossRef]

14. Yamamoto, T.; Yamashita, Y.; Harada, M.; Taoka, N.; Ikeda, K.; Suzuki, K.; Kiso, O.; Sugiyama, N.; Takagi, S. High Performance 60 Nm Gate Length Germanium P-MOSFETs with Ni Germanide Metal Source/Drain. In Proceedings of the 2007 IEEE International Electron Devices Meeting, Washington, DC, USA, 10–12 December 2007; pp. 1041–1043.

15. Xu, J.P.; Zhang, X.F.; Li, C.X.; Lai, P.T.; Chan, C.L. Improved Electrical Properties of Ge P-MOSFET with HfO$_2$ Gate Dielectric by Using TaO$_x$N$_y$ Interlayer. *IEEE Electron Device Lett.* **2008**, *29*, 1155–1158. [CrossRef]

16. Xie, R.; Phung, T.H.; He, W.; Sun, Z.; Yu, M.; Cheng, Z.; Zhu, C. High Mobility High-k/Ge PMOSFETs with 1 Nm EOT—New Concept on Interface Engineering and Interface Characterization. In Proceedings of the 2008 IEEE International Electron Devices Meeting, San Francisco, CA, USA, 15–17 December 2008; pp. 16.2.1–16.2.4.

17. Hashemi, P.; Chern, W.; Lee, H.-S.; Teherani, J.T.; Zhu, Y.; Gonsalvez, J.; Shahidi, G.G.; Hoyt, J.L. Ultrathin Strained-Ge Channel P-MOSFETs with High-K /Metal Gate and Sub-1-Nm Equivalent Oxide Thickness. *IEEE Electron Device Lett.* **2012**, *33*, 943–945. [CrossRef]

18. Zhang, R.; Chern, W.; Yu, X.; Takenaka, M.; Hoyt, J.L.; Takagi, S. High Mobility Strained-Ge PMOSFETs with 0.7-Nm Ultrathin EOT Using Plasma Post Oxidation HfO2/Al2O3/GeOx Gate Stacks and Strain Modulation. In Proceedings of the 2013 IEEE International Electron Devices Meeting, Washington, DC, USA, 9–11 December 2013; pp. 26.1.1–26.1.4.

19. Witters, L.; Arimura, H.; Sebaai, F.; Hikavyy, A.; Milenin, A.P.; Loo, R.; De Keersgieter, A.; Eneman, G.; Schram, T.; Wostyn, K.; et al. Strained Germanium Gate-All-Around PMOS Device Demonstration Using Selective Wire Release Etch Prior to Replacement Metal Gate Deposition. *IEEE Trans. Electron Devices* **2017**, *64*, 4587–4593. [CrossRef]

20. Kuzum, D.; Pethe, A.J.; Krishnamohan, T.; Saraswat, K.C. Ge (100) and (111) N- and P-FETs with High Mobility and Low-T Mobility Characterization. *IEEE Trans. Electron Devices* **2009**, *56*, 648–655. [CrossRef]

21. Zhang, R.; Huang, P.-C.; Lin, J.-C.; Taoka, N.; Takenaka, M.; Takagi, S. High-Mobility Ge p- and n-MOSFETs with 0.7-Nm EOT Using HfO$_2$/Al$_2$O$_3$/GeO$_x$/Ge Gate Stacks Fabricated by Plasma Postoxidation. *IEEE Trans. Electron Devices* **2013**, *60*, 927–934. [CrossRef]

22. Lee, C.H.; Nishimura, T.; Tabata, T.; Lu, C.; Zhang, W.F.; Nagashio, K.; Toriumi, A. Reconsideration of Electron Mobility in Ge N-MOSFETs from Ge Substrate Side—Atomically Flat Surface Formation, Layer-by-Layer Oxidation, and Dissolved Oxygen Extraction. In Proceedings of the 2013 IEEE International Electron Devices Meeting, Washington, DC, USA, 9–11 December 2013; pp. 2.3.1–2.3.4.

23. van Dal, M.J.H.; Duriez, B.; Vellianitis, G.; Doornbos, G.; Oxland, R.; Holland, M.; Afzalian, A.; See, Y.C.; Passlack, M.; Diaz, C.H. Ge N-Channel FinFET with Optimized Gate Stack and Contacts. In Proceedings of the 2014 IEEE International Electron Devices Meeting, San Francisco, CA, USA, 15–17 December 2014; pp. 9.5.1–9.5.4.

24. Lu, C.; Lee, C.H.; Nishimura, T.; Toriumi, A. Yttrium Scandate Thin Film as Alternative High-Permittivity Dielectric for Germanium Gate Stack Formation. *Appl. Phys. Lett.* **2015**, *107*, 072904. [CrossRef]

25. Lin, C.-M.; Chang, H.-C.; Wong, I.-H.; Luo, S.-J.; Liu, C.W.; Hu, C. Interfacial Layer Reduction and High Permittivity Tetragonal ZrO$_2$ on Germanium Reaching Ultrathin 0.39 Nm Equivalent Oxide Thickness. *Appl. Phys. Lett.* **2013**, *102*, 232906. [CrossRef]

26. Rachmady, W.; Agrawal, A.; Sung, S.H.; Dewey, G.; Chouksey, S.; Chu-Kung, B.; Elbaz, G.; Fischer, P.; Huang, C.Y.; Jun, K.; et al. 300 mm Heterogeneous 3D Integration of Record Performance Layer Transfer Germanium PMOS with Silicon NMOS for Low Power High Performance Logic Applications. In Proceedings of the 2019 IEEE International Electron Devices Meeting (IEDM), San Francisco, CA, USA, 7–11 December 2019; pp. 29.7.1–29.7.4.

27. Lin, Y.-W.; Chang, H.-H.; Huang, Y.-H.; Sun, C.-J.; Yan, S.-C.; Lin, S.-W.; Luo, G.-L.; Wu, C.-T.; Wu, Y.-C.; Hou, F.-J. Tightly Stacked 3D Diamond-Shaped Ge Nanowire Gate-All-Around FETs with Superior NFET and PFET Performance. *IEEE Electron Device Lett.* **2021**, *42*, 1727–1730. [CrossRef]
28. Nishimura, T.; Kabuyanagi, S.; Zhang, W.; Lee, C.H.; Yajima, T.; Nagashio, K.; Toriumi, A. Atomically Flat Planarization of Ge(100), (110), and (111) Surfaces in H2 Annealing. *Appl. Phys. Express* **2014**, *7*, 051301. [CrossRef]
29. Morita, Y.; Ota, H.; Masahara, M.; Maeda, T. Impact of H2, O2, and N2 Anneals on Atomic-Scale Surface Flattening for 3-D Ge Channel Architecture. In Proceedings of the 2015 Silicon Nanoelectronics Workshop (SNW), Kyoto, Japan, 14–15 June 2015; pp. 1–2.
30. Nishimura, T.; Takemura, S.; Wang, X.; Shibayama, S.; Yajima, T.; Toriumi, A. Atomically Flat Interface Formation on Ge (111) in Oxidation Process. In Proceedings of the 2018 International Conference on Solid State Devices and Materials, Kyoto, Japan, 9–13 September 2018; pp. 349–350.
31. Datta, S.; Pandey, R.; Agrawal, A.; Gupta, S.K.; Arghavani, R. Impact of Contact and Local Interconnect Scaling on Logic Performance. In Proceedings of the 2014 Symposium on VLSI Technology (VLSI-Technology): Digest of Technical Papers, Honolulu, HI, USA, 9–12 June 2014; p. 16.1.
32. Arghavani, R.; Yang, P.; Ashtiani, K.; Singh, H.; Hemker, D. Low Resistance Contacts to Enable 5 Nm Node Technology: Patterning, Etch, Clean, Metallization and Device Performance. In Proceedings of the 2016 IEEE International Electron Devices Meeting Short Course, San Francisco, CA, USA, 3–7 December 2016.
33. Schroder, D.K. *Semiconductor Material and Device Characterization*, 3rd ed.; Wiley-IEEE Press: Hoboken, NJ, USA, 2015; ISBN 978-0-471-73906-7.
34. Cowley, A.M.; Sze, S.M. Surface States and Barrier Height of Metal-Semiconductor Systems. *J. Appl. Phys.* **1965**, *36*, 3212–3220. [CrossRef]
35. Thanailakis, A.; Northrop, D.C. Metal-Germanium Schottky Barriers. *Solid-State Electron.* **1973**, *16*, 1383–1389. [CrossRef]
36. Marshall, E.D.; Wu, C.S.; Pai, C.S.; Scott, D.M.; Lau, S.S. Metal-Germanium Contacts and Germanide Formation. *MRS Proc.* **1985**, *47*, 161–166. [CrossRef]
37. Dimoulas, A.; Tsipas, P.; Sotiropoulos, A.; Evangelou, E.K. Fermi-Level Pinning and Charge Neutrality Level in Germanium. *Appl. Phys. Lett.* **2006**, *89*, 252110. [CrossRef]
38. Nishimura, T.; Kita, K.; Toriumi, A. Strong Fermi-Level Pinning of Wide Range of Work-Function Metals at Valence Band Edge of Germanium. In Proceedings of the International Conference on Solid State Devices and Materials, Yokohama, Japan, 12–15 September 2006; p. J-6-3.
39. Nishimura, T.; Kita, K.; Toriumi, A. Evidence for Strong Fermi-Level Pinning Due to Metal-Induced Gap States at Metal/Germanium Interface. *Appl. Phys. Lett.* **2007**, *91*, 123123. [CrossRef]
40. Michaelson, H.B. The Work Function of the Elements and Its Periodicity. *J. Appl. Phys.* **1977**, *48*, 4729–4733. [CrossRef]
41. Malov, Y.I.; Onichchenko, A.V.; Mironkova, L.I. Work Function of Rare-Earth Metals. *Phys. Met. Metall.* **1980**, *47*, 195–197.
42. Heine, V. Theory of Surface States. *Phys. Rev.* **1965**, *138*, A1689–A1696. [CrossRef]
43. Louie, S.G.; Cohen, M.L. Electronic Structure of a Metal-Semiconductor Interface. *Phys. Rev. B* **1976**, *13*, 2461–2469. [CrossRef]
44. Tersoff, J. Schottky Barrier Heights and the Continuum of Gap States. *Phys. Rev. Lett.* **1984**, *52*, 465–468. [CrossRef]
45. Mönch, W. Role of Virtual Gap States and Defects in Metal-Semiconductor Contacts. *Phys. Rev. Lett.* **1987**, *58*, 1260–1263. [CrossRef]
46. Mönch, W. Barrier Heights of Real Schottky Contacts Explained by Metal-Induced Gap States and Lateral Inhomogeneities. *J. Vac. Sci. Technol. B* **1999**, *17*, 1867–1876. [CrossRef]
47. Tung, R.T. Chemical Bonding and Fermi Level Pinning at Metal-Semiconductor Interfaces. *Phys. Rev. Lett.* **2000**, *84*, 6078–6081. [CrossRef]
48. Tung, R.T. Formation of an Electric Dipole at Metal-Semiconductor Interfaces. *Phys. Rev. B* **2001**, *64*, 205310. [CrossRef]
49. Spicer, W.E.; Lindau, I.; Skeath, P.; Su, C.Y.; Chye, P. Unified Mechanism for Schottky-Barrier Formation and III-V Oxide Interface States. *Phys. Rev. Lett.* **1980**, *44*, 420–423. [CrossRef]
50. Spicer, W.E.; Lilienthal-Weber, Z.; Weber, E.; Newman, N.; Kendelewicz, T.; Cao, R.; McCants, C.; Mahowald, P.; Miyano, K.; Lindau, I. The Advanced Unified Defect Model for Schottky Barrier Formation. *J. Vac. Sci. Technol. B* **1988**, *6*, 1245–1251. [CrossRef]
51. Hasegawa, H.; Ohno, H. Unified Disorder Induced Gap State Model for Insulator–Semiconductor and Metal–Semiconductor Interfaces. *J. Vac. Sci. Technol. B* **1986**, *4*, 1130–1138. [CrossRef]
52. Hasegawa, H.; Ishii, H.; Koyanagi, K. Formation Mechanism of Schottky Barriers on MBE-Grown GaAs Surfaces Subjected to Various Treatments. *Appl. Surf. Sci.* **1992**, *56–58*, 317–324. [CrossRef]
53. Kurtin, S.; McGill, T.C.; Mead, C.A. Fundamental Transition in the Electronic Nature of Solids. *Phys. Rev. Lett.* **1969**, *22*, 1433–1436. [CrossRef]
54. Hara, S. The Schottky Limit and a Charge Neutrality Level Found on Metal/6H-SiC Interfaces. *Surf. Sci.* **2001**, *494*, L805–L810. [CrossRef]
55. Wang, H.-T.; Jang, S.; Anderson, T.; Chen, J.J.; Kang, B.S.; Ren, F.; Herrero, A.; Gerger, A.M.; Gila, B.P.; Pearton, S.J.; et al. Improved Au Schottky Contacts on GaAs Using Cryogenic Metal Deposition. *J. Vac. Sci. Technol. B* **2006**, *24*, 1799–1802. [CrossRef]
56. Cardona, M.; Christensen, N.E. Acoustic Deformation Potentials and Heterostructure Band Offsets in Semiconductors. *Phys. Rev. B* **1987**, *35*, 6182–6194. [CrossRef]

57. Haesslein, H.; Sielemann, R.; Zistl, C. Vacancies and Self-Interstitials in Germanium Observed by Perturbed Angular Correlation Spectroscopy. *Phys. Rev. Lett.* **1998**, *80*, 2626–2629. [CrossRef]

58. Satoh, M.; Arimoto, K.; Nakagawa, K.; Koh, S.; Sawano, K.; Shiraki, Y.; Usami, N.; Nakajima, K. Acceptorlike Behavior of Defects in SiGe Alloys Grown by Molecular Beam Epitaxy. *Jpn. J. Appl. Phys.* **2008**, *47*, 4630–4633. [CrossRef]

59. Sawano, K.; Hoshi, Y.; Kubo, S.; Arimoto, K.; Yamanaka, J.; Nakagawa, K.; Hamaya, K.; Miyao, M.; Shiraki, Y. Structural and Electrical Properties of Ge(111) Films Grown on Si(111) Substrates and Application to Ge(111)-on-Insulator. *Thin Solid Films* **2016**, *613*, 24–28. [CrossRef]

60. Luo, X.; Nishimura, T.; Yajima, T.; Toriumi, A. Understanding of Fermi Level Pinning at Metal/Germanium Interface Based on Semiconductor Structure. *Appl. Phys. Express* **2020**, *13*, 031003. [CrossRef]

61. Archer, R.J.; Atalla, M.M. Metals Contacts on Cleaved Silicon Surfaces. *Anu. NY Acad. Sci.* **1963**, *101*, 697–708. [CrossRef]

62. Nishimura, T.; Kita, K.; Toriumi, A. Effect of Ultra-Thin Al_2O_3 Insertion on Fermi-Level Pinning at Metal/Ge Interface. In Proceedings of the International Conference on Solid State Devices and Materials, Tsukuba, Japan, 18–21 September 2007; p. A-7-3.

63. Lee, C.H.; Nishimura, T.; Nagashio, K.; Kita, K.; Toriumi, A. High-Electron-Mobility Ge/GeO_2 n-MOSFETs with Two-Step Oxidation. *IEEE Trans. Electron Devices* **2011**, *58*, 1295–1301. [CrossRef]

64. Toriumi, A.; Nishimura, T. Germanium CMOS Potential from Material and Process Perspectives: Be More Positive about Germanium. *Jpn. J. Appl. Phys.* **2017**, *57*, 010101. [CrossRef]

65. Nishimura, T.; Kita, K.; Toriumi, A. A Significant Shift of Schottky Barrier Heights at Strongly Pinned Metal/Germanium Interface by Inserting an Ultra-Thin Insulating Film. *Appl. Phys. Express* **2008**, *1*, 051406. [CrossRef]

66. Zhou, Y.; Ogawa, M.; Han, X.; Wang, K.L. Alleviation of Fermi-Level Pinning Effect on Metal/Germanium Interface by Insertion of an Ultrathin Aluminum Oxide. *Appl. Phys. Lett.* **2008**, *93*, 202105. [CrossRef]

67. Gajula, D.R.; Baine, P.; Modreanu, M.; Hurley, P.K.; Armstrong, B.M.; McNeill, D.W. Fermi Level De-Pinning of Aluminium Contacts to n-Type Germanium Using Thin Atomic Layer Deposited Layers. *Appl. Phys. Lett.* **2014**, *104*, 012102. [CrossRef]

68. Nishimura, T.; Kita, K.; Nagashio, K.; Toriumi, A. Long Range Pinning Interaction in Ultra-Thin Insulator-Inserted Metal/Germanium Junctions. In Proceedings of the 2010 Silicon Nanoelectronics Workshop, Honolulu, HI, USA, 13–14 June 2010; p. 2.3.

69. Lin, J.-Y.J.; Roy, A.M.; Nainani, A.; Sun, Y.; Saraswat, K.C. Increase in Current Density for Metal Contacts to N-Germanium by Inserting TiO_2 Interfacial Layer to Reduce Schottky Barrier Height. *Appl. Phys. Lett.* **2011**, *98*, 092113. [CrossRef]

70. Tsui, B.-Y.; Kao, M.-H. Mechanism of Schottky Barrier Height Modulation by Thin Dielectric Insertion on N-Type Germanium. *Appl. Phys. Lett.* **2013**, *103*, 032104. [CrossRef]

71. Lee, D.; Raghunathan, S.; Wilson, R.J.; Nikonov, D.E.; Saraswat, K.; Wang, S.X. The Influence of Fermi Level Pinning/Depinning on the Schottky Barrier Height and Contact Resistance in Ge/CoFeB and Ge/MgO/CoFeB Structures. *Appl. Phys. Lett.* **2010**, *96*, 052514. [CrossRef]

72. Zhou, Y.; Han, W.; Wang, Y.; Xiu, F.; Zou, J.; Kawakami, R.K.; Wang, K.L. Investigating the Origin of Fermi Level Pinning in Ge Schottky Junctions Using Epitaxially Grown Ultrathin MgO Films. *Appl. Phys. Lett.* **2010**, *96*, 102103. [CrossRef]

73. Jung, H.-W.; Jung, W.-S.; Park, J.-H. Hydrazine-Based Fermi-Level Depinning Process on Metal/Germanium Schottky Junction. *IEEE Electron Device Lett.* **2013**, *34*, 599–601. [CrossRef]

74. Biswas, D.; Biswas, J.; Ghosh, S.; Wood, B.; Lodha, S. Enhanced Thermal Stability of Ti/TiO_2/n-Ge Contacts through Plasma Nitridation of TiO_2 Interfacial Layer. *Appl. Phys. Lett.* **2017**, *110*, 052104. [CrossRef]

75. Lieten, R.R.; Degroote, S.; Kuijk, M.; Borghs, G. Ohmic Contact Formation on N-Type Ge. *Appl. Phys. Lett.* **2008**, *92*, 022106. [CrossRef]

76. Kobayashi, M.; Kinoshita, A.; Saraswat, K.; Wong, H.-S.P.; Nishi, Y. Fermi Level Depinning in Metal/Ge Schottky Junction for Metal Source/Drain Ge Metal-Oxide-Semiconductor Field-Effect-Transistor Application. *J. Appl. Phys.* **2009**, *105*, 023702. [CrossRef]

77. Connelly, D.; Faulkner, C.; Clifton, P.A.; Grupp, D.E. Fermi-Level Depinning for Low-Barrier Schottky Source/Drain Transistors. *Appl. Phys. Lett.* **2006**, *88*, 012105. [CrossRef]

78. Sasaki, S.; Nakayama, T. Defect Distribution and Schottky Barrier at Metal/Ge Interfaces: Role of Metal-Induced Gap States. *Jpn. J. Appl. Phys.* **2016**, *55*, 111302. [CrossRef]

79. Hiraki, A.; Lugujjo, E.; Mayer, J.W. Formation of Silicon Oxide over Gold Layers on Silicon Substrates. *J. Appl. Phys.* **1972**, *43*, 3643–3649. [CrossRef]

80. Lang, N.D.; Kohn, W. Theory of Metal Surfaces: Charge Density and Surface Energy. *Phys. Rev. B* **1970**, *1*, 4555–4568. [CrossRef]

81. Lang, N.D.; Kohn, W. Theory of Metal Surfaces: Work Function. *Phys. Rev. B* **1971**, *3*, 1215–1223. [CrossRef]

82. Kittel, C. *Introduction to Solid State Physics*, 6th ed.; John Wiley & Sons: New York, NY, USA, 1986; ISBN 978-0-471-87474-4.

83. Nava, F.; Tu, K.N.; Thomas, O.; Senateur, J.P.; Madar, R.; Borghesi, A.; Guizzetti, G.; Gottlieb, U.; Laborde, O.; Bisi, O. Electrical and Optical Properties of Silicide Single Crystals and Thin Films. *Mater. Sci. Rep.* **1993**, *9*, 141–200. [CrossRef]

84. Sanderson, R.T. Electronegativity and Bond Energy. *J. Am. Chem. Soc.* **1983**, *105*, 2259–2261. [CrossRef]

85. Gordy, W.; Thomas, W.J.O. Electronegativities of the Elements. *J. Chem. Phys.* **1956**, *24*, 439–444. [CrossRef]

86. Zheng, Z.-W.; Ku, T.-C.; Liu, M.; Chin, A. Ohmic Contact on N-Type Ge Using Yb-Germanide. *Appl. Phys. Lett.* **2012**, *101*, 223501. [CrossRef]

87. Nishimura, T.; Nakatsuka, O.; Akimoto, S.; Takeuchi, W.; Zaima, S. Crystalline Orientation Dependence of Electrical Properties of Mn Germanide/Ge(111) and (001) Schottky Contacts. *Microelectron. Eng.* **2011**, *88*, 605–609. [CrossRef]
88. Senga, K.; Shibayama, S.; Sakashita, M.; Zaima, S.; Nakatsuka, O. Further Reduction of Schottky Barrier Height of Hf-Germanide/n-Ge(001) Contacts by Forming Epitaxial HfGe$_2$. In Proceedings of the 2019 19th International Workshop on Junction Technology (IWJT), Kyoto, Japan, 6–7 June 2019; pp. 1–2.
89. Yamane, K.; Hamaya, K.; Ando, Y.; Enomoto, Y.; Yamamoto, K.; Sadoh, T.; Miyao, M. Effect of Atomically Controlled Interfaces on Fermi-Level Pinning at Metal/Ge Interfaces. *Appl. Phys. Lett.* **2010**, *96*, 162104. [CrossRef]
90. Tu, K.N.; Thompson, R.D.; Tsaur, B.Y. Low Schottky Barrier of Rare-earth Silicide on N-Si. *Appl. Phys. Lett.* **1981**, *38*, 626–628. [CrossRef]
91. Freeouf, J.L.; Woodall, J.M. Schottky Barriers: An Effective Work Function Model. *Appl. Phys. Lett.* **1981**, *39*, 727–729. [CrossRef]
92. Nishimura, T.; Yajima, T.; Toriumi, A. Reexamination of Fermi Level Pinning for Controlling Schottky Barrier Height at Metal/Ge Interface. *Appl. Phys. Express* **2016**, *9*, 081201. [CrossRef]
93. Lin, L.; Guo, Y.; Robertson, J. Metal Silicide Schottky Barriers on Si and Ge Show Weaker Fermi Level Pinning. *Appl. Phys. Lett.* **2012**, *101*, 052110. [CrossRef]
94. Tung, R.T. Schottky-Barrier Formation at Single-Crystal Metal-Semiconductor Interfaces. *Phys. Rev. Lett.* **1984**, *52*, 461–464. [CrossRef]
95. Williams, G.A. Alfven-Wave Propagation in Solid-State Plasmas. I. Bismuth. *Phys. Rev.* **1965**, *139*, A771–A778. [CrossRef]
96. Massalski, T.B.; Murray, J.L.; Bennett, L.H.; Baker, H. *Binary Alloy Phase Diagrams*, 1st ed.; American Society for Metals: Novelty, OH, USA, 1986; ISBN 978-0-87170-261-6.
97. Nishimura, T.; Luo, X.; Matsumoto, S.; Yajima, T.; Toriumi, A. Almost Pinning-Free Bismuth/Ge and /Si Interfaces. *AIP Adv.* **2019**, *9*, 095013. [CrossRef]
98. Shen, P.-C.; Su, C.; Lin, Y.; Chou, A.-S.; Cheng, C.-C.; Park, J.-H.; Chiu, M.-H.; Lu, A.-Y.; Tang, H.-L.; Tavakoli, M.M.; et al. Ultralow Contact Resistance between Semimetal and Monolayer Semiconductors. *Nature* **2021**, *593*, 211–217. [CrossRef]
99. Okada, N.; Uchida, N.; Kanayama, T. Fermi-Level Depinning and Contact Resistance Reduction in Metal/n-Ge Junctions by Insertion of W-Encapsulating Si Cluster Films. *Appl. Phys. Lett.* **2014**, *104*, 062105. [CrossRef]
100. Iyota, M.; Yamamoto, K.; Wang, D.; Yang, H.; Nakashima, H. Ohmic Contact Formation on N-Type Ge by Direct Deposition of TiN. *Appl. Phys. Lett.* **2011**, *98*, 192108. [CrossRef]
101. Wu, Z.; Wang, C.; Huang, W.; Li, C.; Lai, H.; Chen, S. Ohmic Contact Formation of Sputtered TaN on N-Type Ge with Lower Specific Contact Resistivity. *ECS J. Solid State Sci. Technol.* **2012**, *1*, P30–P33. [CrossRef]
102. Seo, Y.; Lee, S.; Baek, S.C.; Hwang, W.S.; Yu, H.-Y.; Lee, S.-H.; Cho, B.J. The Mechanism of Schottky Barrier Modulation of Tantalum Nitride/Ge Contacts. *IEEE Electron Device Lett.* **2015**, *36*, 997–1000. [CrossRef]
103. Suzuki, A.; Asaba, S.; Yokoi, J.; Kato, K.; Kurosawa, M.; Sakashita, M.; Taoka, N.; Nakatsuka, O.; Zaima, S. Reduction of Schottky Barrier Height for N-Type Ge Contact by Using Sn Electrode. *Jpn. J. Appl. Phys.* **2014**, *53*, 04EA06. [CrossRef]
104. Baek, S.C.; Seo, Y.-J.; Oh, J.G.; Albert Park, M.G.; Bong, J.H.; Yoon, S.J.; Seo, M.; Park, S.; Park, B.-G.; Lee, S.-H. Alleviation of Fermi-Level Pinning Effect at Metal/Germanium Interface by the Insertion of Graphene Layers. *Appl. Phys. Lett.* **2014**, *105*, 073508. [CrossRef]
105. Wei, Y.-N.; Hu, X.-G.; Zhang, J.-W.; Tong, B.; Du, J.-H.; Liu, C.; Sun, D.-M.; Liu, C. Fermi-Level Depinning in Metal/Ge Junctions by Inserting a Carbon Nanotube Layer. *Small* **2022**, *18*, 2201840. [CrossRef]
106. Schroder, D.K.; Meier, D.L. Solar Cell Contact Resistance—A Review. *IEEE Trans. Electron Devices* **1984**, *31*, 637–647. [CrossRef]
107. Peng, S.; Jin, Z.; Yao, Y.; Li, L.; Zhang, D.; Shi, J.; Huang, X.; Niu, J.; Zhang, Y.; Yu, G. Metal-Contact-Induced Transition of Electrical Transport in Monolayer MoS2: From Thermally Activated to Variable-Range Hopping. *Adv. Electron. Mater.* **2019**, *5*, 1900042. [CrossRef]
108. Paramahans Manik, P.; Kesh Mishra, R.; Pavan Kishore, V.; Ray, P.; Nainani, A.; Huang, Y.-C.; Abraham, M.C.; Ganguly, U.; Lodha, S. Fermi-Level Unpinning and Low Resistivity in Contacts to n-Type Ge with a Thin ZnO Interfacial Layer. *Appl. Phys. Lett.* **2012**, *101*, 182105. [CrossRef]
109. Yu, H.; Schaekers, M.; Demuynck, S.; Barla, K.; Mocuta, A.; Horiguchi, N.; Collaert, N.; Thean, A.V.-Y.; De Meyer, K. MIS or MS? Source/Drain Contact Scheme Evaluation for 7nm Si CMOS Technology and Beyond. In Proceedings of the 2016 16th International Workshop on Junction Technology (IWJT), Shanghai, China, 9–10 May 2016; pp. 19–24.
110. Manik, P.P.; Lodha, S. Contacts on N-Type Germanium Using Variably Doped Zinc Oxide and Highly Doped Indium Tin Oxide Interfacial Layers. *Appl. Phys. Express* **2015**, *8*, 051302. [CrossRef]
111. Kim, J.-K.; Kim, G.-S.; Nam, H.; Shin, C.; Park, J.-H.; Kim, J.-K.; Cho, B.J.; Saraswat, K.C.; Yu, H.-Y. The Efficacy of Metal-Interfacial Layer-Semiconductor Source/Drain Structure on Sub-10-Nm n-Type Ge FinFET Performances. *IEEE Electron Device Lett.* **2014**, *35*, 1185–1187. [CrossRef]
112. Shayesteh, M.; Huet, K.; Toqué-Tresonne, I.; Negru, R.; Daunt, C.L.M.; Kelly, N.; O'Connell, D.; Yu, R.; Djara, V.; Carolan, P.B.; et al. Atomically Flat Low-Resistive Germanide Contacts Formed by Laser Thermal Anneal. *IEEE Trans. Electron Devices* **2013**, *60*, 2178–2185. [CrossRef]
113. Miyoshi, H.; Ueno, T.; Hirota, Y.; Yamanaka, J.; Arimoto, K.; Nakagawa, K.; Kaitsuka, T. Low Nickel Germanide Contact Resistances by Carrier Activation Enhancement Techniques for Germanium CMOS Application. *Jpn. J. Appl. Phys.* **2014**, *53*, 04EA05. [CrossRef]

114. Nishimura, T.; Toriumi, A. Nishimura, T. (The Univ. of Tokyo, Tokyo, Japan); Toriumi, A. (The Univ. of Tokyo, Tokyo, Japan) Unpublished work, 2016.

115. Hutin, L.; Royer, C.L.; Tabone, C.; Delaye, V.; Nemouchi, F.; Aussenac, F.; Clavelier, L.; Vinet, M. Schottky Barrier Height Extraction in Ohmic Regime: Contacts on Fully Processed GeOI Substrates. *J. Electrochem. Soc.* **2009**, *156*, H522. [CrossRef]

116. Martens, K.; Rooyackers, R.; Firrincieli, A.; Vincent, B.; Loo, R.; De Jaeger, B.; Meuris, M.; Favia, P.; Bender, H.; Douhard, B.; et al. Contact Resistivity and Fermi-Level Pinning in n-Type Ge Contacts with Epitaxial Si-Passivation. *Appl. Phys. Lett.* **2011**, *98*, 013504. [CrossRef]

117. Kim, G.-S.; Kim, J.-K.; Kim, S.-H.; Jo, J.; Shin, C.; Park, J.-H.; Saraswat, K.C.; Yu, H.-Y. Specific Contact Resistivity Reduction Through Ar Plasma-Treated TiO_{2-x} Interfacial Layer to Metal/Ge Contact. *IEEE Electron Device Lett.* **2014**, *35*, 1076–1078. [CrossRef]

Review

Recent Progresses and Perspectives of UV Laser Annealing Technologies for Advanced CMOS Devices

Toshiyuki Tabata *, Fabien Rozé, Louis Thuries, Sébastien Halty, Pierre-Edouard Raynal, Imen Karmous and Karim Huet

Laser Systems & Solutions of Europe (LASSE), 145 Rue Des Caboeufs, 92230 Gennevilliers, France
* Correspondence: toshiyuki.tabata@screen-lasse.com

Abstract: The state-of-the-art CMOS technology has started to adopt three-dimensional (3D) integration approaches, enabling continuous chip density increment and performance improvement, while alleviating difficulties encountered in traditional planar scaling. This new device architecture, in addition to the efforts required for extracting the best material properties, imposes a challenge of reducing the thermal budget of processes to be applied everywhere in CMOS devices, so that conventional processes must be replaced without any compromise to device performance. Ultra-violet laser annealing (UV-LA) is then of prime importance to address such a requirement. First, the strongly limited absorption of UV light into materials allows surface-localized heat source generation. Second, the process timescale typically ranging from nanoseconds (ns) to microseconds (μs) efficiently restricts the heat diffusion in the vertical direction. In a given 3D stack, these specific features allow the actual process temperature to be elevated in the top-tier layer without introducing any drawback in the bottom-tier one. In addition, short-timescale UV-LA may have some advantages in materials engineering, enabling the nonequilibrium control of certain phenomenon such as crystallization, dopant activation, and diffusion. This paper reviews recent progress reported about the application of short-timescale UV-LA to different stages of CMOS integration, highlighting its potential of being a key enabler for next generation 3D-integrated CMOS devices.

Keywords: UV laser annealing; CMOS; 3D integration; thermal budget; FEOL; MOL; BEOL; dopant activation; interconnect; ferroelectricity

Citation: Tabata, T.; Rozé, F.; Thuries, L.; Halty, S.; Raynal, P.-E.; Karmous, I.; Huet, K. Recent Progresses and Perspectives of UV Laser Annealing Technologies for Advanced CMOS Devices. *Electronics* **2022**, *11*, 2636. https://doi.org/10.3390/electronics11172636

Academic Editor: Kiat Seng Yeo

Received: 4 July 2022
Accepted: 20 July 2022
Published: 23 August 2022

Publisher's Note: MDPI stays neutral with regard to jurisdictional claims in published maps and institutional affiliations.

Copyright: © 2022 by the authors. Licensee MDPI, Basel, Switzerland. This article is an open access article distributed under the terms and conditions of the Creative Commons Attribution (CC BY) license (https://creativecommons.org/licenses/by/4.0/).

1. Introduction

Nowadays, sustainable, high-performance, and energy-efficient nanoelectronics drive the further development of CMOS technologies, not only by traditional planar scaling, called "More Moore", but also by opening the integration dimension towards the third axis perpendicular to the silicon (Si) wafer plane. This three-dimensionally (3D) stacked device architecture alleviates the difficulties possibly encountered in the way of pursuing the traditional planar scaling of transistors, while allowing the continuous increment of effective chip density and performance thereby. Here, a "monolithic" or "sequential" approach is talked about, but not a "packaging" one, where interlayer connectivity is dominated by chip bonding alignment accuracy. It also brings another benefit of implementing additional functionalities in CMOS devices, being called "More Than Moore". Furthermore, over the coming end of the current CMOS scaling, "Beyond CMOS" is preconized to determine the best means for continuing technological advancements, having new materials, concepts, and architectures.

Among these three axes of evolution, "More Than Moore" would be the one which has been the most investigated over the last several years, in parallel with the evolution of ultra-violet laser annealing (UV-LA) technologies in the semiconductor industry. To realize 3D-integrated CMOS devices, a new electrically functional Si (or other semiconductor materials) layer must be fabricated directly on the underlayer components, either by wafer bonding [1] or by deposition [2]. Of course, it also requires other steps. For instance, to form a source and drain (S/D), recrystallization and dopant activation are necessary after ion implantation. Otherwise, epitaxy enables in situ doping with good crystal quality. To form a gate stack, dielectric and work function metal deposition, sometimes followed by reliability annealing, are required. Some of them typically need high-temperature processing. Those steps must be accomplished by reducing the thermal budget, which is empirically discussed as a simple combination of temperature (i.e., activation energy) and time (i.e., kinetics) effects, while avoiding any drawback in terms of the performance of each module. Today, as shown in Figure 1, the thermal budget acceptable for 3D-integrated CMOS devices is known as 500 °C for a few hours [1,3,4], and might be further reduced in future developments. UV-LA is then of prime importance to address such requirements for the following reasons. First, the use of a UV light enables strongly limited absorption into materials, and thereby surface-localized heat source generation. Second, the UV-LA process timescale, typically ranging from nanoseconds (ns) to microseconds (µs), efficiently prevents the generated heat from diffusing deeply in the vertical direction. In a given 3D stack, these specific features allow the actual process temperature to be elevated in the top-tier layer without introducing any drawback in the bottom-tier one. In fact, although UV-LA itself has existed for decades, most of the studies of those precedent eras were conducted in a laboratory-scale experiment but not in real industrial-quality devices. This is certainly because there was no UV-LA equipment ready to be implemented in high-volume manufacturing. On the other hand, in the current era, our 300 mm process-compatible UV-LA equipment is commercialized [5], with the capability of UV-LA process simulation [6] to support the UV-LA integration into existing device fabrication flows.

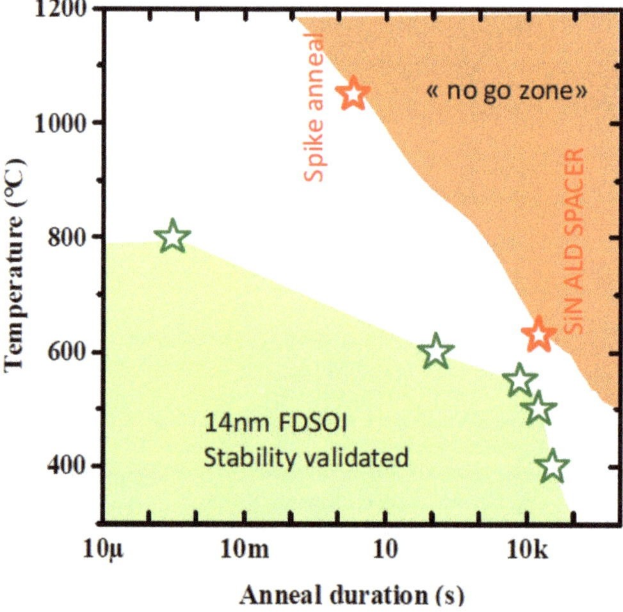

Figure 1. Acceptable thermal budget to preserve the performance of bottom-tier devices. Reprinted with permission from Ref. [3]. 2017, IEEE.

Considering the large freedom of designing applications in 3D integration (e.g., logic-on-logic [3], memory-on-logic [7–9], and sensor-on-logic [3,9]), the thermal budget would have to be managed application-by-application. Furthermore, in some cases components are inserted into interconnect (i.e., the back-end-of-line (BEOL)) layers to enhance the benefits of energy efficiency and look to future neuromorphic computing [10–13]. In such a context, a clear need to freely control the thermal budget is rising. An interesting study is reported in Ref. [14], where the impacts of the buried-oxide (i.e., a Si dioxide (SiO_2) layer separating the top and bottom layers in a 3D stack) thickness and applied thermal budget on the dopant activation in an ion-implanted top and bottom Si layers (i.e., a Si substrate as the bottom layer, whereas an amorphous Si thin layer as the top layer) are systematically investigated. It is noteworthy that a thicker SiO_2 interlayer results in more efficient thermal isolation. Moreover, with a given UV light, laser fluence can be another knob of thermal isolation management. As shown in Figure 2, a more realistic study is conducted using a 28 nm fully depleted Si-on-insulator (FDSOI) CMOS integration, where the impacts of UV-LA on underlying components (i.e., bottom MOS devices and BEOL) are assessed in terms of the performance of transistors and copper (Cu)-based interconnects [4]. The results show no significant degradation in both, confirming the advantage and thermal budget compatibility of UV-LA in 3D integration.

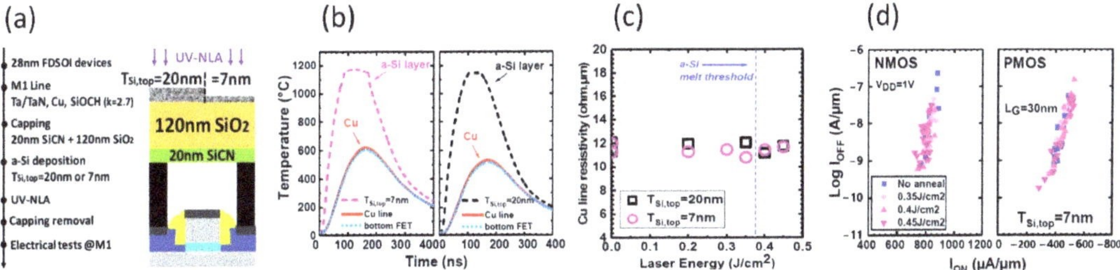

Figure 2. (**a**) Process flow and schematic figure of the 3D structure treated by UV-LA, (**b**) Simulated time–temperature profiles of the applied UV-LA processes, (**c**) Cu line resistivity evaluated before and after UV-LA, (**d**) I_{ON}-I_{OFF} distribution of the bottom n- and p-type MOS devices before and after UV-LA. Reprinted/adapted with permission from Ref. [4]. 2020 IEEE.

In the following sections, we aim at highlighting the potential benefits of short-timescale UV-LA in materials engineering, crossing over diverse applications from the front-end-of-line (FEOL) to BEOL.

2. FEOL Applications

2.1. Reliability Annealing for High-k/SiO$_2$/Si Gate Stacks

In recent planar transistors (e.g., 7 nm node CMOS technology, as shown in Ref. [15]), the SiO_2/Si interface is still utilized in the gate stack (generally a stack of a few nm thick high-permittivity (high-k) hafnium dioxide (HfO_2) film on a Si substrate with a subnanometer-thick SiO_2 interlayer), requiring a lot of efforts to further scale the equivalent oxide thickness (EOT) in a subnanometer region. This gate stack needs to be exposed to a high-temperature annealing process, typically ranging from 800 °C to 1000 °C, for a few seconds in order to ensure temperature instability (BTI) improvement [16–18].

The origin of the BTI is the electrical traps existing at the gate stack interface and within the dielectric films [19]. In the HfO_2/SiO_2/Si stack, there are different BTI sources. One of them is HfO_2 bulk traps, and it is proposed to insert a dipole between the HfO_2 and SiO_2 layers [18] so that the Si channel does not electrically communicate with such defect levels in device operation. Another source is the traps within the interfacial SiO_2/Si system (i.e., the bulk of the thin film and interface), where SiO_2 is often chemically grown to have a very thin film thickness (i.e., ~1 nm or less). Although annealing is necessary to cure or passivate those electrical traps, it must comply with the thermal budget limitations. Recently, low-temperature (from 100 °C to 450 °C) hydrogen-based annealing [20] and plasma treatment [21] have been proposed as an alternative to the conventional high-temperature annealing.

In fact, short timescale UV-LA may also address this problem. A chemically grown SiO_2 thin film is known to have a smaller density than thermally grown films, as well as some impurities remaining inside [22]. Such a situation would also be the case for a subnanometer-thick SiO_2 thin film grown by wet chemical cleaning for a CMOS gate stack. High-temperature processing enabled by UV-LA is then expected to help the removal of the remaining impurities in parallel with the nonequilibrium atomic bonding rearrangement possibly occurring thanks to its short timescale. We have conducted a preliminary study [23] relevant to the potential BTI source annihilation in the wet chemically grown SiO_2/Si system by ns UV-LA (Figure 3), where the effective process temperature is set at 900 °C and the dwell time is in the order of 10^{-7} s. Interestingly, by accumulating the applied ns UV-LA process up to 1000 times, the rearrangement of the Si-O-Si network is observed by attenuated total reflectance (ATR) Fourier transform infrared (FTIR) spectroscopy and X-ray photoelectron spectroscopy (XPS). This coincides with the evolution of the electrical traps captured by a temperature-scan deep-level transient spectroscopy (DLTS). Then, the P_b center-type traps seem annihilated, but it competes with the formation of another type of trap, so-called "tail states" or "U-shaped continuum", which could be associated to Si–O weak bonds formed at the SiO_2/Si interface. It is therefore supposed that UV-LA alone may not be sufficient to perfectly build up the SiO_2/Si system, but a potential of the nonequilibrium SiO_2/Si system engineering is clearly demonstrated.

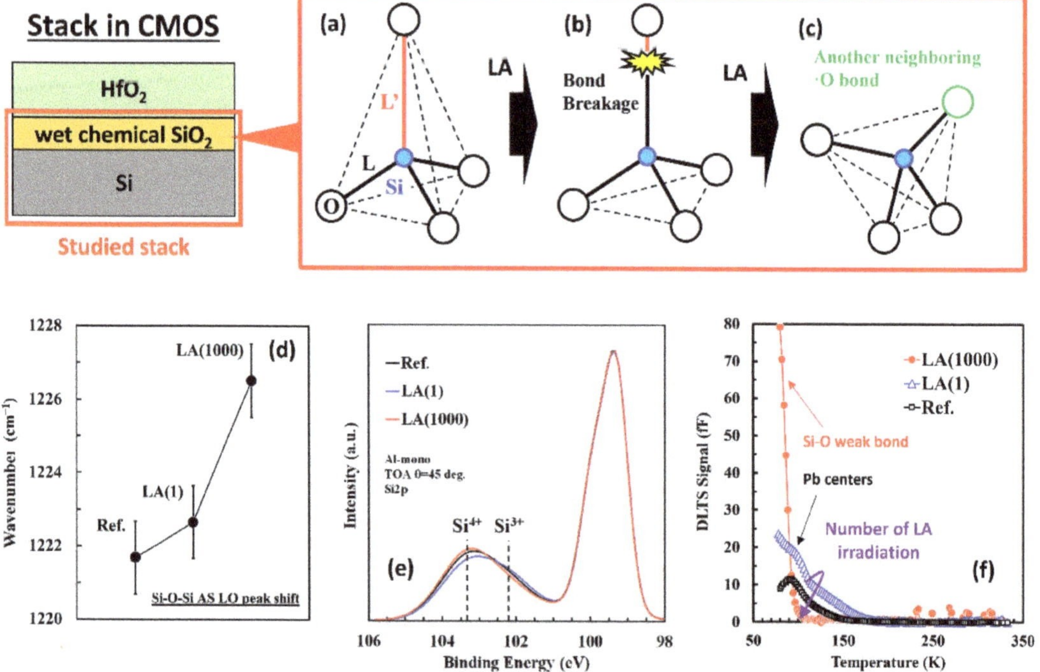

Figure 3. (**a–c**) Schematic figures of a possible local rearrangement procedure of the tetrahedral SiO_4 network. (**d**) Shift of the longitudinal optic (LO) phonon peak of the Si–O–Si asymmetric stretching (AS) vibration as a function of the number of LA irradiation, where LA(1) and LA(1000) stand for 1 and 1000 times irradiation, respectively. (**e**) Normalized Si 2p XPS spectra of the as-grown (Ref.), LA(1), and LA(1000) samples. (**f**) Comparison of the DLTS signals taken for the as-grown (Ref.), LA(1), and LA(1000) samples. Reprinted/adapted with permission from Ref. [23]. 2020, The Japan Society of Applied Physics.

2.2. Channel Doping Engineering to Mitigate Short Channel Effects

The short channel effects (SCE) are known as one of the most critical issues when scaling CMOS devices [24]. To alleviate the SCEs, shallow junction has been conceived, introducing source and drain extension doping beside the channel [25]. Therefore, the control of dopant diffusion in this extension area is critical for device performance. To that end, ns UV-LA may be an ideal solution because, in such a short timescale, dopant diffusion would be strongly limited (or even negligible). Indeed, for conventional dopants such as boron (B), phosphorus (P), and arsenic (As) in Si, it is the case under the Si melting point (i.e., ns UV-LA process does not melt the Si substrate) [26]. When melting the doped Si substrate, although the diffusivity of the dopants strongly enhances (typically towards the order of 10^{-4} cm^2 s^{-1}, as reported in Ref. [27]), their diffusion is limited within the melted region, having a box-like profile [28–31].

A practical example taken in real devices is reported in our previous work [32], where top planar FDSOI transistors having an 11 nm thick Si channel were fabricated (see Figure 4a for their process flow). After the ion implantation in the extension part, ns UV-LA was applied to fully melt the from 5 to 7 nm thick amorphized Si layer (Figure 4b,c) that regrows into the monocrystalline state (Figure 4d,e). After this Si channel regrowth, a raised S/D is properly formed by epitaxy, indicating that ns UV-LA does not seriously degrade the surface morphology of the regrown Si thin layer. In terms of the impact on the SCE, ns UV-LA clearly shows a benefit for suppressing the drain-induced barrier lowering (DIBL) compared to high-temperature spike annealing (Figure 4f), especially with the dopants such as B and P, which are easier to diffuse in Si than As [26].

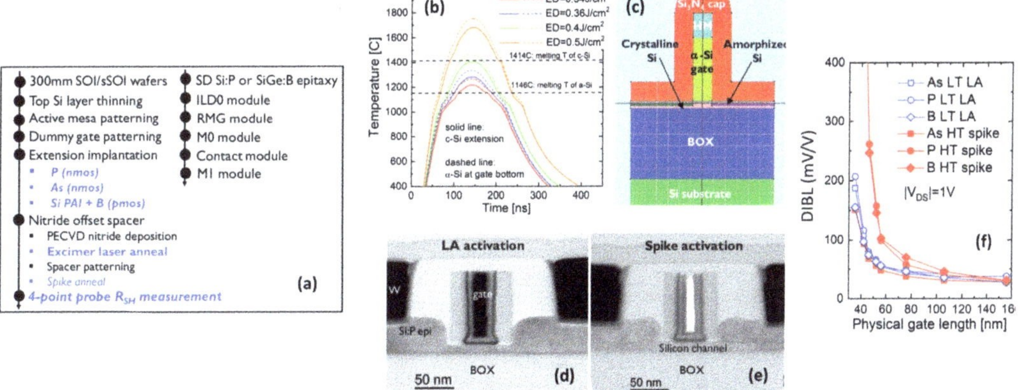

Figure 4. (**a**) Process flow of the FDSOI transistors treated by UV-LA; (**b**) Simulated time–temperature profiles of the applied UV-LA processes; (**c**) Structure used for the simulation; (**d**,**e**) Cross-sectional transmission electron microscopy (TEM) images of the EDSOI transistors after UV-LA and spike annealing; (**f**) DIBL extracted for different physical gate lengths. Reprinted/adapted with permission from Ref. [32]. 2020, IEEE.

3. MOL Applications

The word "middle-of-line (MOL)" stands for the set of processes relevant to the formation of the electrical connection of transistors prior to the BEOL modules. Especially, the low resistive "contact" formation has become one of the biggest challenges in recent years because it is a dominant parasitic factor in scaled transistors. This is a Schottky contact formed by a metal–semiconductor interface. Therefore, to reduce the resistivity, increasing the active carrier concentration in the semiconductor side is supposed to be a strategy of choice. In fact, the required level of specific contact resistivity for future nodes is going to be less than 1×10^{-9} Ω cm^2 [33,34]. To achieve such a low resistivity of the contact, the active carrier concentration needs to be close to 1×10^{21} at./cm^3 [35,36], or even higher [37]. Then, the carrier transport at the Schottky interface should be dominated by the field-emission model [38].

From the viewpoint of the UV-LA processes, there are two approaches. The first one is the liquid phase epitaxial regrowth (LPER), where the doped semiconductor is once melted and epitaxially regrown while activating the dopants. Typically, ns UV-LA enables this approach [34,39–43]. In this article, when discussing LPER, we have a regime so-called "Secondary Melting (SM)" in mind rather than the one so-called "Explosive Melting (EM)". In SM, the melted (i.e., liquid) layer forms at the surface of the semiconductor substrate and simply extends downward [31,44]. On the other hand, in EM, a thin liquid layer (typically a few to 10 nm of thickness [45]) formed at the surface travels in the depth direction, inducing recrystallization into a polycrystalline state [31,44]. The LPER-induced rapid solidification leads to the metastable incorporation of the dopants into the regrown semiconductor crystal. The melting of a doped semiconductor substrate may result in the degradation of the regrown surface morphology [1,40,42], and subsequent steps in a transistor fabrication flow might suffer from it. The second approach is the solid phase epitaxial regrowth (SPER), where the semiconductor layer amorphized by the ion implantation of the dopants can be regrown into the monocrystalline state without melting. Although ns UV-LA can achieve SPER and metastable dopant activation thereby, it requires multiple processes (i.e., more than 20 times the irradiation of the laser pulse) [46]. Therefore, in terms of the productivity, ns UV-LA might not be an optimal option. To address it, extending the process timescale towards μs scale may help [47]. In the previously reported ns UV-LA SPER, the surface morphology does not show serious degradation [46].

3.1. Dopant Activation by Liquid Phase Epitaxial Regrowth (LPER)

A potential advantage of LPER on contact resistivity lowering is that, if a doping element is properly selected, its segregation towards the surface occurs during the solidification of the melted semiconductor material. This may enhance the active carrier concentration near the metal/semiconductor interface.

For that, the segregation coefficient of a doping element ($k = C_S/C_L$, where C_S and C_L stand for the dopant concentrations in the solid (s) and liquid (l) phases at the vicinity of the moving l/s interface) must be less than the unity (i.e., $k < 1$). It should be noted that this k value is a function of the solidification front velocity (V) as shown in Figure 5 [48,49], and increasing V makes k become closer to the unity. For instance, antimony (Sb) in Si (i.e., n-type contact) shows such a surface segregation during ns UV-LA-induced LPER [42], and gallium (Ga) [39–41,43], aluminum (Al) [41], and indium (In) [41] in SiGe (i.e., p-type contact) also do (see Figure 6a,b) as experimental examples.

On the other hand, the activation of these segregated dopants seems not so simple. Firstly, when comparing the ratio of the LPER-induced active carrier concentration to the solid solubility limit of each doping case, that of Al in SiGe is almost unity at different melt conditions. This might be related to the self-compensation of the Al atoms (i.e., electrical deactivation due to lack of a bond between them when their doping concentration becomes extremely high) [50]. Therefore, not every small k dopant may work. Secondly, as shown in Figure 7, the activation also depends on V, drawing a downward convex shape as an overall trend given by the whole red and blue data points. In fact, there is an interesting theoretical prediction about the minimum substitutional concentration (of In in Si) as a function of V [51]. If the nonequilibrium feature of the segregation coefficient (k) is considered (i.e., $k = f(V)$), the relation between this concentration and V is also drawn by a downward convex shape (Figure 8). Hence, increasing V is suggested to enhance the active carrier concentration in LPER. However, it should be noted that it is in a trade-off with the efficiency of surface segregation (i.e., k becomes closer to the unity when increasing V), and a compromise would have to be found when integrating ns UV-LA LPER into real CMOS contact modules.

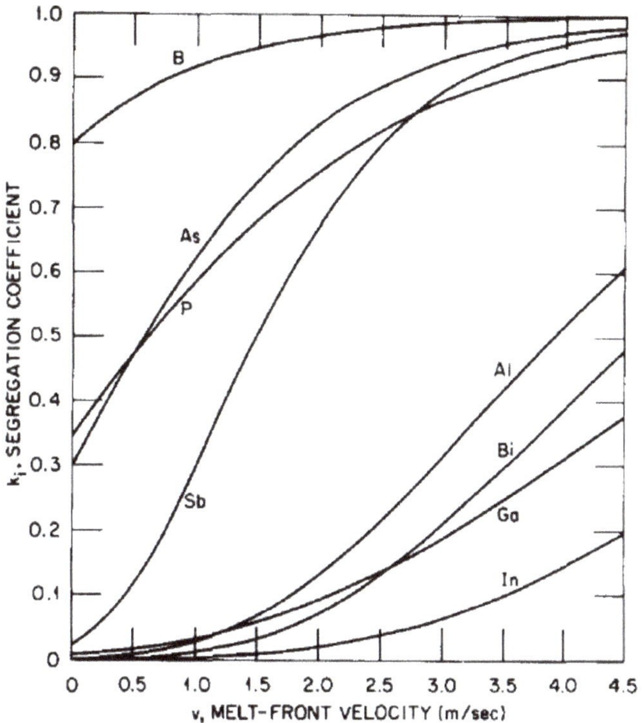

Figure 5. An example of the theoretically predicted dependence of the segregation coefficient (k_i) of dopants in Si on the solidification front velocity (i.e., the melt–front velocity, v). Reprinted with permission from Ref. [48]. 1980, AIP Publishing.

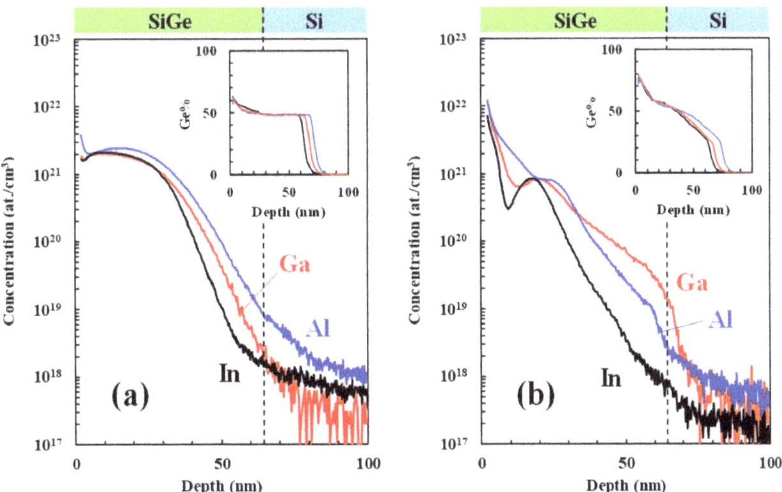

Figure 6. Secondary ion mass spectrometry (SIMS) profiles of Ga, In, and Al taken in the (**a**) as-implanted and (**b**) ns UV-LA-induced full SiGe epilayer melt samples. Reprinted with permission from Ref. [41]. 2019, The Japan Society of Applied Physics.

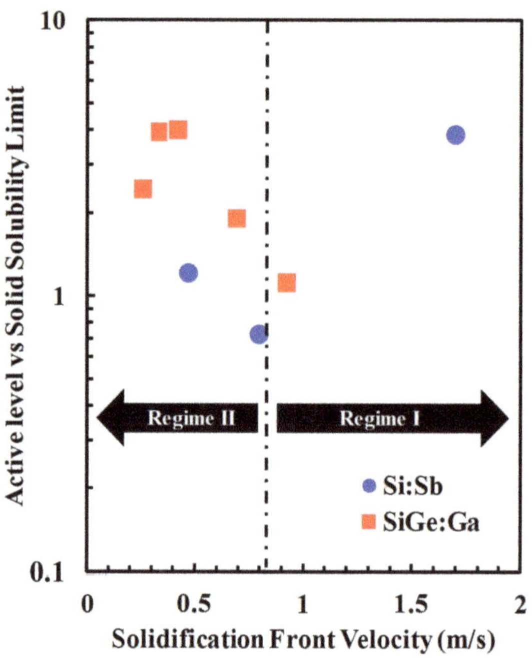

Figure 7. Plots of the degree of surpassing the solid solubility limit as a function of the simulated solidification front velocity (V) for the Sb-implanted Si and Ga-implanted SiGe samples. Reprinted with permission from Ref. [42]. 2020, AIP Publishing.

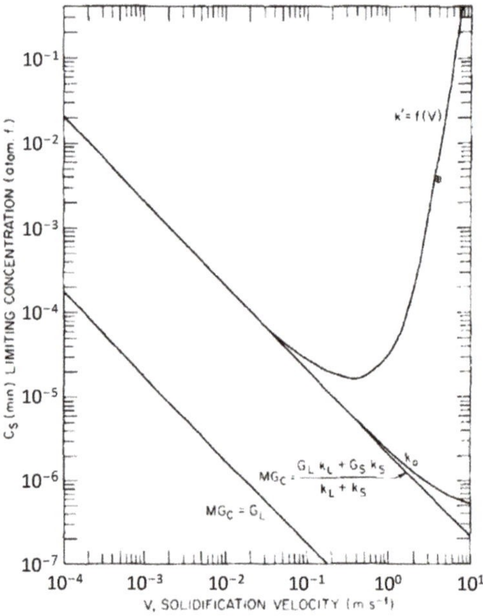

Figure 8. Theoretical prediction of the minimum substitutional concentration (of In in Si) as a function of V. The downward convex shape curve represents the case considering the nonequilibrium feature of the segregation coefficient. Reprinted/adapted with permission from Ref. [51]. 1981, AIP Publishing.

Recently, we have accessed it by using Sb doping for 14 nm node generation FinFET's Si-based contact (Figure 9) [52]. Although there is no electrical data, the segregation of ion-implanted Sb atoms at the top of the fin structure is clearly evidenced (Figure 9d,e, see the position indicated by the arrow "S/D epi top surface"). The V extracted by 3D TCAD simulation is about 4 m/s (Figure 9c). In fact, in real devices, the volume of the S/D parts is negligible compared to the underlying Si substrate, so that heat dissipation becomes fast, and V increases thereby. This value is in a promising range to have a high active carrier concentration, considering the Sb solid solubility limit in Si ($\sim$6.8 $\times$ 10^{19} at./cm^3 [53]) and the expected gain from Figure 7 (more than 10 times at 4 m/s). Furthermore, other promising results are obtained in the p-type contacts of planer FDSOI transistors, receiving high-dose B ion implantation in in situ B-doped SiGe epilayers, then applying ns UV-LA to melt and regrow it [54]. A remaining concern would be crystal defects left after LPER. It is mandatory to melt the ion-implanted region down to the end tail of the dopant profile. Otherwise, the residual point defects and impurities originated from ion implantation precipitate in the form of extended defects in the top of the nonmelted region [30]. Moreover, stacking faults starting from the initial amorphous/crystalline (*a/c*) interface can be observed even at such a full melt condition [43]. They might come either from possible nonuniformity of the as-implanted *a/c* interface or from nonuniform *l/s* interface during UV-LA [55,56].

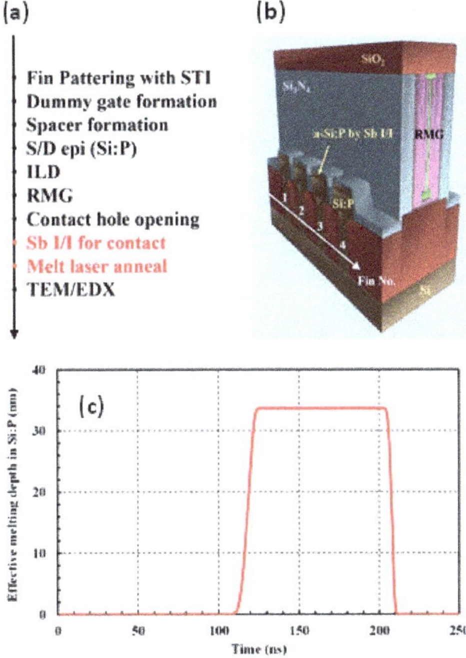

Figure 9. *Cont.*

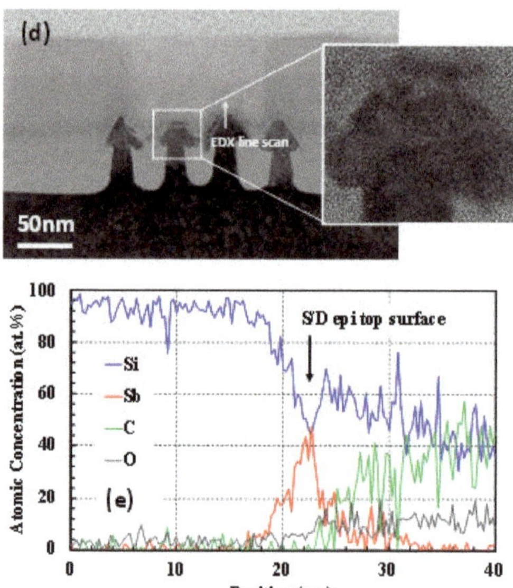

Figure 9. (**a**) Process flow of the FinFET contact modules; (**b**) Structure used for the 3D TCAD simulation; (**c**) Simulated time vs. melting Si fin depth profile during ns UV-LA; (**d**) Cross-sectional TEM image of the Si-based contact module after ns UV-LA; (**e**) Energy-dispersive X-ray spectroscopy line scan taken along the annealed Si fin (the carbon signal comes from the structures surrounding the contact hole). Reprinted/adapted under CC BY 4.0 from Ref. [52]. 2020, T. Tabata et al.

3.2. Dopant Activation by Solid Phase Epitaxial Regrowth (SPER)

As already mentioned above, another way of enabling the metastable activation of the dopants is SPER. In our previously reported ns UV-LA SPER processes [46], multiplying thermally independent laser shots is necessary to fully crystallize a Si layer amorphized by ion implantation (Figure 10a–c). Then, the moving amorphous/crystalline (*a*/*c*) interface maintains a good flatness, resulting in a small root-mean-square value of surface roughness (about 0.10 nm) after SPER completion. An extracted maximum crystallization rate is from 0.8 to 1.8 nm per shot in roughly 10 nm thick amorphous Si layers with different dopants such as B, P, and As (Figure 10d). A rough estimation of film resistivity is also provided based on sheet resistance measurements and an assumption that the conducting layer thickness is equal to the amorphization depth. The calculated film resistivity values imply that the active carrier concentration achieved by ns UV-LA SPER could be higher than that of ns UV-LA LPER.

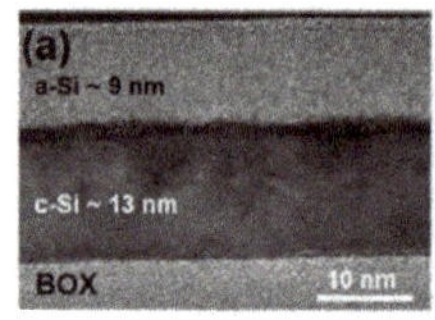

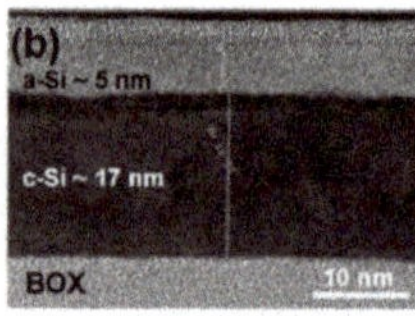

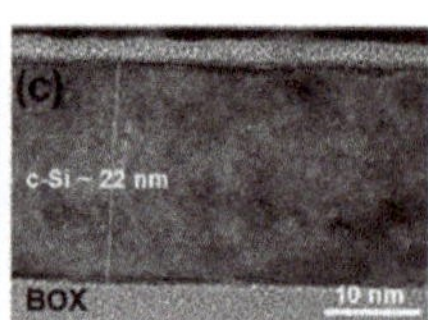

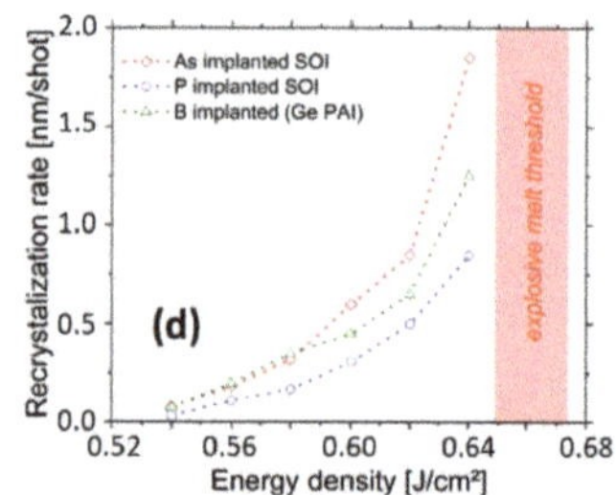

Figure 10. (**a–c**) Progressive SPER with ns UV-LA multiple pulses in a structure of amorphous Si/SiO$_2$/Si substrate. (**d**) Estimated recrystallization rates for each sample as a function of laser energy density. Reprinted/adapted under CC BY 4.0 from Ref. [46]. 2021, P. Acosta Alba et al.

We have reported similar attempts also for µs UV-LA SPER processes [47,57], where the monocrystalline regrowth and flat surface morphology are basically reproduced as in ns UV-LA ones. Interestingly, as shown in Figure 11a, µs UV-LA SPER processes introduce the surface segregation of dopants (indeed, it is known for furnace SPER of As-implanted Si, as reported in Ref. [58]). It is uncertain if our previously reported ns UV-LA processes also induce it. There might be an impact of the SPER rate, which could be different between the reported ns and µs UV-LA processes because of their different ramp-up and cool-down rates. The active carrier concentration measured by the differential Hall effect methodology (DHEM) [59] outperforms 1×10^{21} at./cm^3 near the surface. The thermal stability of these active carriers is investigated by using ns UV-LA as deactivation annealing (Figure 11b). The applied deactivation conditions are those which are reported for copper (Cu) or ruthenium (Ru) based the industrial BEOL interconnect annealing by using ns UV-LA. Up to a submillisecond processing, the maximum sheet resistance degradation is limited to about 5%, encouraging UV-LA integration into different stages of the CMOS devices.

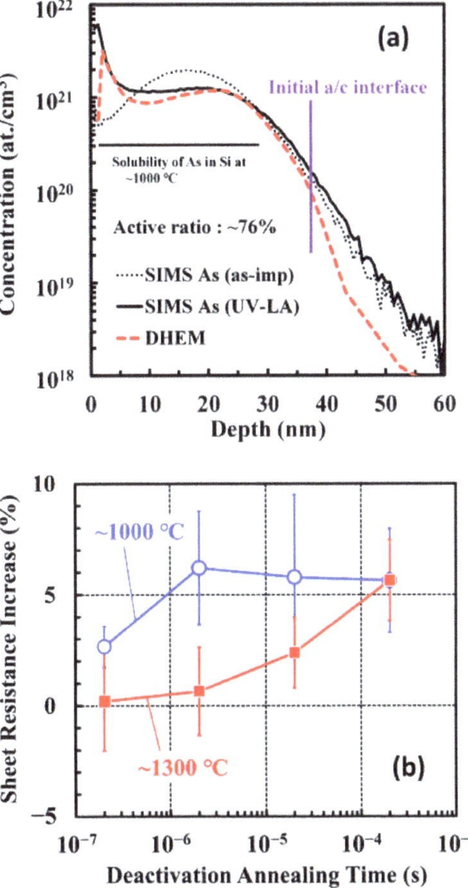

Figure 11. (**a**) SIMS and differential Hall effect methodology (DHEM) profiles taken after µs UV-LA SPER. The initial *a/c* interface position and the As solid solubility in *c*-Si at ~1000 °C reported in Ref. [60] are also indicated. (**b**) Ratio of sheet resistance degradation as a function of the accumulated deactivation annealing time by using the ns UV-LA processes giving different maximum temperatures. Reprinted/adapted under CC BY 4.0 from Ref. [47]. 2022, T. Tabata et al.

4. BEOL Applications

The geometry of the BEOL interconnects (i.e., a trench filled with a highly conductive metal) continuously shrinks. As it limits the growth of metallic grains and results in the increasing density of grain boundaries, electron scattering starts to bring a serious demerit in line resistivity. In fact, there is an exponential relationship between the increase of the line resistivity and the scaling of the cross-sectional area of the BEOL lines (Figure 12) [61–64]. Beyond the 7 nm technology node, alternative metals such as Ru [61,63,65] and cobalt (Co) [61,66,67] start to be considered because of their potential benefit in line resistivity. The figure of merit is determined by a complex combination among bulk resistivity, the cross-sectional area of lines, the mean-free-path of electrons, electro-migration reliability (i.e., melting point of metals), integration compatibility (e.g., availability of raw materials, process uniformity), and the impacts from a barrier and liner. However, some efforts to extend the Cu-based BEOL technologies are found [61,68,69], and indeed Cu-based lines can give lower line resistivity than the alternative metals, even in scaled BEOL modules [61,68]. In addition, it is not realistic to replace all Cu lines in the whole BEOL modules, especially for large interconnects (e.g., semiglobal and global lines [70]).

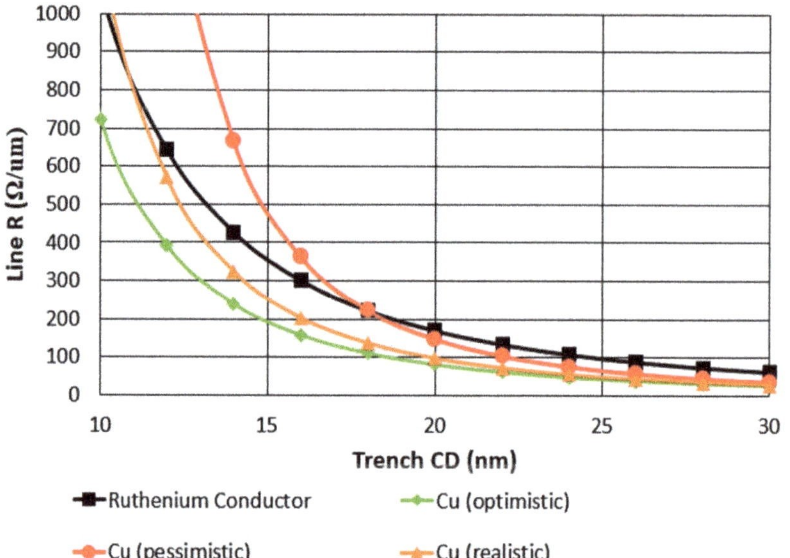

Figure 12. Theoretical trends of line resistance vs. line scale (i.e., trench CD) for Cu- and Ru-based interconnects. Reprinted with permission from Ref. [64]. 2020, American Vacuum Society.

In recent years, the impacts of LA have started to be investigated in real BEOL modules or blanket thin films. Its expected benefit is to enable a bamboo-like structure (i.e., reduction of electron scattering spots) in scaled BEOL lines thanks to the capability of processing wafers at a higher temperature (possibly melting Cu but not Ru) than in typical BEOL furnace annealing (e.g., less than 420 °C up to 1 h [71–74]). As shown in Figures 13 and 14, ns LA (non-UV) indeed shows such a potential [75,76].

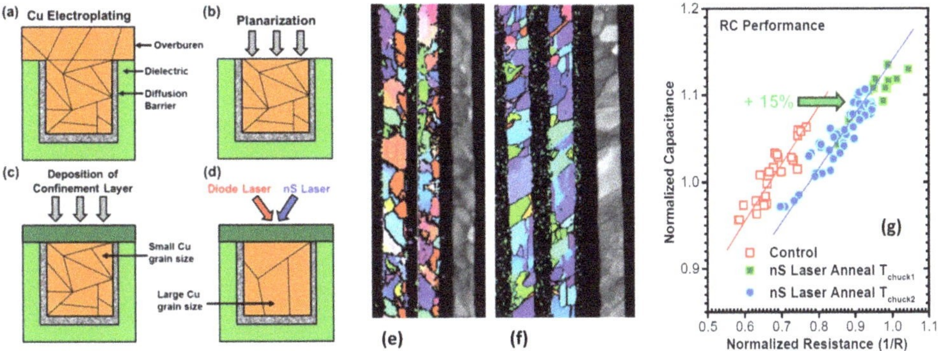

Figure 13. (**a–d**) Schematics of ns LA process flow, (**e,f**) grain analysis performed in the control and Cu melt ns LA lines, (**g**) RC performance measured in the control and Cu melt ns LA lines. Reprinted/adapted with permission from Ref. [75]. 2018, IEEE.

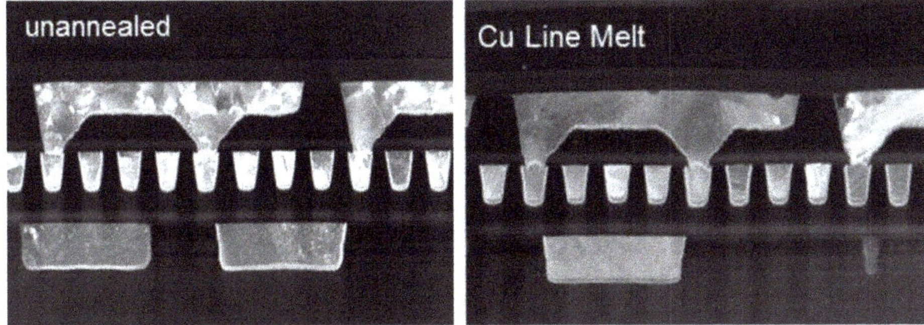

Figure 14. Examples of Cu BEOL lines prior to and after Cu melt LA. Reprinted with permission from Ref. [76]. 2018, The Electrochemical Society (ECS).

4.1. Cu Interconnect

Unfortunately, there is no published work yet about the use of UV-LA in real Cu BEOL modules. However, we have recently conducted some preliminary works by using blanket Cu thin (roughly 50 nm thick) films. Firstly, it has been demonstrated that µs UV-LA enables much greater grain growth than furnace annealing at both Cu submelt and melt conditions. A furnace process at 600 °C typically gives a mean grain size (Av.) of about 100 nm in 50 nm thick Cu films [77,78], whereas it becomes approximately four times and ten times greater in the Cu submelt and melt µs UV-LA, respectively (Figure 15) [79]. Secondly, it has been revealed that even in such a short timescale of process, the structure (i.e., a typical Cu-based BEOL stack of dielectric/Cu/tantalum (Ta)/SiO_2/Si substrate) can be degraded due to interlayer atomic diffusion (Figure 16). Although a process window already exists (e.g., Process A in the Cu submelt shows a 15% reduction of sheet resistance), improving the thermal stability of the Cu-based BEOL structure, for instance, by using an alternative barrier and/or liner (e.g., tantalum nitride (TaN)/Co [68,69], TaN/Ru [68], or tantalum-manganese oxides ($TaMn_xO_y$) [68]) would further extend the merit of using UV-LA.

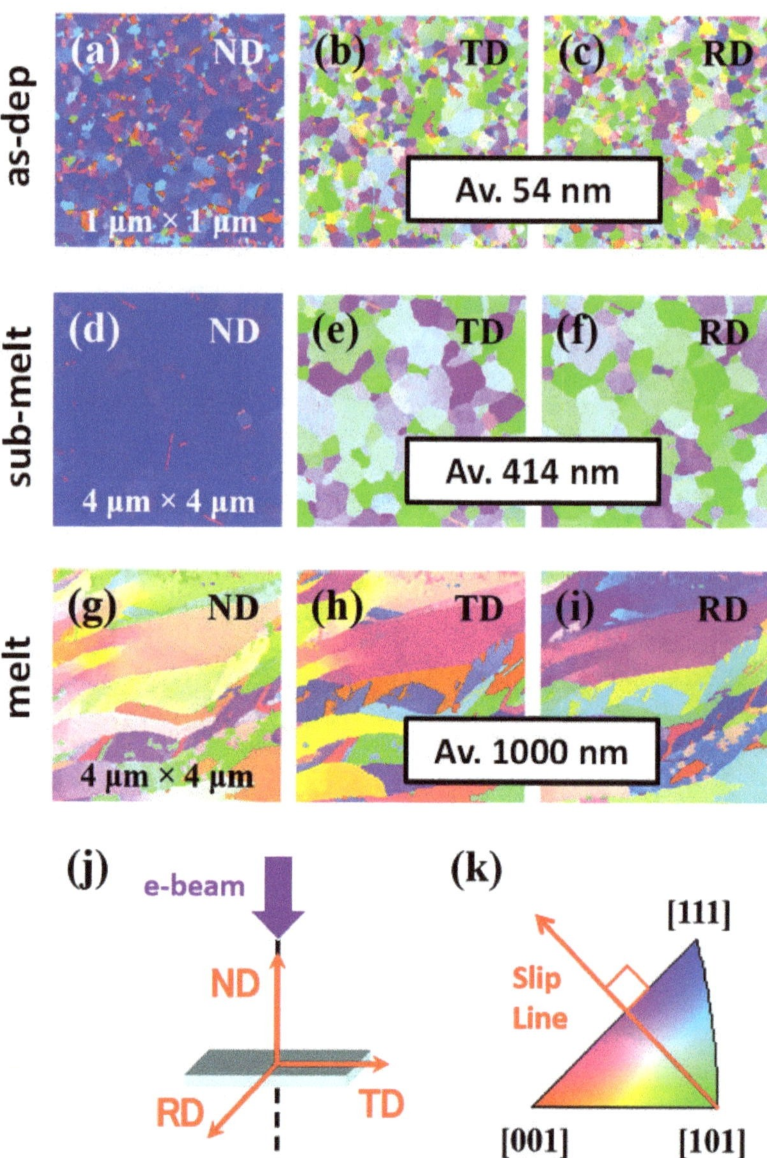

Figure 15. Images of electron diffraction mapping taken for the nonannealed (**a–c**) and annealed Cu thin films (**d–f**) are for a Cu submelt µs UV-LA condition, whereas (**g–i**) are for a Cu melt µs UV-LA condition. As depicted in (**j**), ND, TD, and RD stand for normal direction, transverse direction, and reference direction, respectively. Moreover, a standard triangle of grain orientations is shown in (**k**). Reprinted/adapted with permission from Ref. [79]. 2021, IEEE.

Process	Regime	(i) Cu diffusion into SiO_2	(ii) O diffusion into Cu	(iii) Cu/Ta interface degradation	(iv) Cu surface degradation
A	Sub-melt @800°C				
B	Sub-melt @1000°C	X	X	X	
C	Melt			X	X

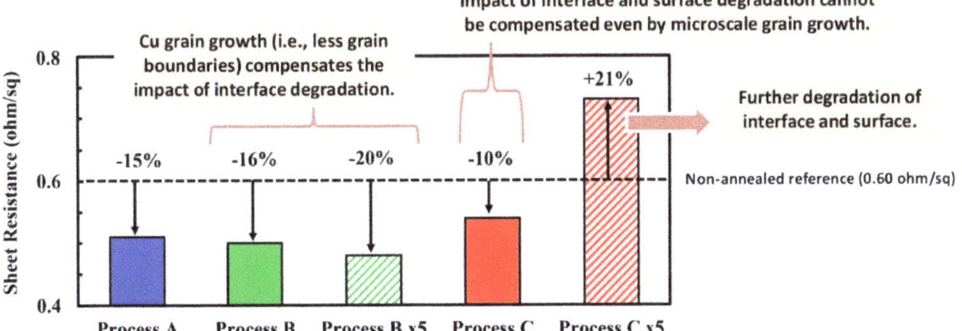

Figure 16. Sheet resistance measured in the Cu-based BEOL blanket structure (dielectric/Cu/Ta/SiO_2/Si substrate) before and after μs UV-LA (Processes A, B, and C). Reprinted/adapted with permission from Ref. [80]. 2022, IEEE.

4.2. Ru Interconnect

Replacement of Cu with an alternative metal in BEOL interconnects is attractive, especially in local lines [70]. One of the listed candidates is Ru, as shown in Figure 17 [81]. Its smaller mean free path of electrons (λ) than Cu reduces electron scattering and compensates a disadvantage of bulk resistivity (ρ_0). Moreover, its higher melting point than Cu is expected to improve electro-migration performance. Furthermore, Ru is already used in industry-like Cu-based BEOL interconnects as a liner material [82]. Although it is not a focus of this review article, Ru-based BEOL processes (e.g., deposition, CMP, and etching) could therefore rapidly mature.

We have previously evaluated the impact of ns UV-LA on the line resistance of Ru lines fabricated by a 21 nm half-pitch dual-damascene process [83]. Then, although the Ru lines are not supposed to be melted because of its much higher (more than two times) melting point than Cu, the line resistance is reduced for up to 25% by multiple ns UV-LA processes compared to the as-deposited Ru lines.

On the other hand, in a way of further scaling in BEOL interconnects, semidamascene processing (also called "subtractive") is emerging for Ru-based BEOL interconnects [84,85]. This new approach might be more adapted for UV-LA because it allows to anneal the metal before line patterning, and thereby may alleviate the challenge of controlling the effects of light pattern interference. In this context, we have also evaluated grain growth in deposited Ru thin films having a similar thickness to the Cu studies [74,83,86], demonstrating effective grain enlargement and the associated sheet resistance drops with multiple ns UV-LA processes.

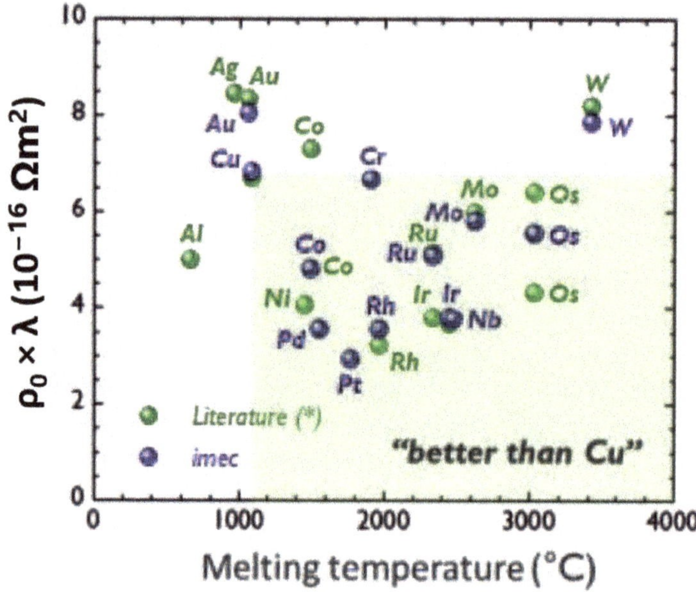

Figure 17. Figure of merit defined as mean free path (λ) × bulk resistivity (ρ_0) vs. melting point for different metals. Reprinted/adjusted with permission from Ref. [81]. 2018, IEEE.

5. "More Than Moore" Applications

Not only in 3D-integrated devices but also in planar ones is the BEOL module the place where the applicable thermal budget is strongly restricted. It often makes material engineering difficult and narrows the range of applications. However, if this restriction is removed, then the diversification of applications will start like cutting a weir. We hereafter present a couple of examples of material engineering in BEOL allowed by short timescale UV-LA.

5.1. Large Poly-Si Grain Formation from Amorphous Si Thin Film

The formation of an electrically active layer (i.e., doped semiconductor with a monocrystalline or polycrystalline state) in BEOL may open a chance for disruptive innovation to boost CMOS performance. However, the maximum process temperature in BEOL is generally much lower than the one enabling solid phase crystallization (SPC) of amorphous Si on SiO_2 (e.g., 600 °C for several hours [87,88]). Therefore, there is a clear need for utilizing short timescale UV-LA, with which both liquid phase crystallization (LPC) and SPC will become accessible.

Recently, FinFET device integration into BEOL has been demonstrated with the aim of reducing chip size and power consumption [2]. In this work, a LA (non-UV) process is performed as a key process for crystallizing amorphous Si via LPC. Such a potential has been also demonstrated by ns UV-LA, as shown in Figure 18 [1]. Si-based photonic device integration in BEOL via ns UV-LA LPC can be also found in the literature [89].

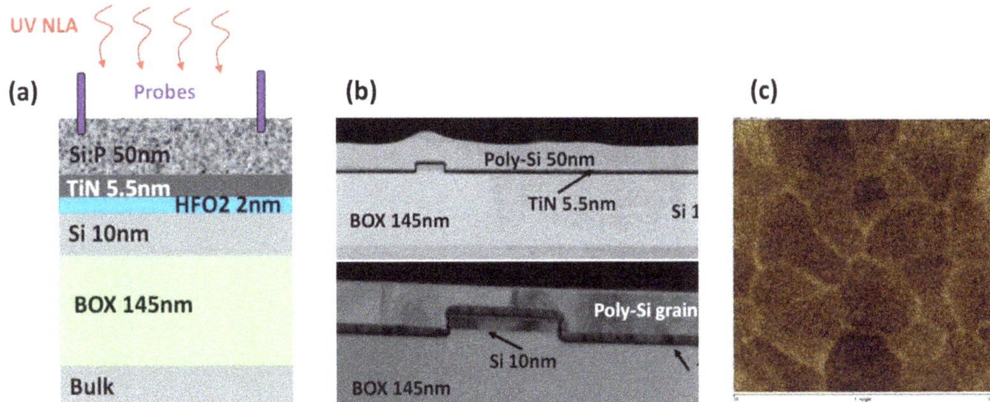

Figure 18. (**a**) Schematic of ns UV-LA in a 3D type structure having an amorphous Si layer on the top, (**b**) cross-sectional TEM images taken in the test structure followed by ns UV-LA and top amorphous Si melting then by CMP to planarize the top surface, (**c**) atomic force microprobe image taken after the CMP planarization. Reprinted/adjusted with permission from Ref. [1]. 2018, IEEE.

It should be noted that melting materials in a device structure may engender drawbacks in subsequent processes. As already shown in Figure 18b, Si LPC clearly presents hillocks on the just-crystallized poly-Si surface, and they must be planarized by CMP to follow the device fabrication process flow. It occurs due to the collision of the solidification front among grains crystallizing from the liquid Si, and the hillocks are formed on the final grain boundaries [90]. Interestingly, in the case of the LPER (i.e., no poly grain boundaries) in Si channel extension, shown in Figure 4d, the regrown Si surface keeps a good flatness and does not impact the subsequent S/D epitaxy. Although the surface tension [91] induced on the top of the liquid Si may induce roughening of the regrown surface [92], a thick (tens of nanometers) dielectric capping layer deposited on the partially amorphized Si channel seems to help to avoid the degradation of the surface morphology. Therefore, one may suppose that a similar capping approach could work also for the Si LPC. However, careful consideration would be necessary, because it is known that a periodic surface pattern (called "wrinkles") emerges and evolves in a stack having an elastic layer on a liquid substrate [93]. Our recent study has revealed that the Si melt induced by ns UV-LA in SiO_2/Si stacks indeed triggers the formation of wrinkles and they grow in a spatial wavelength and height as the maximum melted Si depth increases [94]. As this phenomenon is described by dimensional, mechanical, and viscoelastic parameters of the materials involved in the system, the stack to be treated by ns UV-LA should be carefully designed.

5.2. Hf-Based Ferroelectric Layer Formation

Another application in BEOL which a short timescale UV-LA may enable is the formation of a high relative permittivity or ferroelectric layer (e.g., capacitors integrated into 130 nm CMOS BEOL in Ref. [12]). Hf-based oxides are today well-known as a CMOS technology compatible with ferroelectric thin films [12,95,96]. Their dielectric functionalities (i.e., high relative permittivity or ferroelectricity) rely on crystal phases, and cubic (*c*), tetragonal (*t*), orthorhombic (*o*), and monoclinic (*m*) ones are typical. The ferroelectricity of HfO_2 thin films is related to the *o*-phase (e.g., $Pca2_1$ space group), which emerges during the transition from the *t*-phase to the *m*-phase [97,98]. Therefore, it is critical to control the kinetics of nucleation and crystal phase transformation.

Inspired by previous studies [99,100], we have come up with an idea of shedding a light on it by exploring different process timescales (i.e., dwell time) with ns [101] and μs [102] UV-LA. The expected impact of controlling the dwell time is depicted in Figure 19a. Then, the relative permittivity of the annealed HfO_2 thin film evolves with μs-scale dwell times, as shown in Figure 19b, supporting the proposed idea. The appearance of ferroelectricity is also experimentally evidenced in this study by means of PUND (Positive Up Negative Down)-corrected polarization–voltage (P-V) measurements [103]. When squeezing the dwell timescale toward the ns range, it is expected that HfO_2 phase control comes to rely on nucleation rather than phase transformation. Indeed, repetition of the ns UV-LA process results in a cumulative crystallization of the same phase, as shown in Figure 20. The selectivity of the phase to nucleate in the ns UV-LA seems dependent on the maximum process temperature [12], doping content [12], and stack [101]. As a temperature range of interest is the one near or beyond the Si melting point [12], it is necessary to insert additional layers for efficient UV laser absorption and heat confinement. An example is a stack of metal/HfO_2/metal/SiO_2/Si, as shown in Ref. [12], where the metal layers work for UV laser absorption and the SiO_2 layer does for heat confinement in the upper metal/HfO_2/metal part. Silicon (amorphous, polycrystalline, or monocrystalline) can also be used instead of the metal layers. Some doping elements are known to stabilize the ferroelectric phase in HfO_2 [104,105]. At a given process temperature, the cooling profile may be slightly modulated by the stack and allow partial phase transformation within the nuclei (but their growth with a single ns laser shot would be almost negligible because of limited atomic diffusion in the ns-scale [101]).

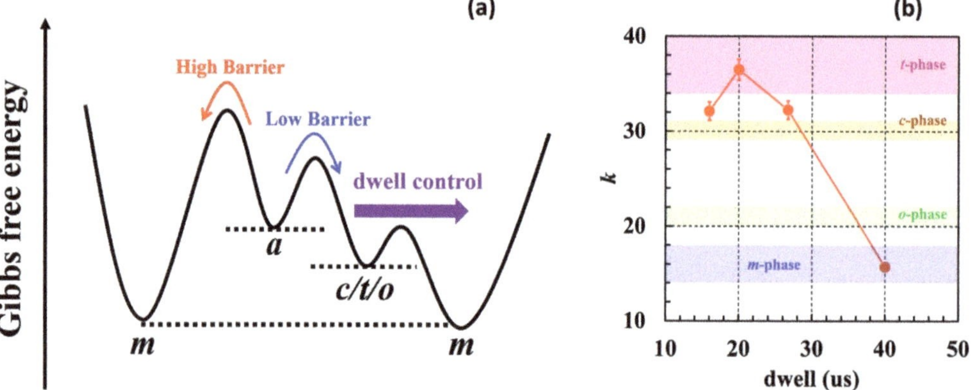

Figure 19. (a) Schematic Gibbs free energy diagram of different HfO_2 crystal phases, where the character "*a*" stands for the amorphous state, "*c*" for the *c*-phase, "*t*" for the *t*-phase, "*o*" for the *o*-phase, and "*m*" for the *m*-phase, respectively. (b) Relative permittivity (*k*) values of the 10 nm thick HfO_2 films as a function of the UV-LA dwell time. The theoretically predicted *k*-value range of each phase is shown together. Reprinted/adjusted with permission from Ref. [102]. 2021, The Japan Society of Applied Physics.

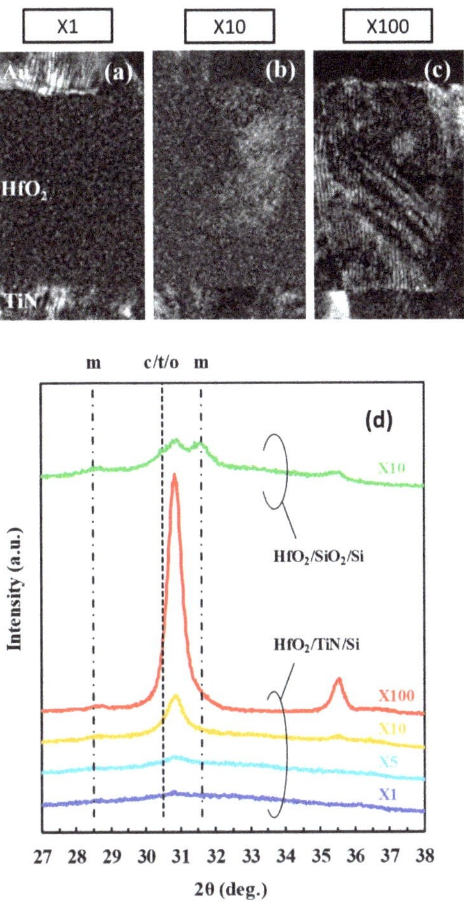

Figure 20. (**a–c**) Cross-sectional dark-field TEM images and (**d**) XRD patterns taken on 50 nm thick HfO_2/TiN/Si samples after single or multiple ns UV-LA processing. Reprinted/adjusted with permission from Ref. [101]. 2020, The Japan Society of Applied Physics.

6. Conclusions

In this review, recent progresses of ns and μs UV-LA technologies have been reviewed, focusing on applications relevant to CMOS devices. The selectivity of heating in the vertical direction that short timescale UV-LA provides is highlighted as a key feature for 3D integration. The presented applications vary from FEOL to BEOL, indicating the high potential of integrating UV-LA processes into different stages of a CMOS fabrication flow. From the viewpoint of materials science, short timescale UV-LA opens new fields of research, especially related to its nonequilibrium aspect. Although the reported range of process timescale from ns to μs is long enough to be at thermal equilibrium [106], it is still chemically and mechanically out of equilibrium. Therefore, phenomena such as atomic bonding rearrangement, diffusion, activation, crystallization (including grain growth), and stress relaxation show interesting behaviors. From them, module-level major improvements to boost the entire CMOS performance may be achieved. Even if there is still a lot of effort required to integrate the presented UV-LA processes into the industrial CMOS devices, the technological maturity of short-timescale UV-LA is steadily progressing.

Author Contributions: Writing—original draft preparation, T.T.; writing—review and editing, F.R., L.T., S.H., P.-E.R., I.K. and K.H. All authors have read and agreed to the published version of the manuscript.

Funding: This project has received funding from the ECSEL Joint Undertaking (JU) under grant agreement no. 875999. The JU receives support from the European Union's Horizon 2020 research and innovation programme, and the Netherlands, Belgium, Germany, France, Austria, Hungary, United Kingdom, Romania, and Israel.

Institutional Review Board Statement: Not applicable.

Informed Consent Statement: Not applicable.

Data Availability Statement: Not applicable.

Acknowledgments: The authors thank their former colleagues, Joris Aubin and Fulvio Mazzamuto, for their contribution to the works cited in this review article, as well as all other LASSE employees for their contribution to the development of ns and μs UV-LA platforms.

Conflicts of Interest: The authors declare no conflict of interest.

References

1. Brunet, L.; Fenouillet-Beranger, C.; Batude, P.; Beaurepaire, S.; Ponthenier, F.; Rambal, N.; Mazzocchi, V.; Pin, J.-B.; Acosta-Alba, P.; Kerdilès, S.; et al. Breakthroughs in 3D Sequential technology. In Proceedings of the 2018 IEEE International Electron Devices Meeting (IEDM), San Francisco, CA, USA, 1–5 December 2018; pp. 7.2.1–7.2.4. [CrossRef]
2. Hsieh, P.-Y.; Chang, Y.-J.; Chen, P.-J.; Chen, C.-L.; Yang, C.-C.; Huang, P.-T.; Chen, Y.-J.; Shen, C.-M.; Liu, Y.-W.; Huang, C.-C.; et al. Monolithic 3D BEOL FinFET switch arrays using location-controlled-grain technique in voltage regulator with better FOM than 2D regulators. In Proceedings of the 2019 IEEE International Electron Devices Meeting (IEDM), San Francisco, CA, USA, 7–11 December 2019; pp. 3.1.1–3.1.4. [CrossRef]
3. Batude, P.; Brunet, L.; Fenouillet-Beranger, C.; Andrieu, F.; Colinge, J.-P.; Lattard, D.; Vianello, E.; Thuries, S.; Billoint, O.; Vivet, P.; et al. 3D Sequential Integration: Application-driven technological achievements and guidelines. In Proceedings of the 2017 IEEE International Electron Devices Meeting (IEDM), San Francisco, CA, USA, 2–6 December 2017; pp. 3.1.1–3.1.4. [CrossRef]
4. Cavalcante, C.; Fenouillet-Beranger, C.; Batude, P.; Garros, X.; Federspiel, X.; Lacord, J.; Kerdilès, S.; Royet, A.S.; Acosta-Alba, P.; Rozeau, O.; et al. 28nm FDSOI CMOS Technology (FEOL and BEOL) Thermal Stability for 3D Sequential Integration: Yield and Reliability Analysis. In Proceedings of the 2020 IEEE Symposium on VLSI Technology, Honolulu, HI, USA, 16–19 June 2020; pp. 1–2. [CrossRef]
5. Huet, K.; Mazzamuto, F.; Tabata, T.; Toqué-Tresonne, I.; Mori, Y. Doping of semiconductor devices by Laser Thermal Annealing. *Mater. Sci. Semicond. Process.* **2017**, *62*, 92–102. [CrossRef]
6. Lombardo, S.F.; Fisicaro, G.; Deretzis, I.; La Magna, A.; Curver, B.; Lespinasse, B.; Huet, K. Theoretical study of the laser annealing process in FinFET structures. *Appl. Surf. Sci.* **2019**, *467–468*, 666–672. [CrossRef]
7. Shen, C.-H.; Shieh, J.-M.; Wu, T.-T.; Huang, W.-H.; Yang, C.-C.; Wan, C.-J.; Lin, C.-D.; Wang, H.-H.; Chen, B.-Y.; Huang, G.-W.; et al. Monolithic 3D chip integrated with 500ns NVM, 3ps logic circuits and SRAM. In Proceedings of the 2013 IEEE International Electron Devices Meeting, Washington, DC, USA, 9–11 December 2013; pp. 9.3.1–9.3.4. [CrossRef]
8. Shulaker, M.M.; Wu, T.F.; Pal, A.; Zhao, L.; Nishi, Y.; Saraswat, K.; Wong, H.-S.P.; Mitra, S. Monolithic 3D integration of logic and memory: Carbon nanotube FETs, resistive RAM, and silicon FETs. In Proceedings of the 2014 IEEE International Electron Devices Meeting, San Francisco, CA, USA, 15–17 December 2014; pp. 27.4.1–27.4.4. [CrossRef]
9. Wu, T.-T.; Shen, C.-H.; Shieh, J.-M.; Huang, W.-H.; Wang, H.-H.; Hsueh, F.-K.; Chen, H.-C.; Yang, C.-C.; Hsieh, T.-Y.; Chen, B.-Y.; et al. Low-cost and TSV-free monolithic 3D-IC with heterogeneous integration of logic, memory and sensor analogy circuitry for Internet of Things. In Proceedings of the 2015 IEEE International Electron Devices Meeting (IEDM), Washington, DC, USA, 7–9 December 2015; pp. 25.4.1–25.4.4. [CrossRef]
10. Yang, C.-C.; Hsieh, T.-Y.; Huang, P.-T.; Chen, K.-N.; Wu, W.-C.; Chen, S.-W.; Chang, C.-H.; Shen, C.-H.; Shieh, J.-M.; Hu, C.; et al. Location-controlled-grain Technique for Monolithic 3D BEOL FinFET Circuits. In Proceedings of the 2018 IEEE International Electron Devices Meeting (IEDM), San Francisco, CA, USA, 1–5 December 2018; pp. 11.3.1–11.3.4. [CrossRef]
11. Francois, T.; Grenouillet, L.; Coignus, J.; Blaise, P.; Carabasse, C.; Vaxelaire, N.; Magis, T.; Aussenac, F.; Loup, V.; Pellissier, C.; et al. Demonstration of BEOL-compatible ferroelectric Hf0.5Zr0.5O2 scaled FeRAM co-integrated with 130 nm CMOS for embedded NVM applications. In Proceedings of the 2019 IEEE International Electron Devices Meeting (IEDM), San Francisco, CA, USA, 7–11 December 2019; pp. 15.7.1–15.7.4. [CrossRef]

12. Grenouillet, L.; Francois, T.; Coignus, J.; Kerdilès, S.; Vaxelaire, N.; Carabasse, C.; Mehmood, F.; Chevalliez, S.; Pellissier, C.; Triozon, F.; et al. Nanosecond Laser Anneal (NLA) for Si-Implanted HfO_2 Ferroelectric Memories Integrated in Back-End of Line (BEOL). In Proceedings of the 2020 IEEE Symposium on VLSI Technology, Honolulu, HI, USA, 16–19 June 2020; pp. 1–2. [CrossRef]

13. Srimani, T.; Hills, G.; Bishop, M.; Lau, C.; Kanhaiya, P.; Ho, R.; Amer, A.; Chao, M.; Yu, A.; Wright, A.; et al. Heterogeneous Integration of BEOL Logic and Memory in a Commercial Foundry: Multi-Tier Complementary Carbon Nanotube Logic and Resistive RAM at a 130 nm node. In Proceedings of the 2020 IEEE Symposium on VLSI Technology, Honolulu, HI, USA, 16–19 June 2020; pp. 1–2. [CrossRef]

14. Morin, P.; Tabata, T.; Rozé, F.; Saib, M.; Thielens, H.; Thuries, L.; Huet, K.; Mazzamuto, F. Impact of the buried oxide thickness in UV laser heated 3D stacks. In Proceedings of the 2021 Solid State Devices and Materials (SSDM), Virtual, 6–9 September 2021; p. A-6-03.

15. Ragnarsson, L.-Å.; Dekkers, H.; Matagne, P.; Schram, T.; Conard, T.; Horiguchi, N.; Thean, A.V.-Y. Zero-thickness multi work function solutions for N7 bulk FinFETs. In Proceedings of the 2016 IEEE Symposium on VLSI Technology, Honolulu, HI, USA, 14–16 June 2016; pp. 1–2. [CrossRef]

16. Bury, E.; Kaczer, B.; Arimura, H.; Toledano Luque, M.; Ragnarsson, L.Å.; Roussel, P.; Veloso, A.; Chew, S.A.; Togo, M.; Schram, T.; et al. Reliability in gate first and gate last ultra-thin-EOT gate stacks assessed with CV-eMSM BTI characterization. In Proceedings of the 2013 IEEE International Reliability Physics Symposium (IRPS), Monterey, CA, USA, 14–18 April 2013; pp. GD.3.1–GD.3.5. [CrossRef]

17. Rzepa, G.; Franco, J.; Subirats, A.; Jech, M.; Chasin, A.; Grill, A.; Waltl, M.; Knobloch, T.; Stampfer, B.; Chiarella, T.; et al. Efficient physical defect model applied to PBTI in high-κ stacks. In Proceedings of the 2017 IEEE International Reliability Physics Symposium (IRPS), Monterey, CA, USA, 2–6 April 2017; pp. XT-11.1–XT-11.6. [CrossRef]

18. Franco, J.; Wu, Z.; Rzepa, G.; Vandooren, A.; Arimura, H.; Ragnarsson, L.-Å.; Hellings, G.; Brus, S.; Cott, D.; De Heyn, V.; et al. BTI Reliability Improvement Strategies in Low Thermal Budget Gate Stacks for 3D Sequential Integration. In Proceedings of the 2018 IEEE International Electron Devices Meeting (IEDM), San Francisco, CA, USA, 1–5 December 2018; pp. 34.2.1–34.2.4. [CrossRef]

19. Denais, M.; Huard, V.; Parthasarathy, C.; Ribes, G.; Perrier, F.; Revil, N.; Bravaix, A. Interface trap generation and hole trapping under NBTI and PBTI in advanced CMOS technology with a 2-nm gate oxide. *IEEE Trans. Device Mater. Reliab.* **2004**, *4*, 715–722. [CrossRef]

20. Franco, J.; Arimura, H.; de Marneffe, J.-F.; Wu, Z.; Vandooren, A.; Ragnarsson, L.-Å.; Dentoni Litta, E.; Horiguchi, N.; Croes, K.; Linten, D.; et al. Low-temperature atomic and molecular hydrogen anneals for enhanced chemical SiO_2 IL quality in low thermal budget RMG stacks. In Proceedings of the 2021 IEEE International Electron Devices Meeting (IEDM), San Francisco, CA, USA, 11–16 December 2021; pp. 31.4.1–31.4.4. [CrossRef]

21. Franco, J.; de Marneffe, J.-F.; Vandooren, A.; Arimura, H.; Ragnarsson, L.-Å.; Claes, D.; Dentoni Litta, E.; Horiguchi, N.; Croes, K.; Linten, D.; et al. Low Temperature Atomic Hydrogen Treatment for Superior NBTI Reliability—Demonstration and Modeling across SiO_2 IL Thicknesses from 1.8 to 0.6 nm for I/O and Core Logic. In Proceedings of the 2021 Symposium on VLSI Technology, Kyoto, Japan, 13–19 June 2021; pp. 1–2.

22. Sometani, M.; Hasunuma, R.; Ogino, M.; Kuribayashi, H.; Sugahara, Y.; Uedono, A.; Yamabe, K. Variation of Chemical Vapor Deposited SiO_2 Density Due to Generation and Shrinkage of Open Space During Thermal Annealing. *Jpn. J. Appl. Phys.* **2012**, *51*, 021101. [CrossRef]

23. Tabata, T.; Inoue, K.; Yoshida, Y.; Takahashi, H. Non-equilibrium engineering of chemically grown SiO_2/Si by UV nanosecond pulsed laser annealing from the viewpoint of bias temperature instability sources. *Appl. Phys. Express* **2020**, *14*, 011003. [CrossRef]

24. Khanna, V.K. Short-Channel Effects in MOSFETs. In *Integrated Nanoelectronics. NanoScience and Technology*, 1st ed.; Springer: New Delhi, India, 2016; pp. 73–93. [CrossRef]

25. Yau, L.D. A simple theory to predict the threshold voltage of short-channel IGFET's. *Solid-State Electron.* **1974**, *17*, 1059–1063. [CrossRef]

26. Bracht, H. Advanced dopant and self-diffusion studies in silicon. *Nucl. Instrum. Methods Phys. Res. Sect. B Beam Interact. Mater. At.* **2006**, *253*, 105–112. [CrossRef]

27. Kodera, H. Diffusion Coefficients of Impurities in Silicon Melt. *Jpn. J. Appl. Phys.* **1963**, *2*, 212–219. [CrossRef]

28. Huet, K.; Boniface, C.; Fisicaro, G.; Desse, F.; Variam, N.; Erokhin, Y.; La Magna, A.; Privitera, V.; Schuhmacher, M.; Besaucele, H.; et al. Experimental and theoretical analysis of dopant activation in double implanted silicon by pulsed laser thermal annealing. In Proceedings of the 17th International Conference on Advanced Thermal Processing of Semiconductors, Albany, NY, USA, 29 September–2 October 2009; pp. 1–16. [CrossRef]

29. Venturini, J. Laser Thermal Annealing: Enabling ultra-low thermal budget processes for 3D junctions formation and devices. In Proceedings of the 12th International Workshop on Junction Technology, Shanghai, China, 14–15 May 2012; pp. 57–62. [CrossRef]

30. Qiu, Y.; Cristiano, F.; Huet, K.; Mazzamuto, F.; Fisicaro, G.; La Magna, A.; Quillec, M.; Cherkashin, N.; Wang, H.; Duguay, S.; et al. Extended Defects Formation in Nanosecond Laser-Annealed Ion Implanted Silicon. *Nano Lett.* **2014**, *14*, 1769–1775. [CrossRef]

31. Lombardo, S.F.; Lombardo, S.F.; Boninelli, S.; Cristiano, F.; Deretzis, I.; Grimaldi, M.G.; Huet, K.; Napolitani, E.; La Magna, A. Phase field model of the nanoscale evolution during the explosive crystallization phenomenon. *J. Appl. Phys.* **2018**, *123*, 105105. [CrossRef]

32. Vandooren, A.; Wu, Z.; Parihar, N.; Franco, J.; Parvais, B.; Matagne, P.; Debruyn, H.; Mannaert, G.; Devriendt, K.; Teugels, L.; et al. 3D Sequential Low Temperature Top Tier Devices using Dopant Activation with Excimer Laser Anneal and Strained Silicon as Performance Boosters. In Proceedings of the 2020 IEEE Symposium on VLSI Technology, Honolulu, HI, USA, 16–19 June 2020; pp. 1–2. [CrossRef]

33. Ni, C.-N.; Rao, K.V.; Khaja, F.; Sharma, S.; Tang, S.; Chen, J.J.; Hollar, K.E.; Breil, N.; Li, X.; Jin, M.; et al. Ultra-low NMOS contact resistivity using a novel plasma-based DSS implant and laser anneal for post 7 nm nodes. In Proceedings of the 2016 IEEE Symposium on VLSI Technology, Honolulu, HI, USA, 14–16 June 2016; pp. 1–2. [CrossRef]

34. Van Dal, M.J.H.; Vellianitis, G.; Doornbos, G.; Duriez, B.; Holland, M.C.; Vasen, T.; Afzalian, A.; Chen, E.; Su, S.K.; Chen, T.K.; et al. Ge CMOS gate stack and contact development for Vertically Stacked Lateral Nanowire FETs. In Proceedings of the 2018 IEEE International Electron Devices Meeting (IEDM), San Francisco, CA, USA, 1–5 December 2018; pp. 21.1.1–21.1.4. [CrossRef]

35. Yu, H.; Schaekers, M.; Hikavyy, A.; Rosseel, E.; Peter, A.; Hollar, K.; Khaja, F.A.; Aderhold, W.; Date, L.; Mayur, A.J.; et al. Ultralow-resistivity CMOS contact scheme with pre-contact amorphization plus Ti (germano-)silicidation. In Proceedings of the 2016 IEEE Symposium on VLSI Technology, Honolulu, HI, USA, 14–16 June 2016; pp. 1–2. [CrossRef]

36. Wang, L.-L.; Yu, H.; Schaekers, M.; Everaert, J.-L.; Franquet, A.; Douhard, B.; Date, L.; del Agua Borniquel, J.; Hollar, K.; Khaja, F.A.; et al. Comprehensive study of Ga activation in Si, SiGe and Ge with 5×10^{-10} $\Omega \cdot cm^2$ contact resistivity achieved on Ga doped Ge using nanosecond laser activation. In Proceedings of the 2017 IEEE International Electron Devices Meeting (IEDM), San Francisco, CA, USA, 2–6 December 2017; pp. 22.4.1–22.4.4. [CrossRef]

37. Niimi, H.; Liu, Z.; Gluschenkov, O.; Mochizuki, S.; Fronheiser, J.; Li, J.; Demarest, J.; Zhang, C.; Liu, B.; Yang, J.; et al. Sub-10^{-9} $\Omega \cdot cm^2$ n-Type Contact Resistivity for FinFET Technology. *IEEE Electron Device Lett.* **2016**, *37*, 1371–1374. [CrossRef]

38. Schroder, D.K. Contact Resistance and Schottky Barriers. In *Semiconductor Material and Device Characterization*, 3rd ed.; John Wiley & Sons, Inc.: Hoboken, NJ, USA, 2006; pp. 127–184. [CrossRef]

39. Everaert, J.-L.; Schaekers, M.; Yu, H.; Wang, L.-L.; Hikavyy, A.; Date, L.; del Agua Borniquel, J.; Hollar, K.; Khaja, F.A.; Aderhold, W.; et al. Sub-10^{-9} $\Omega \cdot cm^2$ contact resistivity on p-SiGe achieved by Ga doping and nanosecond laser activation. In Proceedings of the 2017 Symposium on VLSI Technology, Kyoto, Japan, 5–8 June 2017; pp. T214–T215. [CrossRef]

40. Tabata, T.; Aubin, J.; Huet, K.; Mazzamuto, F. Segregation and activation of Ga in high Ge content SiGe by UV melt laser anneal. *J. Appl. Phys.* **2019**, *125*, 215702. [CrossRef]

41. Tabata, T.; Huet, K.; Mazzamuto, F.; La Magna, A. Surface segregated Ga, In, and Al activation in high Ge content SiGe during UV melt laser induced non-equilibrium solidification. *Jpn. J. Appl. Phys.* **2020**, *58*, 120911. [CrossRef]

42. Tabata, T.; Raynal, P.-E.; Huet, K.; Everaert, J.-L. Segregation and activation of Sb implanted in Si by UV nanosecond-laser-anneal-induced non-equilibrium solidification. *J. Appl. Phys.* **2020**, *127*, 135701. [CrossRef]

43. Tabata, T.; Aubin, J.; Mazzamuto, F. Multilayered highly-active dopant distribution by UV nanosecond melt laser annealing in Ga and B co-implanted high Ge content SiGe:B epilayers. *Jpn. J. Appl. Phys.* **2020**, *59*, 050903. [CrossRef]

44. Tabata, T.; Huet, K.; Rozé, F.; Mazzamuto, F.; Sermage, B.; Kopalidis, P.; Roh, D. Dopant Redistribution and Activation in Ga Ion-Implanted High Ge Content SiGe by Explosive Crystallization during UV Nanosecond Pulsed Laser Annealing. *ECS J. Solid State Sci. Technol.* **2021**, *10*, 023005. [CrossRef]

45. Albenze, E.J.; Thompson, M.O.; Clancy, P. Atomistic computer simulation of explosive crystallization in pure silicon and germanium. *Phys. Rev. B* **2004**, *70*, 094110. [CrossRef]

46. Acosta Alba, P.; Aubin, J.; Perrot, S.; Mazzamuto, F.; Grenier, A.; Kerdilès, S. Solid phase recrystallization induced by multi-pulse nanosecond laser annealing. *Appl. Surf. Sci. Adv.* **2021**, *3*, 100053. [CrossRef]

47. Tabata, T.; Rozé, F.; Thuries, L.; Halty, S.; Raynal, P.-E.; Huet, K.; Mazzamuto, F.; Joshi, A.; Basol, B.M.; Acosta Alba, P. Microsecond non-melt UV laser annealing for future 3D-stacked CMOS. *Appl. Phys. Express* **2022**, *15*, 061002. [CrossRef]

48. Wood, R.F. Model for nonequilibrium segregation during pulsed laser annealing. *Appl. Phys. Lett.* **1980**, *37*, 302. [CrossRef]

49. Galenko, P. Solute trapping and diffusionless solidification in a binary system. *Phys. Rev. E* **2007**, *76*, 031606. [CrossRef]

50. Poulton, J.T.L.; Bowler, D.R. An Ab Initio Study of Aluminium self-compensation in Bulk Silicon. *arXiv* **2019**, arXiv:1907.03636. [CrossRef]

51. Narayan, J. Interface instability and cell formation in ion-implanted and laser-annealed silicon. *J. Appl. Phys.* **1981**, *52*, 1289. [CrossRef]

52. Tabata, T.; Curvers, B.; Huet, K.; Chew, S.A.; Everaert, J.-L.; Horiguchi, N. 3D Simulation for Melt Laser Anneal Integration in FinFET's Contact. *IEEE J. Electron Devices Soc.* **2020**, *8*, 1323–1327. [CrossRef]

53. Trumbore, F.A. Solid solubilities of impurity elements in germanium and silicon. *Bell Syst. Tech. J.* **1960**, *39*, 205–233. [CrossRef]

54. Vandooren, A.; Tabata, T.; Eyben, P.; Roseel, E.; Hikavyy, A.; Huet, K.; Mazzamuto, F.; Dentoni Litta, E.; Horiguchi, N. Potential benefits of S/D HDD activation by melt laser annealing in 3D-inte-grated top-tier FDSOI FETs. In Proceedings of the 2021 Solid State Devices and Materials (SSDM), Virtual, 6–9 September 2021; p. A-6-02.

55. Dagault, L.; Acosta-Alba, P.; Kerdilès, S.; Barnes, J.P.; Hartmann, J.M.; Gergaud, P.; Nguyen, T.T.; Grenier, A.; Papon, A.M.; Bernier, N.; et al. Impact of UV Nanosecond Laser Annealing on Composition and Strain of Undoped $Si_{0.8}Ge_{0.2}$ Epitaxial Layers. *ECS J. Solid State Sci. Technol.* **2019**, *8*, P202. [CrossRef]

56. Dagault, L.; Kerdilès, S.; Acosta Alba, P.; Hartmann, J.-M.; Barnes, J.-P.; Gergaud, P.; Scheid, E.; Cristiano, F. Investigation of recrystallization and stress relaxation in nanosecond laser annealed $Si_{1-x}Ge_x$/Si epilayers. *Appl. Surf. Sci.* **2020**, *527*, 146752. [CrossRef]

57. Tabata, T.; Roze, F.; Alba, P.A.; Halty, S.; Raynal, P.-E.; Karmous, I.; Kerdiles, S.; Mazzamuto, F. Solid Phase Recrystallization in Arsenic Ion-Implanted Silicon-On-Insulator by Microsecond UV Laser Annealing. *arXiv* **2022**, arXiv:2204.12167. [CrossRef]

58. Hopstaken, M.J.P.; Tamminga, Y.; Verheijen, M.A.; Duffy, R.; Venezia, V.C.; Heringa, A. Effects of crystalline regrowth on dopant profiles in preamorphized silicon. *Appl. Surf. Sci.* **2004**, *231–232*, 688–692. [CrossRef]

59. Joshi, A.; Basol, B.M. Sub-nm Near-Surface Activation Profiling for Highly Doped Si and Ge Using Differential Hall Effect Metrology (DHEM). *ECS Trans.* **2020**, *98*, 405. [CrossRef]

60. Lietoila, A.; Gibbons, J.F.; Sigmon, T.W. The solid solubility and thermal behavior of metastable concentrations of As in Si. *Appl. Phys. Lett.* **1980**, *36*, 765. [CrossRef]

61. Nogami, T.; Zhang, X.; Kelly, J.; Briggs, B.; You, H.; Patlolla, R.; Huang, H.; McLaughlin, P.; Lee, J.; Shobha, H.; et al. Comparison of key fine-line BEOL metallization schemes for beyond 7 nm node. In Proceedings of the 2017 Symposium on VLSI Technology, Kyoto, Japan, 5–8 June 2017; pp. T148–T149. [CrossRef]

62. Nogami, T. Overview of interconnect technology for 7nm node and beyond-New materials and technologies to extend Cu and to enable alternative conductors. In Proceedings of the 2019 Electron Devices Technology and Manufacturing Conference (EDTM), Singapore, 12–15 March 2019; pp. 38–40. [CrossRef]

63. Nogami, T.; Gluschenkov, O.; Sulehria, Y.; Nguyen, S.; Huang, H.; Lanzillo, N.A.; DeSilva, A.; Mignot, Y.; Church, J.; Lee, J.; et al. Advanced BEOL Interconnects. In Proceedings of the 2020 IEEE International Interconnect Technology Conference (IITC), San Jose, CA, USA, 5–8 October 2020; pp. 1–3. [CrossRef]

64. Simon, A.; van der Straten, O.; Lanzillo, N.A.; Yang, C.-C.; Nogami, T.; Edelstein, D.C. Role of high aspect-ratio thin-film metal deposition in Cu back-end-of-line technology. *J. Vac. Sci. Technol. A* **2020**, *38*, 053402. [CrossRef]

65. Murdoch, G.; Tokei, Z.; Paolillo, S.; Varela Pedreira, O.; Vanstreels, K.; Wilson, C.J. Semidamascene Interconnects for 2nm node and Beyond. In Proceedings of the 2020 IEEE International Interconnect Technology Conference (IITC), San Jose, CA, USA, 5–8 October 2020; pp. 4–6. [CrossRef]

66. Hu, C.-K.; Kelly, J.; Chen, J.H.-C.; Huang, H.; Ostrovski, Y.; Patlolla, R.; Peethala, B.; Adusumilli, P.; Spooner, T.; Gignac, L.M.; et al. Electromigration and resistivity in on-chip Cu, Co and Ru damascene nanowires. In Proceedings of the 2017 IEEE International Interconnect Technology Conference (IITC), Hsinchu, Taiwan, 16–18 May 2017; pp. 1–3. [CrossRef]

67. Bhosale, P.; Parikh, S.; Lanzillo, N.; Tao, R.; Nogami, T.; Gage, M.; Shaviv, R.; Huang, H.; Simon, A.; Stolfi, M.; et al. Composite Interconnects for High-Performance Computing beyond the 7 nm Node. In Proceedings of the 2020 IEEE Symposium on VLSI Technology, Honolulu, HI, USA, 16–19 June 2020; pp. 1–2. [CrossRef]

68. Nogami, T.; Briggs, B.D.; Korkmaz, S.; Chae, M.; Penny, C.; Li, J.; Wang, W.; McLaughlin, P.S.; Kane, T.; Parks, C.; et al. Through-Cobalt Self Forming Barrier (tCoSFB) for Cu/ULK BEOL: A novel concept for advanced technology nodes. In Proceedings of the 2015 IEEE International Electron Devices Meeting (IEDM), Washington, DC, USA, 7–9 December 2015; pp. 8.1.1–8.1.4. [CrossRef]

69. Lanzillo, N.A.; Yang, C.-C.; Motoyama, K.; Huang, H.; Cheng, K.; Maniscalco, J.; Van Der Straten, O.; Penny, C.; Standaert, T.; Choi, K. Exploring the Limits of Cobalt Liner Thickness in Advanced Copper Interconnects. *IEEE Electron Device Lett.* **2019**, *40*, 1804–1807. [CrossRef]

70. Ezz-Eldin, R.; El-Moursy, M.A.; Hamed, H.F.A. Interconnection. In *Analysis and Design of Networks-on-Chip Under High Process Variation*, 1st ed.; Springer: Berlin/Heidelberg, Germany, 2015; pp. 45–56. [CrossRef]

71. Yang, C.-C.; Witt, C.; Wang, P.-C.; Edelstein, D.; Rosenberg, R. Stress control during thermal annealing of copper interconnects. *Appl. Phys. Lett.* **2011**, *98*, 051911. [CrossRef]

72. Yang, C.-C.; Spooner, T.; McLaughlin, P.; Hu, C.K.; Huang, H.; Mignot, Y.; Ali, M.; Lian, G.; Quon, R.; Standaert, T.; et al. Microstructure modulation for resistance reduction in copper interconnects. In Proceedings of the 2017 IEEE International Interconnect Technology Conference (IITC), Hsinchu, Taiwan, 16–18 May 2017; pp. 1–3. [CrossRef]

73. Dutta, S.; Moors, K.; Vandemaele, M.; Adelmann, C. Finite Size Effects in Highly Scaled Ruthenium Interconnects. *IEEE Electron Device Lett.* **2018**, *39*, 268–271. [CrossRef]

74. Sil, D.; Sulehria, Y.; Gluschenkov, O.; Nogami, T.; Cornell, R.; Simon, A.; Li, J.; Demarest, J.; Haran, B.; Lavoie, C.; et al. Impact of Nanosecond Laser Anneal on PVD Ru Films. In Proceedings of the 2021 IEEE International Interconnect Technology Conference (IITC), Kyoto, Japan, 6–9 July 2021; pp. 1–3. [CrossRef]

75. Lee, R.T.P.; Petrov, N.; Kassim, J.; Gribelyuk, M.; Yang, J.; Cao, L.; Yeap, K.B.; Shen, T.; Zainuddin, A.N.; Chandrashekar, A.; et al. Nanosecond Laser Anneal for BEOL Performance Boost in Advanced FinFETs. In Proceedings of the 2018 IEEE Symposium on VLSI Technology, Honolulu, HI, USA, 18–22 June 2018; pp. 61–62. [CrossRef]

76. Gluschenkov, O.; Jagannathan, H. Laser Annealing in CMOS Manufacturing. *ECS Trans.* **2018**, *85*, 11. [CrossRef]

77. Sun, T.; Yao, B.; Warren, A.P.; Barmak, K.; Toney, M.F.; Peale, R.E.; Coffey, K.R. Surface and grain-boundary scattering in nanometric Cu films. *Phys. Rev. B* **2010**, *81*, 155454. [CrossRef]

78. Chawla, J.S.; Gstrein, F.; O'Brien, K.P.; Clarke, J.S.; Gall, D. Electron scattering at surfaces and grain boundaries in Cu thin films and wires. *Phys. Rev. B* **2011**, *84*, 235423. [CrossRef]

79. Tabata, T.; Raynal, P.-E.; Rozé, F.; Halty, S.; Thuries, L.; Cristiano, F.; Scheid, E.; Mazzamuto, F. Copper Large-Scale Grain Growth by UV Nanosecond Pulsed Laser Annealing. In Proceedings of the 2021 IEEE International Interconnect Technology Conference (IITC), Kyoto, Japan, 6–9 July 2021; pp. 1–3. [CrossRef]

80. Demoulin, R.; Daubriac, R.; Thuries, L.; Scheid, E.; Rozé, F.; Cristiano, F.; Tabata, T.; Mazzamuto, F. Failure Mode Analysis in Microsecond UV Laser Annealing of Cu Thin Films. In Proceedings of the 2022 IEEE International Interconnect Technology Conference (IITC), San Jose, CA, USA, 27–30 June 2022; p. 5.5.

81. Croes, K.; Adelmann, C.; Wilson, C.J.; Zahedmanesh, H.; Varela Pedreira, O.; Wu, C.; Leśniewska, A.; Oprins, H.; Beyne, S.; Ciofi, I.; et al. Interconnect metals beyond copper: Reliability challenges and opportunities. In Proceedings of the 2018 IEEE International Electron Devices Meeting (IEDM), San Francisco, CA, USA, 1–5 December 2018; pp. 5.3.1–5.3.4. [CrossRef]

82. Motoyama, K.; van der Straten, O.; Maniscalco, J.; Huang, H.; Kim, Y.B.; Choi, J.K.; Lee, J.H.; Hu, C.-K.; McLaughlin, P.; Standaert, T.; et al. Ru Liner Scaling with ALD TaN Barrier Process for Low Resistance 7 nm Cu Interconnects and Beyond. In Proceedings of the 2018 IEEE International Interconnect Technology Conference (IITC), Santa Clara, CA, USA, 4–7 June 2018; pp. 40–42. [CrossRef]

83. Jourdan, N.; Rozé, F.; Tabata, T.; Lariviere, S.; Contino, A.; Mazzamuto, F.; Zsolt, T. UV nanosecond laser annealing for Ru interconnects. In Proceedings of the 2020 IEEE International Interconnect Technology Conference (IITC), San Jose, CA, USA, 5–8 October 2020; pp. 163–165. [CrossRef]

84. Wan, D.; Paolillo, S.; Rassoul, N.; Kutrzeba Kotowska, B.; Blanco, V.; Adelmann, C.; Lazzarino, F.; Ercken, M.; Murdoch, G.; Bömmels, J.; et al. Subtractive Etch of Ruthenium for Sub-5nm Interconnect. In Proceedings of the 2018 IEEE International Interconnect Technology Conference (IITC), Santa Clara, CA, USA, 4–7 June 2018; pp. 10–12. [CrossRef]

85. Na, M.H.; Jang, D.; Baert, R.; Sarkar, S.; Patli, S.; Zografos, O.; Chehab, B.; Spessot, A.; Sisto, G.; Schuddinck, P.; et al. Disruptive Technology Elements, and Rapid and Accurate Block-Level Performance Evaluation for 3nm and Beyond. In Proceedings of the 5th IEEE Electron Devices Technology & Manufacturing Conference (EDTM), Chengdu, China, 8–11 April 2021; pp. 1–3. [CrossRef]

86. Nogami, T.; Gluschenkov, O.; Sulehria, Y.; Nguyen, S.; Peethala, B.; Huang, H.; Shobha, H.; Lanzillo, N.; Patlolla, R.; Sil, D.; et al. Advanced BEOL Materials, Processes, and Integration to Reduce Line Resistance of Damascene Cu, Co, and Subtractive Ru Interconnects. In Proceedings of the 2022 Symposium on VLSI Technology, Honolulu, HI, USA, 13–17 June 2022; p. TFS2-1.

87. Shimizu, T.; Ishihara, S. Effect of SiO_2 Surface Treatment on the Solid-Phase Crystallization of Amorphous Silicon Films. *J. Electrochem. Soc.* **1995**, *142*, 298. [CrossRef]

88. Ryu, M.-K.; Hwang, S.-M.; Kim, T.-H.; Kim, K.-B. The effect of surface nucleation on the evolution of crystalline microstructure during solid phase crystallization of amorphous Si films on SiO_2. *Appl. Phys. Lett.* **1997**, *71*, 3063. [CrossRef]

89. Lee, Y.H.D.; Lipson, M. Back-End Deposited Silicon Photonics for Monolithic Integration on CMOS. *IEEE J. Sel. Top. Quantum Electron.* **2013**, *19*, 8200207. [CrossRef]

90. He, M.; Metselaar, W.; Beenakker, K. ⟨100⟩-textured self-assembled square-shaped polycrystalline silicon grains by multiple shot excimer laser crystallization. *J. Appl. Phys.* **2006**, *100*, 083103. [CrossRef]

91. Eustathopoulos, N.; Drevet, B. Surface tension of liquid silicon: High or low value? *J. Cryst. Growth* **2013**, *371*, 77–83. [CrossRef]

92. Choi, D.; Shin, J. Study of phosphorus-doped Si annealed by a multi-wavelength laser. *Results Phys.* **2022**, *38*, 105632. [CrossRef]

93. Huang, R.; Im, S.H. Dynamics of wrinkle growth and coarsening in stressed thin films. *Phys. Rev. E* **2006**, *74*, 026214. [CrossRef]

94. Karmous, I.; Rozé, F.; Raynal, P.-E.; Huet, K.; Acosta-Alba, P.; Tabata, T.; Kerdilès, S. Wrinkles Emerging in SiO_2/Si Stack during UV Nanosecond Laser Anneal. *ECS Trans.* **2021**, *102*, 125–137. [CrossRef]

95. Böscke, T.S.; Müller, J.; Bräuhaus, D.; Schröder, U.; Böttger, U. Ferroelectricity in hafnium oxide: CMOS compatible ferroelectric field effect transistors. In Proceedings of the 2011 International Electron Devices Meeting, Washington, DC, USA, 5–7 December 2011; pp. 24.5.1–24.5.4. [CrossRef]

96. Müller, J.; Schröder, U.; Böscke, T.S.; Müller, I.; Böttger, U.; Wilde, L.; Sundqvist, J.; Lemberger, M.; Kücher, P.; Mikolajick, T.; et al. Ferroelectricity in yttrium-doped hafnium oxide. *J. Appl. Phys.* **2011**, *110*, 114113. [CrossRef]

97. Huan, T.D.; Sharma, V.; Rossetti, G.A.; Ramprasad, R. Pathways towards ferroelectricity in hafnia. *Phys. Rev. B* **2014**, *90*, 064111. [CrossRef]

98. Park, M.H.; Schenk, T.; Fancher, C.M.; Grimley, E.D.; Zhou, C.; Richter, C.; LeBeau, J.M.; Jones, J.L.; Mikolajick, T.; Schroeder, U. A comprehensive study on the structural evolution of HfO_2 thin films doped with various dopants. *J. Mater. Chem. C* **2017**, *5*, 4677–4690. [CrossRef]

99. Nakajima, Y.; Kita, K.; Nishimura, T.; Nagashio, K.; Toriumi, A. Phase transformation kinetics of HfO_2 polymorphs in ultra-thin region. In Proceedings of the 2011 Symposium on VLSI Technology, Kyoto, Japan, 14–16 June 2011; pp. 84–85.

100. Mori, Y.; Nishimura, T.; Yajima, T.; Migita, S.; Toriumi, A. Impacts of doped element on ferroelectric phase stabilization in HfO_2 through non-equilibrium PDA. In Proceedings of the 2018 Solid State Devices and Materials (SSDM), Tokyo, Japan, 9–13 September 2018; p. PS-10-18.

101. Tabata, T. Nucleation and crystal growth in HfO_2 thin films by UV nanosecond pulsed laser annealing. *Appl. Phys. Express* **2019**, *13*, 015509. [CrossRef]

102. Tabata, T.; Halty, S.; Rozé, F.; Huet, K.; Mazzamuto, F. Non-doped HfO_2 crystallization controlled by dwell time in laser annealing. *Appl. Phys. Express* **2021**, *14*, 115503. [CrossRef]

103. Fina, I.; Fàbrega, L.; Langenberg, E.; Martí, X.; Sánchez, F.; Varela, M.; Fontcuberta, J. Nonferroelectric contributions to the hysteresis cycles in manganite thin films: A comparative study of measurement techniques. *J. Appl. Phys.* **2011**, *109*, 074105. [CrossRef]

104. Xu, L.; Shibayama, S.; Izukashi, K.; Nishimura, T.; Yajima, T.; Migita, S.; Toriumi, A. General relationship for cation and anion doping effects on ferroelectric HfO_2 formation. In Proceedings of the 2016 IEEE International Electron Devices Meeting (IEDM), San Francisco, CA, USA, 3–7 December 2016; pp. 25.2.1–25.2.4. [CrossRef]
105. Xu, L.; Nishimura, T.; Shibayama, S.; Yajima, T.; Migita, S.; Toriumi, A. Kinetic pathway of the ferroelectric phase formation in doped HfO_2 films. *J. Appl. Phys.* **2017**, *122*, 124104. [CrossRef]
106. Sundaram, S.K.; Mazur, E. Inducing and probing non-thermal transitions in semiconductors using femtosecond laser pulses. *Nat. Mater.* **2002**, *1*, 217–224. [CrossRef]

electronics

MDPI

Review

Heterogeneous and Monolithic 3D Integration Technology for Mixed-Signal ICs

Jaeyong Jeong [1,†], Dae-Myeong Geum [2,†] and SangHyeon Kim [1,2,*]

[1] School of Electrical Engineering, Korea Advanced Institute of Science and Technology (KAIST), Daejeon 34141, Korea
[2] Information and Electronics Research Institute, Korea Advanced Institute of Science and Technology (KAIST), Daejeon 34141, Korea
* Correspondence: shkim.ee@kaist.ac.kr
† These authors contributed equally to this work.

Abstract: For next-generation system-on-chips (SoCs) in diverse applications (RF, sensor, display, etc.) which require high-performance, small form factors, and low power consumption, heterogeneous and monolithic 3D (M3D) integration employing advanced Si CMOS technology has been intriguing. To realize the M3D-based systems, it is important to take into account the relationship between the top and bottom devices in terms of thermal budget, electrical coupling, and operability when using different materials and various processes during integration and sequential fabrication. In this paper, from this perspective, we present our recent progress of III-V devices on Si bottom devices/circuits for providing informative guidelines in RF and imaging devices. Successful fabrication of the high-performance InGaAs high electron mobility transistors (HEMTs) on the bottom ICs, with a high unity current gain cutoff frequency (f_T) and unity power gain cutoff frequency (f_{MAX}) was accomplished without substrate noise. Furthermore, the insertion of an intermediate metal plate between the top and bottom devices reduced the thermal interaction. Furthermore, the InGaAs photodetectors (PDs) were monolithically integrated on Si bottom devices without thermal damage due to low process temperature. Based on the integrated devices, we successfully evaluated the device scalability using sequential fabrication and basic readout functions of integrated circuits.

Keywords: heterogeneous integration; monolithic 3D; sequential 3D; wafer bonding; RF application; image sensor; mixed-signal IC; system-on-chip

Citation: Jeong, J.; Geum, D.-M.; Kim, S. Heterogeneous and Monolithic 3D Integration Technology for Mixed-Signal ICs. *Electronics* **2022**, *11*, 3013. https://doi.org/10.3390/electronics11193013

Academic Editor: Andrea Boni

Received: 5 August 2022
Accepted: 14 September 2022
Published: 22 September 2022

Publisher's Note: MDPI stays neutral with regard to jurisdictional claims in published maps and institutional affiliations.

Copyright: © 2022 by the authors. Licensee MDPI, Basel, Switzerland. This article is an open access article distributed under the terms and conditions of the Creative Commons Attribution (CC BY) license (https://creativecommons.org/licenses/by/4.0/).

1. Introduction

Recently, 3-dimensional (3D) integration technology has been actively investigated to overcome the disadvantages of conventional 2D integration in highly-dense device systems, such as interconnection delay and high-power consumption [1–3]. Indeed, various mixed-signal IC chips using though Si vias (TSVs) based 3D chip stacking technology have been reported in communication, image sensors, etc. [4]. For image sensors, in order to solve the image distortion for moving targets, a three-layer stacked structure consisting of photodiodes on dynamic random-access memory (DRAM) on a logic circuit demonstrated the improved frame speed rate [5]. Similarly, in communication, by using Si-interposer, and TSV, RF system-on-chips (SoCs) have been demonstrated [6–9]. However, while 3D integration technology has improved the mixed-signal IC performance, form factor, large via size, alignment accuracy, and via densities still need to be improved.

As a result, heterogeneous and monolithic 3D (M3D) integration has been extensively studied in order to maximize the benefits of 3D integration in terms of low power consumption, interconnection delay, and via densities. M3D-based RF transistors on Si CMOS ICs, high-resolution microdisplays on Si CMOS driving circuits, and imaging systems on MOSFETs and neuromorphic devices have been demonstrated [10–19]. Furthermore, the aforementioned studies demonstrated the heterogeneous integration of different materials of III-V compound semiconductors with Si- and Ge-based bottom devices, indicating better flexibility in process design and performance improvements, as shown in Figure 1. Although the heterogeneous integration method has many advantages, thermal management on bottom devices, physical coupling, and other factors should be considered first for establishing the M3D-based device architectures.

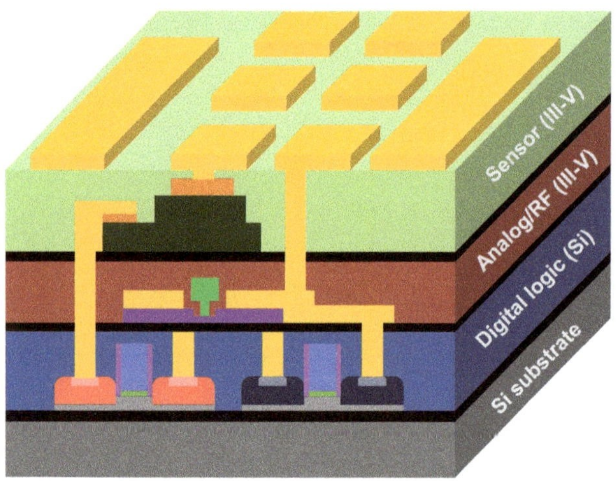

Figure 1. Conceptual illustration of heterogeneous and monolithic 3D integrated system including digital logic (blue layer, Si technology), analog/RF (red layer, III-V technology), and sensor (green layer, III-V technology).

Thus, in this paper, we present our recent research on the M3D integration of RF devices on Si CMOS circuits and InGaAs photodetectors on Si bottom FETs for realizing future M3D-based mixed signal systems. First, we discuss the improved RF device performances, noise de-coupling, and thermal coupling on Si CMOS ICs, which should be considered in future M3D-based 6G systems. Secondly, to implement high-resolution III-V image sensors, the thermal damage on bottom devices during the top device process, device scalability, and basic functionality of readout operation will be discussed in imaging devices.

2. RF System-on-Chip (SoC)

Thanks to the excellent performance improvement through scaling of Si CMOS, not only digital but also analog/RF domains were able to be successfully implemented using Si CMOS technology [20–22]. However, it is expected that next-generation wireless communication will use a frequency band of 100–300 GHz, which is challenging to implement a high-performance RF chip with only Si CMOS technology. As a result, the heterogeneous and monolithic 3D integration emerged as a promising solution to overcome the limited RF performance of Si CMOS technology and utilize the advantages of Si CMOS technology on the digital side [23–26]. Here, we discuss the heterogeneous and M3D integrated RF SoC, which contributes to optimizing performance both on digital and analog/RF and enhances the functionality.

2.1. Why Heterogeneous and Monolithic 3D Integration for RF SoC?

A lot of research has been done on RF devices using III-V HEMT, III-V HBT, SiGe HBT, and Si CMOS technology over a long period. Although the RF performance of various technologies has been improved with the scaling down of technology nodes, the III-V-based devices exhibit outstanding RF performance when compared with the other technologies due to their high electron mobility. The figure of merit (FOM) is used to estimate the performance and capabilities of various technologies. The main FOMs of RF devices are the unity current gain cutoff frequency (f_T) and unity power gain cutoff frequency (f_{MAX}). Figure 2 shows f_T and f_{MAX} of the state-of-the-art III-V HEMT, III-V HBT, SiGe HBT, and Si CMOS. The f_T and f_{MAX} represent the frequency at which the transistor provides a unity gain. The transistor cannot provide gain at frequencies higher than f_T and f_{MAX}.

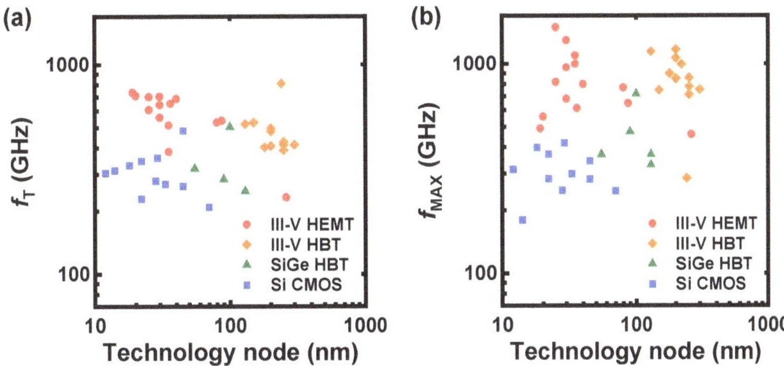

Figure 2. The (**a**) f_T and (**b**) f_{MAX} versus technology node of the state-of-the-art RF transistors.

The III-V HEMT, III-V HBT, SiGe HBT, and Si CMOS have made significant progress in the goal of increasing f_T and f_{MAX} to process more high-frequency signals. In the case of Si-based technology, the RF performance has continued to increase into the range of 300–500 GHz by technology node scaling [20–22]. For CMOS technology, as the gate length of Si MOSFET is extremely reduced, advanced structures, such as FinFETs or gate-all-around (GAA) FETs, become essential. However, these structures cause increasing parasitic capacitance [20]. Therefore, Si CMOS technology has reached the limit in which f_T and f_{MAX} do not increase even if the gate length is scaled. The SiGe HBT technology that can be integrated with Si CMOS has reached f_T of 500 and f_{MAX} of 700 [27]. Even though SiGe HBT exhibits higher f_T and f_{MAX} than that of Si CMOS, it is not enough f_T and f_{MAX} to successfully implement 6G.

On the other hand, the transistors exhibiting f_T above 700 GHz and f_{MAX} above 1.5 THz have been demonstrated with III-V-based devices, especially InGaAs-based transistors, which has never been achieved in other solid-state transistors [28–30]. Furthermore, the III-V-based RF device is the only technology option at this moment that shows a cutoff frequency (f_T and f_{MAX}) over 700 GHz, appropriate to amplify the signal between 100 and 300 GHz for future wireless communication [31]. However, despite their excellent RF performances, the III-V technology has been restricted for a long time because the III-V material cannot be simply integrated with Si CMOS technology. Integrating with Si CMOS technology is important because the RF chip needs not only RF circuits, such as low noise amplifier (LNA) and power amplifier (PA) but also the digital processor.

Therefore, to overcome the limitation of Si CMOS in high frequency and take advantage of highly advanced Si CMOS technology in digital, the heterogeneous integration of III-V and Si is required, as shown in Figure 3. In conventional heterogeneous integration, the III-V technology-based analog/RF circuits and Si CMOS technology-based digital circuits are fabricated respectively and integrated through packaging. However, the packaging technology has disadvantages, such as a long connection distance, high loss, high power consumption, and a large form factor. Furthermore, considering the development direction of the current wireless communication technology, massive multiple-input multiple-output (MIMO) and dense device integration are very important features one must take into account for hardware development [32,33]. For the application of Massive MIMO using a small antenna array, the requirement for the fine pitch size is increasing, and it is urgent to develop a technology that can achieve the micrometer level or less pitch size of I/O pins required for communication between the modules. According to the above trend, the packaging-based integration would be facing this technical difficulty, whereas the monolithic 3D integration process would be applicable for the finer pitch size as the integration density has been proven in other various applications (ex. Logic). Therefore, heterogeneous and monolithic 3D integration is essential for future wireless communication.

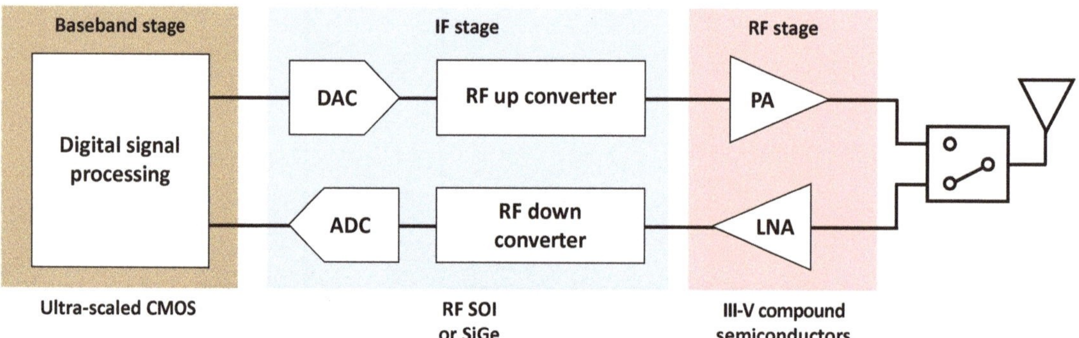

Figure 3. The strategy of heterogeneous integration in RF chips to optimize the performance in both digital and analog/RF circuits.

2.2. Heterogenous and Monolithic 3D Integration of III-V-Based RF Devices on Si CMOS

For future next-generation wireless communication, the co-integration of III-V-based RF devices and Si CMOS digital circuits is positively necessary, as shown in Figure 4a. We have successfully implemented the heterogeneous and monolithic 3D integration of III-V-based RF devices on Si CMOS by direct wafer bonding [10]. The cross-sectional SEM image is shown in Figure 4b. The III-V layers were uniformly bonded on Si CMOS. The top RF devices are based on InGaAs HEMT, and the bottom Si CMOS is a standard MOSFET structure with SiO_2 gate dielectric and poly-Si gate. The heterogeneous and monolithic 3D integration process flow consists of (1) bottom Si CMOS fabrication with 180 nm standard CMOS technology, (2) back end of line (BEOL) process, (3) III-V heterostructure transfer by wafer bonding, (4) top device fabrication, (5) interconnect between top and bottom. Such monolithic 3D integration by direct wafer bonding enables tight integration of different technologies. The process temperature after the front end of the line (FEOL) of Si CMOS is very low at 300 °C or less [10,34]. Figure 5 shows the impact of monolithic 3D integration on the performance of bottom Si CMOS. The transistor characteristics, such as subthreshold swing, on current, off current, and threshold voltage, did not change after monolithic 3D integration, as shown in Figure 5a. Not only the characteristics of the individual transistor but also the function composed of several transistors showed almost no change in characteristics, as shown in Figure 5b. The thermal budget of 3D integration and top device fabrication is 300 °C, which is suitable for monolithic 3D integration.

We showed the basic concept of heterogeneous and monolithic 3D integrated III-V-based RF devices on Si CMOS circuits, as shown in Figure 6a. The top III-V-based RF devices consist of f_T-oriented InGaAs HEMTs, which are designed to optimize the f_T characteristic and f_{MAX}-oriented InGaAs HEMTs, which are engineered to maximize the f_{MAX} characteristic and bottom Si CMOS circuits are composed of the analog-to-digital converters (ADCs) and digital-to-analog converters (DACs). The f_T-oriented InGaAs HEMT has a narrow head, and short gate stem, and the f_{MAX}-oriented InGaAs HEMT features a wide head and short gate stem. The top RF device can amplify the high-frequency signals of hundreds of GHz because the III-V-based RF devices exhibit outstanding f_T and f_{MAX}, as shown in Figure 6b. At the same time, the bottom Si CMOS circuits of ADC and DAC can convert analog and digital signals to digital and analog signals, as shown in Figure 6c–h. A key feature of our heterogeneous and monolithic 3D integrated III-V-based RF devices on Si CMOS circuits is the ability to process analog/RF signals and digital signals with excellent performance at the same chip, utilizing different technologies (III-V and Si CMOS technology).

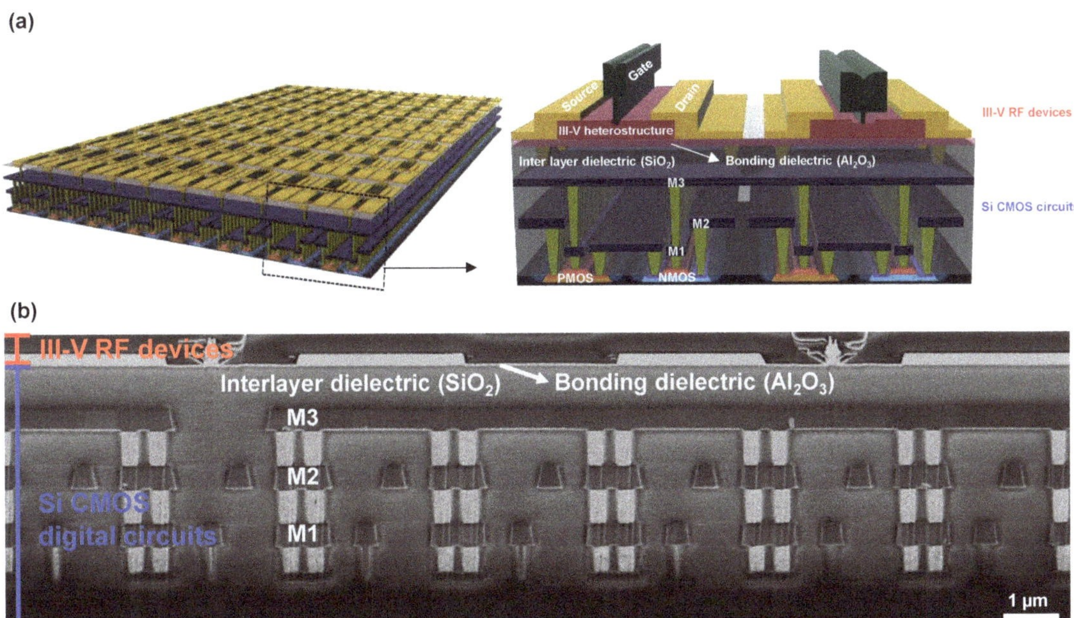

Figure 4. (**a**) Schematic of heterogeneous and monolithic 3D integrated III-V-based RF devices in Si CMOS. (**b**) Cross-sectional STEM image of top transistor based on InGaAs HEMT. Reprinted with permission from Ref. [10]. 2022, ACS Nano.

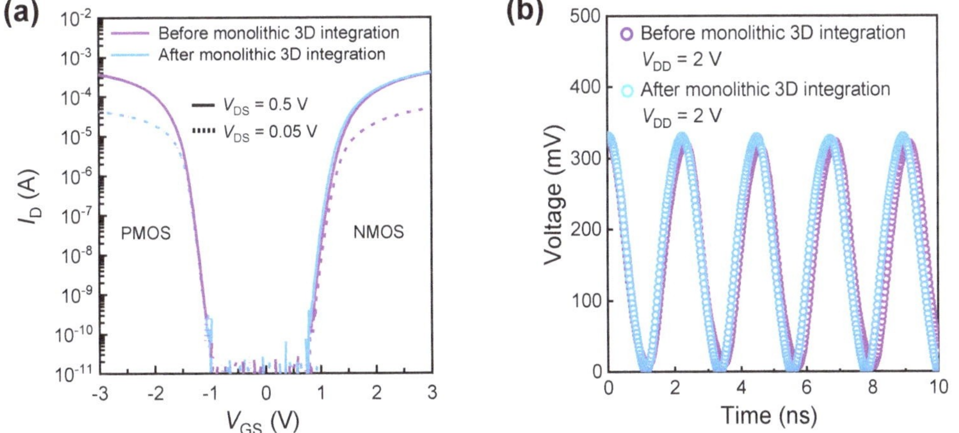

Figure 5. (**a**) Transfer characteristics of bottom single Si MOSFET before and after monolithic 3D integration. (**b**) Output waveform of VCO based on a ring oscillator composed of multiple Si MOSFETs. Reprinted with permission from Ref. [10]. 2022, ACS Nano.

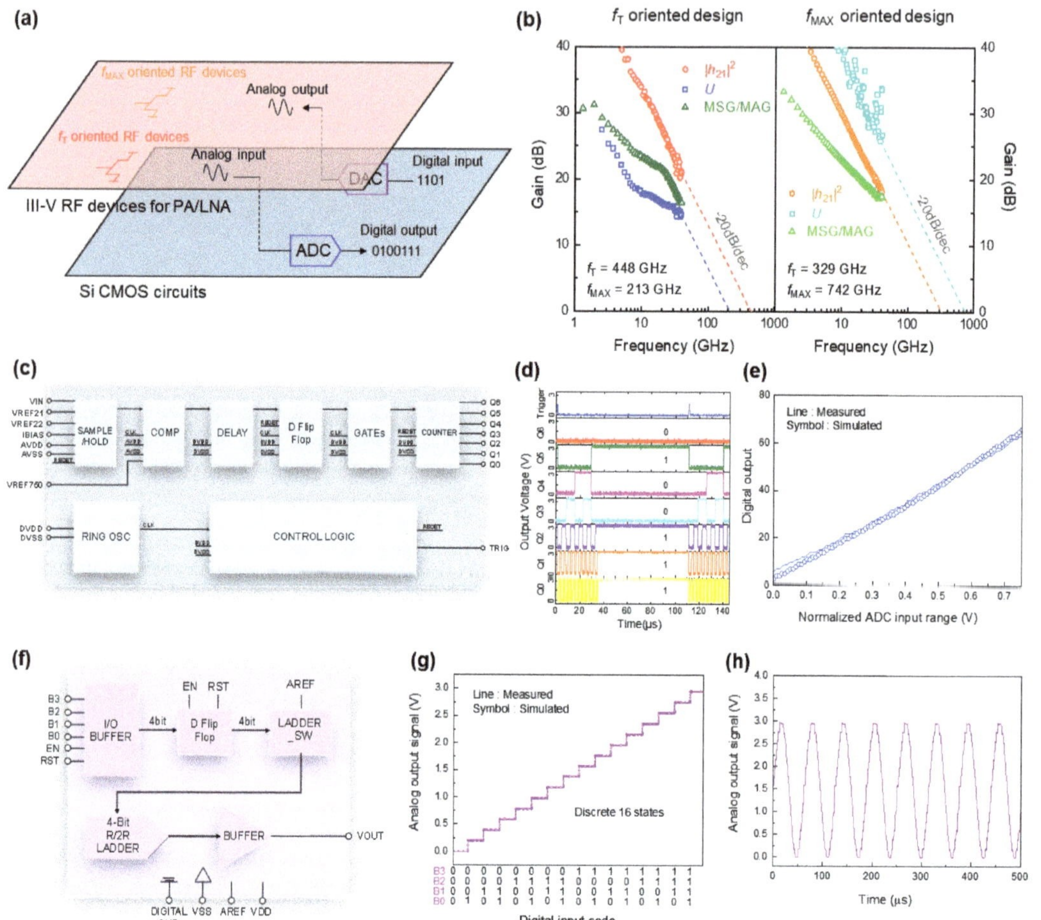

Figure 6. (**a**) Conceptual schematic of heterogeneous and monolithic 3D integrated III-V-based RF devices on Si CMOS. (**b**) Gain plot of f_T and f_{MAX}-oriented top InGaAs HEMTs. (**c**) The block diagram of ADC. (**d**) Time domain output voltage of ADC. (**e**) Digital output of ADC. (**f**) The block diagram of DAC. (**g**) Analog output of DAC. (**h**) Sinewave output of DAC. Reprinted with permission from Ref. [10]. 2022, ACS Nano.

2.3. Issues on Monolithic 3D Integrated Analog/RF-Digital Mixed-Signal IC

2.3.1. Substrate Digital Interference

One of the critical issues in analog/RF-digital mixed-signal integrated circuits is electromagnetic coupling [35,36]. The monolithic integration of analog/RF circuits and digital circuits in the same chip induces the crosstalk between them. Therefore, the appropriate shielding technology is essential. Representatively, there is substrate digital interference in mixed-signal IC. The substrate digital interference, which is caused by switching any digital circuits, can be spread across the shared substrate and degrade the performance of analog/RF circuits when the digital circuits and analog/RF circuits are integrated on the same substrate. Therefore, in the traditional Si-based analog/RF-digital mixed-signal circuits, shielding technology, such as guard ring, deep N-well, deep trench isolation, and through silicon via (TSV), has been necessarily used [37,38]. However, these approaches require many additional process steps and occupy an additional area, imposing critical costs and area penalties. On the other hand, in the case of monolithic 3D integrated analog/RF-digital mixed-signal IC, the digital circuits and analog/RF circuits do not share the same substrate. These two different types of circuits are separated by an interlayer dielectric due to the inherent nature of the floating thin body structure of top devices. Therefore, it is expected that the digital interference from bottom digital circuits can be shielded and has no effect on top analog/RF circuits when an appropriate thickness of interlayer dielectric and BEOL lines are introduced.

To investigate the effect of substrate coupled digital interference in both conventional and monolithic 3D mixed-signal IC, we used a 31-stage ring oscillator in digital circuits as a signal source and a single transistor to sense the substrate digital interference in analog/RF circuits, as shown in Figure 7a. The schematics of the conventional and monolithic 3D mixed-signal systems are shown in Figure 7b,c. The top RF devices and bottom Si CMOS BEOL are separated with a 600-nm-thick interlayer dielectric. The results of substrate digital inference analysis are shown in Figure 7d,e. In the case of M3D mixed-signal IC, the bottom digital interference cannot propagate to the top circuits through the substrate because the digital and analog/RF circuits do not share the same substrate, whereas conventional 2D integrated transistors with the digital circuits strongly feel the digital interference when the digital circuits operate as in Figure 7d. This is another strong advantage of the M3D integration in mixed-signal IC.

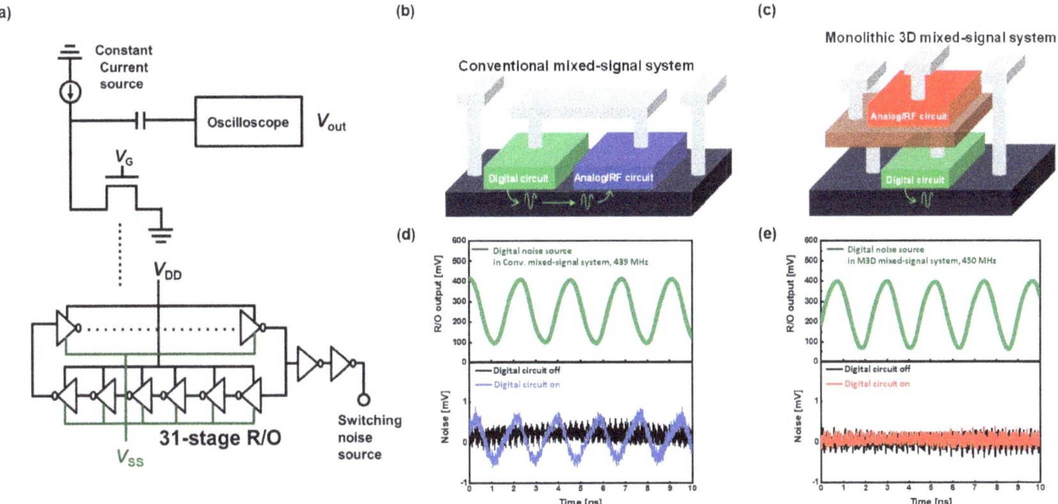

Figure 7. (**a**) Substrate digital interference measurement setup. The schematic of (**b**) conventional and (**c**) monolithic 3D integrated digital-analog/RF mixed-signal IC. The result of substrate digital interference analysis in (**d**) conventional and (**e**) monolithic 3D integrated digital-analog/RF mixed-signal IC. Reprinted with permission from Ref. [11]. 2021, JSAP.

2.3.2. Self-Heating

Another critical issue in monolithic 3D systems is the self-heating of top devices, which can affect the device's performance. As reported by many research groups, the top devices are thermally isolated by an interlayer dielectric which has a low thermal conductivity, thereby, the performance and reliability can be degraded [39,40]. In the monolithic 3D system, the self-heating of the top device is one of the limiting factors in high performance, hence, many groups have reported on the self-heating effect in the M3D systems [39,40]. We presented one of the possible solutions to relaxing self-heating by introducing the metal plate, which has a high thermal conductivity at the backside of the top RF devices, as shown in Figure 8a [41,42]. The embedded metal can act as an additional heat dissipation path because the metal has high thermal conductivity. We used high-resolution thermoreflectance microscopy (TRM) to investigate the self-heating during the transistor operation [41,43]. Since variation in relative reflectivity ($\Delta R/R$) has a linear relationship with variation in surface temperature (ΔT), the temperature of the device can be determined by detecting the change in reflectivity(R). The real-time thermal distribution of 3D integrated InGaAs-based RF transistors with different structures is shown in Figure 8b–e.

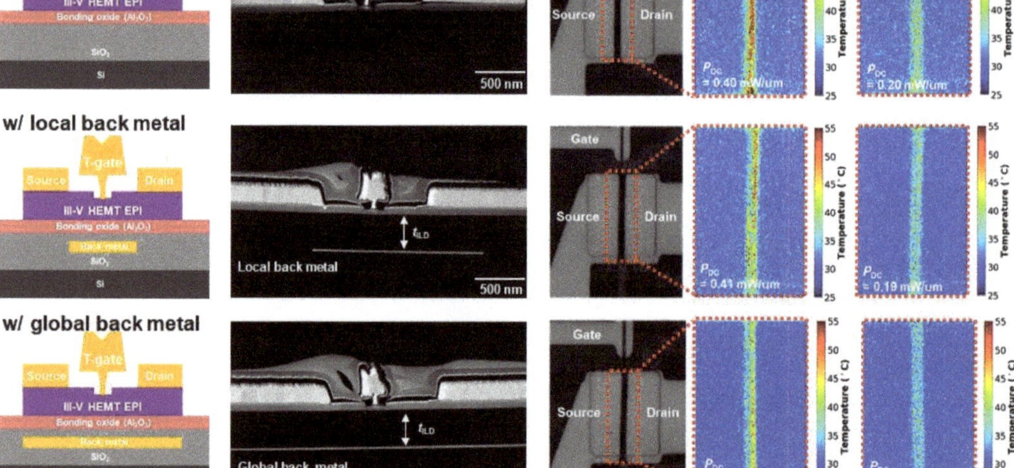

Figure 8. (**a**) Conceptual schematic of heterogeneous and monolithic 3D integrated III-V-based RF device on Si CMOS with back metal to relax the self-heating of top devices. The arrows represent the heat spread. The (**b**) schematics, (**c**) cross-sectional images, (**d**) CCD, and (**e**) thermal images of 3D integrated InGaAs HEMTs with different structures. Reprinted with permission from Ref. [41]. 2022, IEEE.

The 3D integrated InGaAs HEMTs with back metal show reduced self-heating characteristics compared to the 3D integrated InGaAs HEMTs without back metal because the back metal with high thermal conductivity offers an additional heat dissipation path, resulting in low static device temperature. The self-heating relaxing effect of the top devices increased as the size of the back metal increased, and the distance from the top device to the back metal decreased. However, the additional back parasitic capacitance is caused by the back metal. Moreover, when the metal is introduced near the top devices to relax the self-heating, the effect of back capacitance becomes more important, requiring careful device design to mitigate this trade-off relationship. The influence of the back metal on the RF performance of top devices was investigated [41]. However, a more systematic investigation of the influence of the back metal on the top RF devices in the M3D platform will be needed for successful heterogeneous and monolithic 3D integrated RF SoC.

2.4. Benchmarking of the Monolithic 3D Integrated RF Transistors

Many research groups have developed monolithic 3D integrated RF transistors. Figure 9 and Table 1 show the RF performance and information of the process to compare the state-of-the-art 3D integrated RF transistors [10,12–14,44–48]. The most important metric is cutoff frequency because it determines the maximum frequency at which the device can be used. As shown in Figure 9 the III-V-based RF devices, especially InGaAs-based RF transistors, obviously outperform Si transistors. This would motivate the heterogeneous and monolithic 3D integration in future wireless communication. Furthermore, the RF performance versus power consumption of the monolithic 3D integrated InGaAs on Si and monolithic 3D integrated Si on Si are shown in Figure 9. The monolithic 3D integrated InGaAs-based RF transistors on Si show higher RF performance at the same power consumption or less power consumption at the same RF performance than monolithic 3D integrated Si-based RF transistors on Si. This characteristic is also very important to be used for cryogenic LNA toward scalable quantum computing [49]. One drawback would be the fact that InGaAs-based devices are still difficult to be grown on large substrates, such as 200 or 300 mm sizes, resulting in a size mismatch to Si CMOS technology. The GaN-based devices overcome this problem by growing the GaN on Si wafers, but their RF performances fall short of InGaAs-based devices. Even in the case of InGaAs, it is thought that this problem can also be solved by introducing the growth of III-V on Si substrate or die to wafer bonding [50–52].

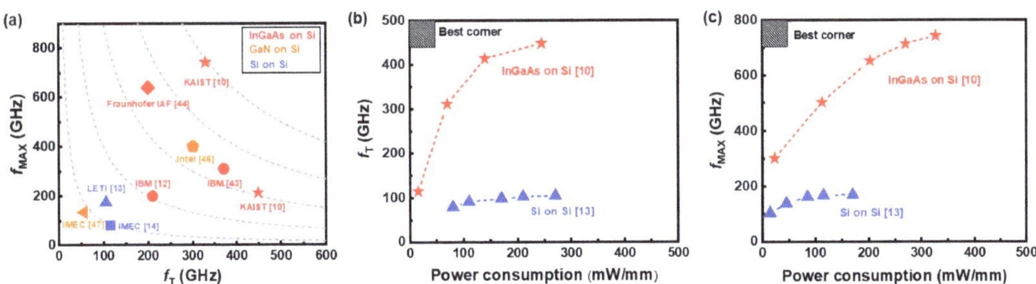

Figure 9. (a) The important FOMs of f_T and f_{MAX} are shown with different types of state-of-the-art monolithic 3D RF transistors. The (b) f_T and (c) f_{MAX} versus power consumption of InGaAs on Si and Si on Si. The best performances of each technology were used for comparison.

Table 1. Comparison of M3D RF transistors.

	RF Device	Tprocess of RF Device	L_G	f_T/f_{MAX}	$\sqrt{f_T \cdot f_{MAX}}$	Integration with Si CMOS
KAIST [10]	InGaAs HEMT	≤300 °C	80 nm (f_{MAX} oriented design)	329/742 GHz	494 GHz	Yes
			140 nm (f_T oriented design)	448/213 GHz	309 GHz	

Table 1. *Cont.*

	RF Device	**Tprocess of RF Device**	L_G	f_T/f_{MAX}	$\sqrt{f_T \cdot f_{MAX}}$	**Integration with Si CMOS**
IBM [43]	InGaAs MOSFET	$\leq$500 °C	20 nm	380/310 GHz	343 GHz	No
IBM [12]	InGaAs MOSFET	$\leq$500 °C	25 nm	210/200 GHz	205 GHz	Yes
Fraunhofer IAF [44]	InGaAs MOSHEMT	$\leq$300 °C	20 nm	640/200 GHz	358 GHz	No
NUS [45]	InGaAs MOSHEMT	$\leq$350 °C	150 nm	60/- GHz	-	Yes
Intel [46]	GaN MOSHEMT	-	30 nm	300/400 GHz	346 GHz	Yes
IMEC [47]	GaN HEMT	-	110 nm	56/135 GHz	87 GHz	No
IMEC [14]	Si MOSFET	$\leq$525 °C	80 nm	80/115 GHz	96 GHz	No
LETI [13]	Si MOSFET	$\leq$500 °C	45 nm	105/175 GHz	136 GHz	No

3. M3D Integration for High-Resolution III-V Image Sensors

Other than RF-SoCs, image sensors are one of the important applications for obtaining visual information in an autonomous car, time-of-flight sensors, and industrial surveillance systems [53–55]. Recently, 3-dimensional packaging using through-Si vias (TSVs) technology has been used to develop high-speed and high-resolution image sensors based on visible light absorption [56]. It is because there are several advantages, such as small chip size, high fill factor, and high bandwidth, whereas the conventional planar integration method, which places pixel region and pixel transistors and circuit section on the same plane, reduces the pixel area and an increases chip size, as shown in Figure 10 [5,57].

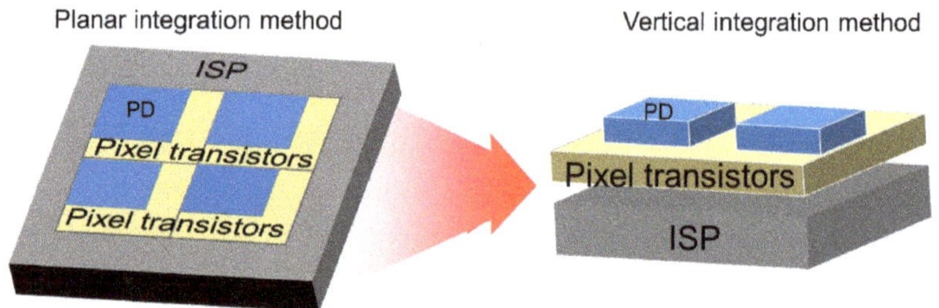

Figure 10. The development trend for fabricating image sensors from conventional planar integration method to newly 3D vertical integration including pixel area, pixel transistors, and image signal processor (ISP).

Although packaging technology has greatly accelerated the performance of image sensors, there is still room for improvement by full 3D integration with pixel transistors, etc. Therefore, ultimately, M3D integration can be a viable approach for future image sensors requiring much more functionalities. Beyond the visible light, detection of the short-wavelength infrared (SWIR) light has become increasingly important, which is mostly accomplished with the III-V material-based photodetectors (PDs) [55,56]. The packaging-based hybrid integration method currently used in III-V device integration with Si ICs has obvious limitations in terms of the resolution of several μm [56]. Thus, the benefits of M3D integration can be maximized in fabricating III-V image sensors because III-V materials can be processed at low temperatures, even a room temperature, and can be separated from donor substrates, making them very suitable for sequential fabrication [58,59]. Moreover,

III-V thin film-based pixels can provide broadband detection capabilities with simple structure design, such as TMD/III-V heterojunction, Metal/III-V Schottky junction, III-V with optical mode resonance structure, etc. [60–62].

Here, we demonstrated the M3D integration of III-V-based PDs on SOI pixel transistors using CMOS-compatible fabrication processes, such as wafer-bonding and low-temperature sequential fabrication processes.

3.1. InGaAs Photodetectors on SOI-MOSFETs by Using Monolithic Integration and Sequential Fabrication Process

Figure 11 depicts the entire fabrication process flow for M3D InGaAs PDs on SOI-MOSFETs. SOI-MOSFETs have 50-nm Si channel thickness, 365-nm-thick BOX, and TiN/HfAlOx gate stack with a conventional InGaAs PD structure. Additionally, a chemical-mechanical polishing (CMP) process was used to obtain a flat surface for wafer bonding by depositing RF-sputtered SiO_2 inter-layer dielectric (ILD) on SOI-MOSFET. The InGaAs PD structure was then bonded to SOI-MOSFETs by utilizing an intermediate layer of 40-nm-thick Al_2O_3 layer with oxygen plasma treatment. Then, the sequential fabrication process of InGaAs PD on SOI-MOSFET with various mesa dimensions was performed, indicating pixel pitch scalability of InGaAs PDs. We used the room-temperature process for M3D integration here, even without the contact annealing process of PDs. Finally, as shown in Figure 11h, the interconnection between InGaAs PD cathode and the source region of SOI-MOSFET was performed for 1 pixel/1-pixel transistor operation.

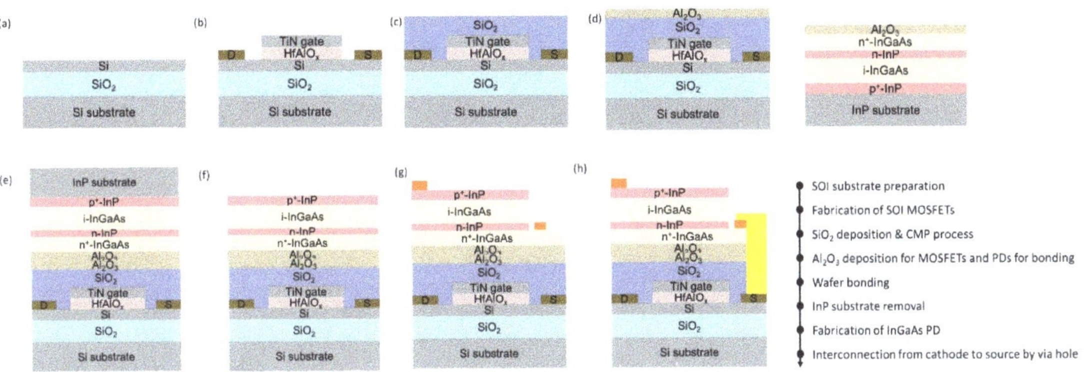

Figure 11. The fabrication process of vertically integrated InGaAs PDs on SOI-MOSFETs. (**a**) SOI substrate preparation. (**b**) Fabrication of SOI MOSFETs with TiN gate and HfAlOx. (**c**) SiO2 deposition and CMP process for planarization. (**d**) Al2O3 deposition for MOSFETs and PDs for wafer bonding. (**e**) Wafer bonding with plasma treatment. (**f**) InP substrate removal by using HCl solution. (**g**) Sequentially processed InGaAs PDs. (**h**) Interconnection from cathode of PDs to sources of MOSFETs by filling via holes. Reprinted with permission from Ref. [15]. 2020, IEEE.

Figure 12a shows a low-magnification transmission electron microscopy (TEM) image of the M3D-integrated InGaAs PD on the SOI-MOSFET structure. Bottom SOI-MOSFET and top InGaAs PD layers were clearly seen to be free of voids and dislocations, indicating integration stability with only a low process temperature. Furthermore, the inset photograph of Figure 12a exhibits the actual fabrication of InGaAs PD on the 2 cm × 2 cm SOI-MOSFET chip with various dimensions of PD devices. These devices were formed by using a sequential fabrication process on the SOI-MOSFETs, implying that pixel pitch scaling could be facilitated by a lithographic alignment due to the elimination of a mechanical alignment used in TSVs and hybrid integration methods. In addition, energy dispersive X-ray (EDX) spectroscopy analysis was performed along with the red arrow, as shown in Figure 12a, to evaluate the bonding interface. The resulting EDX line profile was shown in Figure 12b for various atoms. We observed the formation of abrupt interfaces between different materials of InGaAs/Al$_2$O$_3$ and Al$_2$O$_3$/SiO$_2$, suggesting our fabrication steps did not induce any interdiffusion of elements due to the low-temperature process.

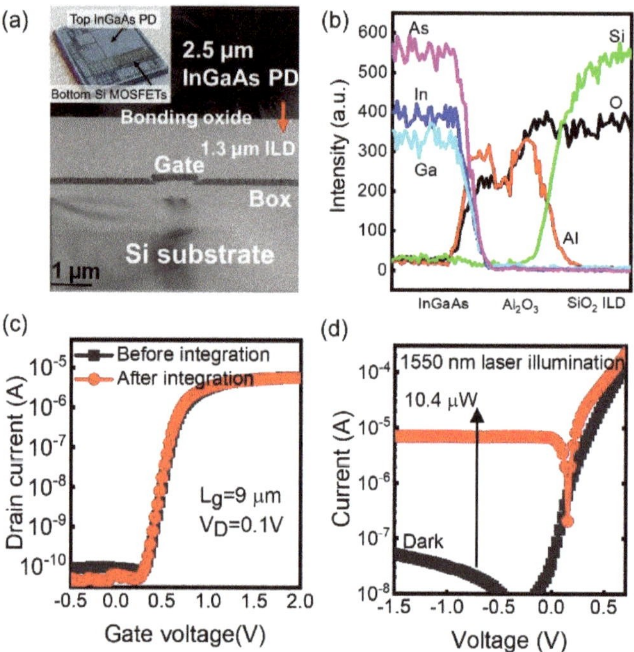

Figure 12. (**a**) Cross-sectional low-magnified TEM image of InGaAs PDs on SOI-MOSFETs with SiO$_2$ ILD layer and inset photograph for an actual chip. The red arrow is the EDX scan line. (**b**) EDX line profile of various atoms of In, Ga, As, Si, O, and Al along to the red arrow in Figure 11a. (**c**) The transfer curves of SOI-MOSFET with 9-µm gate length before and after monolithic 3D integration. (**d**) Current-voltage characteristics of InGaAs PD with a 1550-nm laser under dark/light conditions. The current is increased after laser illumination. Reprinted with permission from Ref. [15]. 2022, IEEE.

Then, we performed the electrical and optical characterization of each device, as shown in Figure 12c,d, to confirm device performances of InGaAs PDs and SOI-MOSFETs before and after the M3D integration process. Figure 12c illustrates the transfer curves of the 9-µm gate length SOI-MOSFET with and without the M3D integration process. This evaluation is highly significant because the fabrication process of top layer devices can affect the bottom devices during the thermal processes and plasma process, etc. In our devices, there is a negligible difference in the subthreshold swing (S.S), on/off ratio, and saturation current of SOI-MOSFETs before and after M3D integration. It was confirmed that the top InGaAs devices process and integration process do not degrade the performance of the bottom devices. Moreover, the integrated InGaAs PD device has a good current ratio at ±1.5 V, corresponding to approximately 10^4 under dark conditions. Figure 12d shows a clear photoresponse with 0.7 A/W of responsivity without anti-reflection coating (ARC) obtained with 1550-nm laser illumination at the surface. From these results, top and bottom device qualities would be well preserved during the thin-film transfer process, and the device fabrication process that indicated the proposed M3D integration could be used to fabricate an actual M3D image sensor with a high resolution and a small chip size.

3.2. Evaluation of Fabricated InGaAs Photodetectors on SOI-MOSFETs for Mimicking IC Operation

Finally, we investigated the readout function for mimicking the actual ROIC operation using M3D integrated InGaAs PDs on SOI-MOSFETs. In actual ROIC operation, Figure 13a illustrates the fundamental building block of typical circuit configurations, such as direct injection and source-follower per detector [63]. Although IC operation mechanisms vary in many configurations, the basic element close to PDs is a simple combination of 1 PD and 1 Tr. Thus, we evaluated the fundamental function of IC operation in M3D integrated InGaAs PDs on SOI-MOSFETs (1PD and 1Tr structure). The readout voltage (V_{out}) measurement was performed for the top InGaAs PD layer under the illumination condition of a 1550 nm laser. Figure 13b shows the resulting V_{out} as a function of the light intensity by varying the gate bias (V_{GS}) from 0.425 V to 1 V, where 0.425 V of V_{GS} is in the subthreshold regime. This is due to the SOI-MOSFETs' ability to function as a charge transfer gate in direct injection mode, where photo-generated carriers are directly injected via the source on the output stage. V_{out} gradually increased as V_{GS} increased from 0.4 V to 1 V, which could be attributed to the division of resistance between Tr and PD caused by the Tr resistance decrease. Furthermore, in fixed V_{GS}, increasing light intensity from 0.1 to 10 μW resulted in the V_{out} decrease due to PD resistance decrease. These results suggested the successful mixed-signal operation of the direct injection mechanism of ROIC by using M3D integrated PD and Tr structure. Recently, we demonstrated an M3D integrated MicroLED display on CMOS driver IC with a low-temperature process [17]. This integration process could be directly applied to integrated PD on ROIC in the future.

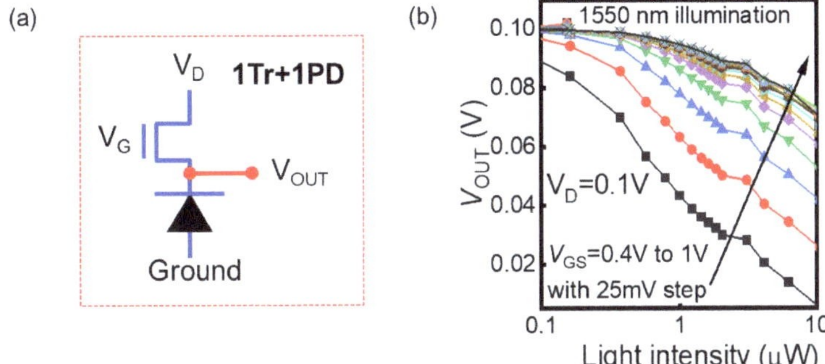

Figure 13. (**a**) The measured unit cell for readout operation for the fabricated InGaAs PD on SOI-MOSFET for mimicking the direct injection operation. (**b**) Electrical responses of the measured unit cell with various gate biases and light intensity on top InGaAs PDs. Reprinted with permission from Ref. [15]. 2022, IEEE.

4. Conclusions

We presented the M3D integration-based InGaAs HEMTs and InGaAs PDs on Si-based bottom devices/ICs for next-generation RF and image sensor applications. For the RF applications, we demonstrated the heterogenous and monolithic 3D integration of III-V RF devices on Si CMOS circuits, which enable performance optimization in both RF and digital integrated circuits. Furthermore, we discussed the issues and solutions of substrate digital interference and self-heating in terms of the M3D integrated RF platform. For the image sensor applications, to overcome the inherent limitation of hybrid integration technology for fabricating high-resolution SWIR image sensors, the integrated 1 PD and 1 Tr devices successfully exhibited good PD performances after the layer transfer and fabrication process without bottom Si device degradations. Photolithographic alignment allowed dimension scaling, which can scale the pixel pitch to high-resolution (<1 μm). Furthermore, the basic IC operation changing from light signals to electrical signals was achieved. These results strongly suggested that the heterogeneous and monolithic 3D integration will provide high performance and multi-functionality in various applications, such as RF and image sensors.

Author Contributions: Conceptualization, methodology, analysis, J.J. and D.-M.G.; writing—original draft preparation; S.K. supervision of this paper. All authors have read and agreed to the published version of the manuscript.

Funding: This research was funded by the System Semiconductor Development Program funded by Gyeonggi-do, NRF of Korea (2022M3F3A2A01065057, 2022M3I8A107725711, 2020M3F3A2A02082450, 2020M3F3A2A01082329), NNFC OI LAB Project, and BrainKorea 21 (BK21) FOUR.

Conflicts of Interest: The authors declare no conflict of interest.

References

1. Mallik, A.; Vandooren, A.; Witters, L.; Walke, A.; Franco, J.; Sherazi, Y.; Weckx, P.; Yakimets, D.; Bardon, M.; Parvais, B.; et al. The Impact of Sequential-3D Integration on Semiconductor Scaling Roadmap. In Proceedings of the 2018 IEEE International Electron Devices Meeting (IEDM), San Francisco, CA, USA, 1–5 December 2018. [CrossRef]

2. Batude, P.; Brunet, L.; Fenouillet-Beranger, C.; Andrieu, F.; Colinge, J.P.; Lattard, D.; Vianello, E.; Thuries, S.; Billoint, O.; Vivet, P.; et al. 3D Sequential Integration: Application-Driven Technological Achievements and Guidelines. In Proceedings of the 2018 IEEE International Electron Devices Meeting (IEDM), San Francisco, CA, USA, 1–5 December 2018. [CrossRef]

3. Shulaker, M.M.; Hills, G.; Park, R.S.; Howe, R.T.; Saraswat, K.; Wong, H.S.P.; Mitra, S. Three-Dimensional Integration of Nanotechnologies for Computing and Data Storage on a Single Chip. *Nature* **2017**, *547*, 74–78. [CrossRef] [PubMed]

4. Motoyoshi, M. Through-Silicon Via (TSV). *Proc. IEEE* **2009**, *97*, 43–48. [CrossRef]

5. Haruta, T.; Nakajima, T.; Hashizume, J.; Umebayashi, T.; Takahashi, H.; Taniguchi, K.; Kuroda, M.; Sumihiro, H.; Enoki, K.; Yamasaki, T.; et al. A 1/2.3inch 20Mpixel 3-Layer Stacked CMOS Image Sensor with DRAM. In Proceedings of the 2017 IEEE International Solid-State Circuits Conference (ISSCC), San Francisco, CA, USA, 5–9 February 2017. [CrossRef]

6. Estrada, J.A.; Lasser, G.; Pinto, M.; Herrault, F.; Popovic, Z. Metal-Embedded Chip Assembly Processing for Enhanced RF Circuit Performance. *IEEE Trans. Microw. Theory Tech.* **2019**, *67*, 3537–3546. [CrossRef]

7. Herrault, F.; Wong, J.; Ramos, I.; Tai, H.; King, M. Chiplets in Wafers (CiW) Process Design Kit and Demonstration of High-Frequency Circuits with GaN Chiplets in Silicon Interposers. In Proceedings of the 2021 IEEE 71st Electronic Components and Technology Conference (ECTC), San Diego, CA, USA, 1–4 June 2021. [CrossRef]

8. Weimann, N.; Hossain, M.; Krozer, V.; Heinrich, W.; Lisker, M.; Mai, A. Tight Focous Toward the Future: Tight Material Combination for Millimeter-Wave RF Power Applications: InP HBT SiGe BiCMOS Heterogeneous Wafer-Level Integration. *IEEE Microw. Mag.* **2017**, *18*, 74–82. [CrossRef]

9. Urteaga, M.; Carter, A.; Griffith, Z.; Pierson, R.; Bergman, J.; Arias, A.; Rowell, P.; Hacker, J.; Brar, B.; Rodwell, M.J.W. THz Bandwidth InP HBT Technologies and Heterogeneous Integration with Si CMOS. In Proceedings of the 2016 IEEE Bipolar/BiCMOS Circuits and Technology Meeting (BCTM), New Brunswick, NJ, USA, 25–27 September 2016. [CrossRef]

10. Jeong, J.; Kim, S.K.; Kim, J.; Geum, D.M.; Kim, D.; Jo, E.; Jeong, H.; Park, J.; Jang, J.H.; Choi, S.; et al. Heterogeneous and Monolithic 3D Integration of III-V-Based Radio Frequency Devices on Si CMOS Circuits. *ACS Nano* **2022**, *16*, 9031–9040. [CrossRef] [PubMed]

11. Jeong, J.; Kim, S.K.; Kim, J.; Geum, D.; Park, J.; Jang, J.; Kim, S. High-Performance InGaAs-On-Insulator HEMTs on Si CMOS for Substrate Coupling Noise-Free Monolithic 3D Mixed-Signal IC. In Proceedings of the 2021 IEEE Symposium on VLSI Technology, Kyoto, Japan, 13–19 June 2021.

12. Caimi, D.; Tiwari, P.; Sousa, M.; Moselund, K.E.; Zota, C.B. Heterogeneous Integration of III-V Materials by Direct Wafer Bonding for High-Performance Electronics and Optoelectronics. *IEEE Trans. Electron Devices* **2021**, *68*, 3149–3156. [CrossRef]

13. Frutuoso, T.M.; Sideris, P.; Lugo-Alvarez, J.; Garros, X.; Brunet, L.; Fenouille-Beranger, C.; Batude, P.; Theodorou, C.; Ferrari, P.; Gaillard, F. RF Performance of Devices Processed in Low-Temperature Sequential Integration. *IEEE Trans. Electron Devices* **2021**, *68*, 3157–3162. [CrossRef]

14. Vandooren, A.; Franco, J.; Parvais, B.; Wu, Z.; Witters, L.; Walke, A.; Li, W.; Peng, L.; Desphande, V.; Bufler, F.M.; et al. 3D Sequential Stacked Planar Devices on 300 Mm Wafers Featuring Replacement Metal Gate Junction-Less Top Devices Processed at 525 °C with Improved Reliability. In Proceedings of the 2018 IEEE Symposium on VLSI Technology, Honolulu, HI, USA, 18–22 June 2018. [CrossRef]

15. Geum, D.M.; Kim, S.K.; Lee, S.; Lim, D.; Kim, H.J.; Choi, C.H.; Kim, S.H. Monolithic 3D Integration of InGaAs Photodetectors on Si MOSFETs Using Sequential Fabrication Process. *IEEE Electron Device Lett.* **2020**, *41*, 433–436. [CrossRef]

16. Kim, S.K.; Geum, D.M.; Lim, H.R.; Han, J.; Kim, H.; Jeong, Y.; Kim, S.H. Photo-Responsible Synapse Using Ge Synaptic Transistors and GaAs Photodetectors. *IEEE Electron Device Lett.* **2020**, *41*, 605–608. [CrossRef]

17. Park, J.; Geum, D.M.; Baek, W.; Shieh, J.; Kim, S. Monolithic 3D sequential integration realizing 1600-PPI red micro-LED display on Si CMOS driver IC. In Proceedings of the 2022 IEEE Symposium on VLSI Technology and Circuits (VLSI Technology and Circuits), Honolulu, HI, USA, 12–17 June 2022. [CrossRef]

18. Kim, S.K.; Jeong, Y.; Bidenko, P.; Lim, H.R.; Jeon, Y.R.; Kim, H.; Lee, Y.J.; Geum, D.M.; Han, J.; Choi, C.; et al. 3D Stackable Synaptic Transistor for 3D Integrated Artificial Neural Networks. *ACS Appl. Mater. Interfaces* **2020**, *12*, 7372–7380. [CrossRef]

19. Han, J.K.; Sim, J.; Geum, D.M.; Kim, S.K.; Yu, J.M.; Kim, J.; Kim, S.; Choi, Y.K. 3D Stackable Broadband Photoresponsive InGaAs Biristor Neuron for a Neuromorphic Visual System with Near 1 v Operation. In Proceedings of the 2021 IEEE International Electron Devices Meeting (IEDM), San Francisco, CA, USA, 11–16 December 2021. [CrossRef]

20. Lee, H.J.; Rami, S.; Ravikumar, S.; Neeli, V.; Phoa, K.; Sell, B.; Zhang, Y. Intel 22nm FinFET (22FFL) Process Technology for RF and Mm Wave Applications and Circuit Design Optimization for FinFET Technology. In Proceedings of the 2018 IEEE International Electron Devices Meeting (IEDM), San Francisco, CA, USA, 1–5 December 2018. [CrossRef]

21. Kane, O.M.; Lucci, L.; Scheiblin, P.; Lepilliet, S.; Danneville, F. 22nm Ultra-Thin Body and Buried Oxide FDSOI RF Noise Performance. In Proceedings of the 2019 IEEE Radio Frequency Integrated Circuits Symposium (RFIC), Boston, MA, USA, 2–4 June 2019. [CrossRef]

22. Van Der Voorn, P.; Agostinelli, M.; Choi, S.J.; Curello, G.; Deshpande, H.; El-Tanani, M.A.; Hafez, W.; Jalan, U.; Janbay, L.; Kang, M.; et al. A 32nm Low Power RF CMOS SOC Technology Featuring High-k/Metal Gate. In Proceedings of the 2010 Symposium on VLSI Technology, Honolulu, HI, USA, 15–17 June 2010. [CrossRef]

23. Vandooren, A.; Parvais, B.; Witters, L.; Walke, A.; Vais, A.; Merckling, C.; Lin, D.; Waldron, N.; Wambacq, P.; Mocuta, D.; et al. 3D Technologies for Analog/RF Applications. In Proceedings of the IEEE SOI-3D-Subthreshold Microelectronics Technology Unified Conference (S3S), Burlingame, CA, USA, 16–19 October 2017. [CrossRef]

24. Collaert, N.; Alian, A.; Banerjee, A.; Chauhan, V.; Elkashlan, R.Y.; Hsu, B.; Ingels, M.; Khaled, A.; Kodandarama, K.V.; Kunert, B.; et al. From 5G to 6G: Will Compound Semiconductors Make the Difference? In Proceedings of the 2020 IEEE 15th International Conference Solid-State Integration Circuit Technol. ICSICT, Kunning, China, 3–6 November 2020. [CrossRef]

25. Parvais, B.; Vais, A.; Yadav, S.; Mols, Y.; Vermeersch, B.; Kodanarama, K.V.; Boccardi, G.; Kunert, B.; Collaert, N. III-V on a Si Platform for the next Generations of Communication Systems. In Proceedings of the 2022 6th IEEE Electron Devices Technology and Manufacturing Conference (EDTM), Oita, Japan, 6–9 March 2022. [CrossRef]

26. Green, D.S.; Dohrman, C.L.; Demmin, J.; Chang, T.H. Path to 3D Heterogeneous Integration. In Proceedings of the 2015 International 3D Systems Integration Conference, Sendai, Japan, 31 August–2 September 2015. [CrossRef]

27. Heinemann, B.; Rucker, H.; Barth, R.; Barwolf, F.; Drews, J.; Fischer, G.G.; Fox, A.; Fursenko, O.; Grabolla, T.; Herzel, F.; et al. SiGe HBT with Fx/Fmax of 505 GHz/720 GHz. In Proceedings of the 2016 IEEE International Electron Devices Meeting (IEDM), San Francisco, CA, USA, 3–7 December 2016. [CrossRef]

28. Lai, R.; Mei, X.B.; Deal, W.R.; Yoshida, W.; Kim, Y.M.; Liu, P.H.; Lee, J.; Uyeda, J.; Radisic, V.; Lange, M.; et al. Sub 50 nm InP HEMT Device with Fmax Greater than 1 THz. In Proceedings of the 2007 IEEE International Electron Devices Meeting, Washington, DC, USA, 10–12 December 2007. [CrossRef]

29. Mei, X.; Yoshida, W.; Lange, M.; Lee, J.; Zhou, J.; Liu, P.H.; Leong, K.; Zamora, A.; Padilla, J.; Sarkozy, S.; et al. First Demonstration of Amplification at 1 THz Using 25-nm InP High Electron Mobility Transistor Process. *IEEE Electron Device Lett.* **2015**, *36*, 327–329. [CrossRef]

30. Jo, H.B.; Yun, S.W.; Kim, J.G.; Baek, J.M.; Lee, I.G.; Kim, D.H.; Kim, T.W.; Kim, S.K.; Yun, J.; Kim, T.; et al. Sub-30-nm In0.8Ga0.2As Composite-Channel High-Electron-Mobility Transistors with Record High-Frequency Characteristics. *IEEE Trans. Electron Devices* **2021**, *68*, 2010–2016. [CrossRef]

31. Thome, F.; Heinz, F.; Leuther, A. InGaAs MOSHEMT W-Band LNAs on Silicon and Gallium Arsenide Substrates. *IEEE Microw. Wirel. Compon. Lett.* **2020**, *30*, 1089–1092. [CrossRef]

32. Larsson, E.G.; Edfors, O.; Tufvesson, F.; Marzetta, T.L. Massive MIMO for next Generation Wireless Systems. *IEEE Commun. Mag.* **2014**, *52*, 186–195. [CrossRef]

33. Lu, L.; Li, G.Y.; Swindlehurst, A.L.; Ashikhmin, A.; Zhang, R. An Overview of Massive MIMO: Benefits and Challenges. *IEEE J. Sel. Top. Signal Process.* **2014**, *8*, 742–758. [CrossRef]

34. Jeong, J.; Kim, S.K.; Kim, J.; Geum, D.M.; Park, J.; Jang, J.H.; Kim, S. Stackable InGaAs-on-Insulator HEMTs for Monolithic 3-D Integration. *IEEE Trans. Electron Devices* **2021**, *68*, 2205–2211. [CrossRef]

35. Su, D.K.; Loinaz, M.J.; Masui, S.; Wooley, B.A. Experimental Results and Modeling Techniques for Substrate Noise in Mixed-Signal Integrated Circuits. *IEEE J. Solid-State Circuits* **1993**, *28*, 420–429. [CrossRef]

36. Xu, M.; Su, D.K.; Shaeffer, D.K.; Lee, T.H.; Wooley, B.A. Measuring and Modeling the Effects of Substrate Noise on the LNA for a CMOS GPS Receiver. *IEEE J. Solid-State Circuits* **2001**, *36*, 473–485. [CrossRef]

37. Kosaka, D.; Nagata, M.; Hiraoka, Y.; Imanoshi, I.; Maeda, M.; Murasaka, Y.; Iwata, A. Isolation Strategy Against Substrate Coupling in CMOS Mixed-Signal/RF Circuits. In Proceedings of the 2005 Symposium on VLSI Circuits, Kyoto, Japan, 16–18 July 2005. [CrossRef]

38. Uemura, S.; Hiraoka, Y.; Kai, T.; Dosho, S. Isolation Techniques Against Substrate Noise Coupling Utilizing Through Silicon Via (TSV) Process for RF/Mixed-Signal SoCs. *IEEE J. Solid-State Circuits* **2012**, *47*, 810–816. [CrossRef]

39. Shin, S.H.; Kim, S.H.; Kim, S.; Wu, H.; Ye, P.D.; Alam, M.A. Substrate and Layout Engineering to Suppress Self-Heating in Floating Body Transistors. In Proceedings of the 2016 IEEE International Electron Devices Meeting (IEDM), San Francisco, CA, USA, 3–7 December 2016. [CrossRef]

40. Triantopoulos, K.; Casse, M.; Brunet, L.; Batude, P.; Fenouillet-Beranger, C.; Mathieu, B.; Vinet, M.; Ghibaudo, G.; Reimbold, G. Thermal effects in 3D sequential technology. In Proceedings of the 2017 IEEE International Electron Devices Meeting (IEDM), San Francisco, CA, USA, 2–6 December 2017. [CrossRef]

41. Jeong, J.; Kim, S.K.; Kim, J.; Geum, D.M.; Kim, S. Monolithic 3D Integrated InGaAs HEMTs on Si for Next-Generation Communication: Record {MAX} and Relaxed Self-Heating of Top Devices by a Novel M3D Structure. In Proceedings of the 2021 IEEE International Electron Devices Meeting (IEDM), San Francisco, CA, USA, 11–16 December 2021. [CrossRef]

42. Jeong, J.; Kim, S.K.; Kim, J.; Geum, D.-M.; Kim, S. Heat Management in Monolithic 3D RF Platform. In Proceedings of the 2022 6th IEEE Electron Devices Technology & Manufacturing Conference (EDTM), Oita, Japan, 6–9 March 2022. [CrossRef]

43. Shim, J.; Lim, J.; Geum, D.M.; You, J.B.; Yoon, H.; Kim, J.P.; Baek, W.J.; Han, J.H.; Kim, S. TiOx/Ti/TiOx Tri-Layer Film-Based Waveguide Bolometric Detector for On-Chip Si Photonic Sensor. *IEEE Trans. Electron Devices* **2022**, *69*, 2151–2158. [CrossRef]

44. Zota, C.B.; Convertino, C.; Baumgartner, Y.; Sousa, M.; Caimi, D.; Czornomaz, L. High Performance Quantum Well InGaAs-On-Si MOSFETs with Sub-20 nm Gate Length for RF Applications. In Proceedings of the 2018 IEEE International Electron Devices Meeting (IEDM), San Francisco, CA, USA, 1–5 December 2018. [CrossRef]

45. Tessmann, A.; Leuther, A.; Heinz, F.; Bernhardt, F.; John, L.; Massler, H.; Czornomaz, L.; Merkle, T. 20-nm In0.8Ga0.2As MOSHEMT MMIC Technology on Silicon. *IEEE J. Solid-State Circuits* **2019**, *54*, 2411–2418. [CrossRef]

46. Yadav, S.; Kumar, A.; Nguyen, X.S.; Lee, K.H.; Liu, Z.; Xing, W.; Masudy-Panah, S.; Lee, K.; Tan, C.S.; Fitzgerald, E.A.; et al. High Mobility $In_{0.30}Ga_{0.70}As$ MOSHEMTs on Low Threading Dislocation Density 200 mm Si Substrates: A Technology Enabler Towards Heterogenous Integration of Low Noise and Medium Power Amplfiers with Si CMOS. In Proceedings of the 2017 IEEE International Electron Devices Meeting (IEDM), San Francisco, CA, USA, 2–6 December 2017. [CrossRef]

47. Then, H.W.; Radosavljevic, M.; Koirala, P.; Thomas, N.; Nair, N.; Ban, I.; Talukdar, T.; Nordeen, P.; Ghosh, S.; Bader, S.; et al. Advanced Scaling of Enhancement Mode High-K Gallium Nitride-on-300mm-Si(111) Transistor and 3D Layer Transfer GaN-Silicon Finfet CMOS Integration. In Proceedings of the 2021 IEEE International Electron Devices Meeting (IEDM), San Francisco, CA, USA, 11–16 December 2021. [CrossRef]

48. Parvais, B.; Alian, A.; Peralagu, U.; Rodriguez, R.; Yadav, S.; Khaled, A.; Elkashlan, R.Y.; Putcha, V.; Sibaja-Hernandez, A.; Zhao, M.; et al. GaN-on-Si Mm-Wave RF Devices Integrated in a 200mm CMOS Compatible 3-Level Cu BEOL. In Proceedings of the 2020 IEEE International Electron Devices Meeting (IEDM), San Francisco, CA, USA, 12–18 December 2020. [CrossRef]

49. Jeong, J.; Kim, S.K.; Kim, J.; Geum, D.; Lee, J.; Park, S.; Kim, S. 3D Stackable Cryogenic InGaAs HEMTs for Heterogeneous and Monolithic 3D Integrated Highly Scalable Quantum Computing Systems. In Proceedings of the 2022 IEEE Symposium on VLSI Technology and Circuits (VLSI Technology and Circuits), Honolulu, HI, USA, 12–17 June 2022. [CrossRef]

50. Kim, S.H.; Ikku, Y.; Yokoyama, M.; Nakane, R.; Li, J.; Kao, Y.C.; Takenaka, M.; Takagi, S. High Performance InGaAs-on-Insulator MOSFETs on Si by Novel Direct Wafer Bonding Technology Applicable to Large Wafer Size. In Proceedings of the 2014 Symposium on VLSI Technology (VLSI-Technology), Honolulu, HI, USA, 9–12 June 2014. [CrossRef]

51. Uhrmann, T.; Burggraf, J.; Eibelhuber, M. Heterogeneous Integration by Collective Die-to-Wafer Bonding. In Proceedings of the 2018 International Wafer Level Packaging Conference (IWLPC), San Hose, CA, USA, 23–25 October 2018. [CrossRef]

52. Lee, S.; Kim, S.K.; Han, J.H.; Song, J.D.; Jun, D.H.; Kim, S.H. Epitaxial Lift-Off Technology for Large Size III-V-on-Insulator Substrate. *IEEE Electron Device Lett.* **2019**, *40*, 1732–1735. [CrossRef]

53. Choi, B.S.; Kim, S.H.; Lee, J.; Chang, S.; Park, J.; Lee, S.J.; Shin, J.K. CMOS Image Sensor Using Pixel Aperture Technique for Single-Chip 2D and 3D Imaging. In Proceedings of the 2017 IEEE SENSORS, Glasgow, UK, 29 October–1 November 2017. [CrossRef]

54. Lange, R.; Seitz, P. Solid-State Time-of-Flight Range Camera. *IEEE J. Quantum Electron.* **2001**, *37*, 390–397. [CrossRef]

55. Geum, D.M.; Kim, S.; Kim, S.K.; Kang, S.; Kyhm, J.; Song, J.; Choi, W.J.; Yoon, E. Monolithic Integration of Visible GaAs and Near-Infrared InGaAs for Multicolor Photodetectors by Using toward High-Resolution Imaging Systems. *Sci. Rep.* **2019**, *9*, 18661. [CrossRef] [PubMed]

56. Manda, S.; Zaizen, Y.; Hirano, T.; Iwamoto, H.; Matsumoto, R.; Saito, S.; Maruyama, S.; Minari, H.; Hirano, T.; Takachi, T.; et al. High-Definition Visible-SWIR InGaAs Image Sensor Using Cu-Cu Bonding of III-V to Silicon Wafer. In Proceedings of the 2019 IEEE International Electron Devices Meeting (IEDM), San Francisco, CA, USA, 7–11 December 2019. [CrossRef]

57. Xu, K. Integrated Silicon Directly Modulated Light Source Using P-Well in Standard CMOS Technology. *IEEE Sens. J.* **2016**, *16*, 6184–6191. [CrossRef]

58. Kim, S.H.; Kim, S.K.; Shim, J.P.; Geum, D.M.; Ju, G.; Kim, H.S.; Lim, H.J.; Lim, H.R.; Han, J.H.; Lee, S.; et al. Heterogeneous Integration Toward a Monolithic 3-D Chip Enabled by III-V and Ge Materials. *IEEE J. Electron Devices Soc.* **2018**, *6*, 579–587. [CrossRef]

59. Kim, S.K.; Shim, J.P.; Geum, D.M.; Kim, C.Z.; Kim, H.S.; Song, J.D.; Choi, S.J.; Kim, D.H.; Choi, W.J.; Kim, H.J.; et al. Fabrication of InGaAs-on-Insulator Substrates Using Direct Wafer-Bonding and Epitaxial Lift-Off Techniques. *IEEE Trans. Electron Devices* **2017**, *64*, 3601–3608. [CrossRef]

60. Jang, J.H.; Geum, D.M.; Kim, S.H. Broadband Au/n-GaSb Schottky photodetector array with a spectral range from 300 nm to 1700 nm. *Opt. Express* **2021**, *29*, 38894–38903. [CrossRef]

61. Geum, D.M.; Kim, S.; Khym, J.; Lim, J.; Kim, S.; Ahn, S.Y.; Kim, T.S.; Kang, K.; Kim, S.H. Arrayed MoS_2–$In_{0.53}Ga_{0.47}As$ van der Waals Heterostructure for High-Speed and Broadband Detection from Visible to Shortwave-Infrared Light. *Small* **2021**, *17*, 2007357. [CrossRef]

62. Geum, D.M.; Lim, J.; Jang, J.; Ahn, S.Y.; Kim, S.K.; Shim, J.; Kim, B.H.; Park, J.H.; Baek, W.J.; Jeong, J.; et al. A sub-micron-thick InGaAs broadband (400-1700 nm) photodetectors with a high external quantum efficiency (>70%). In Proceedings of the 2022 IEEE Symposium on VLSI Technology and Circuits (VLSI Technology and Circuits), Honolulu, HI, USA, 12–17 June 2022. [CrossRef]

63. Bielecki, Z. Readout electronics for optical detectors. *OPTO-Electron. Rev.* **2004**, *12*, 129–137.

Review

Recent Trends in Copper Metallization

Hyung-Woo Kim

China Nanhu Academy of Electronics and Information Technology, Jiaxing 314001, China; kimhyungwoo@cnaeit.com

Abstract: The Cu/low-k damascene process was introduced to alleviate the increase in the RC delay of Al/SiO_2 interconnects, but now that the technology generation has reached $1\times$ nm or lower, a number of limitations have become apparent. Due to the integration limit of low-k materials, the increase in the RC delay due to scaling can only be suppressed through metallization. As a result, various metallization methods have been proposed, including traditional barrier/liner thickness scaling, and new materials and integration schemes have been developed. This paper introduces these methods and summarizes the recent trends in metallization. It also includes a brief introduction to the Cu damascene process, an explanation of why the low-k approach faces limitations, and a discussion of the measures of reliability (electromigration and time-dependent dielectric breakdown) that are essential for all validation schemes.

Keywords: BEOL; interconnect; copper; low-k; metallization; scaling

Citation: Kim, H.-W. Recent Trends in Copper Metallization. *Electronics* **2022**, *11*, 2914. https://doi.org/10.3390/electronics11182914

Academic Editor: Changhwan Shin

Received: 15 August 2022
Accepted: 9 September 2022
Published: 14 September 2022

Publisher's Note: MDPI stays neutral with regard to jurisdictional claims in published maps and institutional affiliations.

Copyright: © 2022 by the author. Licensee MDPI, Basel, Switzerland. This article is an open access article distributed under the terms and conditions of the Creative Commons Attribution (CC BY) license (https://creativecommons.org/licenses/by/4.0/).

1. Introduction

Continuous developments in integrated chip scaling have improved circuit density and chip performance in recent decades. Moore's law, which states that the number of transistors on a microchip doubles every two years, is now accepted as an empirical law. These advances have been achieved not only with simple reductions in dimensions; they have also been spurred by new patterning approaches, innovative device architectures, tool improvements, design–technology co-optimization, and the integration of new materials [1,2].

The increasing number of transistors on an integrated chip usually leads to an increase in the complexity of the interconnections. This can be resolved by increasing the number of vertical stacking levels based on hierarchical wiring schemes for effective design allocation [3,4]. A hierarchical wiring system is commonly constructed by vertically distributing various metal levels with different minimum pitches. Depending on their purpose, these interconnects can be divided into three groups: local, intermediate, or global interconnects. **Local interconnects:** The minimum metal pitch is applied to the levels closest to the transistor. They adopt state-of-the-art processes and technologies with low-k dielectrics and metallization to minimize the resistance–capacitance (RC) delay, which is most affected by patterning.

Intermediate interconnects: They connect primitive cells or signal transmission, such as the system clock. Because the congestion is lower than for local interconnects, a relaxed metal pitch and an increase in thickness are allowed.

Global interconnects: These are the wiring levels at the top of the integrated chip and are used to minimize the power transmission and voltage drop. They typically have a thick metal layer and a relaxed pitch. However, they are subject to additional requirements related to connections with the outside of the chip, i.e., the packaging.

The purpose of scaling is to aggregate more transistors within the same unit area while improving the performance of the semiconductor chip [5]. When the dimensions of a transistor are reduced, the transit time of the carriers passing through the transistor channel decreases. However, the interconnect performance is degraded due to an increase in the RC delay, which is the product of the conductor resistance (R) and the dielectric capacitance (C). R and C are calculated using Equations (1) and (2), respectively:

$$R = \rho \, \frac{L}{WT} \tag{1}$$

$$C = k\frac{LT}{S} \tag{2}$$

where ρ is the metal resistivity and k is the dielectric constant. L, W, and T are the length, width, and thickness of the metal lines, respectively, while S is the spacing between the metal lines. Based on these equations, it can be seen that the interconnect RC delay is determined by the physical dimensions and constituent material of the wiring.

W and S are determined by the physical scaling in terms of the minimum pitch. However, T and H (where H is the via height, the space between M_x and M_{x+1}) can be optimized during the fabrication process. Therefore, attempts have been made to improve the RC delay by optimizing the physical dimensions via process integration based on the given material and the target electrical and reliability specifications. Another consideration is the metal and dielectric insulators in the interconnects. Traditionally, aluminum (Al) and SiO_2 have been used as metal and dielectric insulators, respectively, since the introduction of 3 μm technology. However, as shown in Figure 1, when the technology generation reached 0.25 μm, the interconnect RC delay began to exceed the gate delay, leading to a bottleneck for integrated chip performance [6].

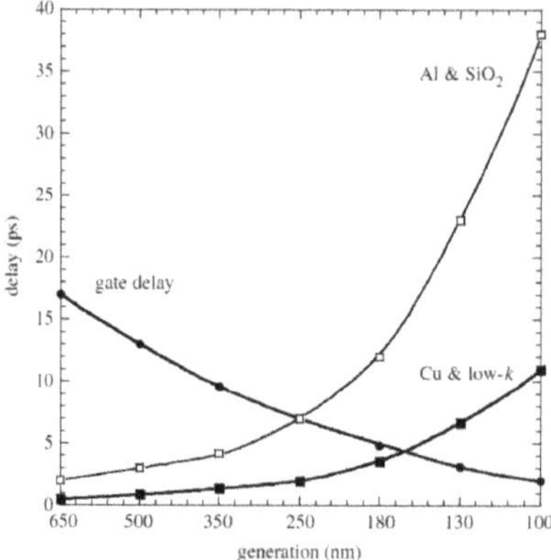

Figure 1. Gate and interconnect delay in accordance with technology generations (ITRS '99) [6].

In order to address this increase in the RC delay, new materials to replace Al and SiO_2 have been sought. Due to their low resistivity, copper (Cu), gold (Au), and silver (Ag) have been considered as replacements for Al, with Cu in particular showing lower resistivity and producing better electromigration (EM) performance [7]. In addition, porous SiCOH (pSiCOH) has been developed from fluorinated silicon glass (FSG) and SiCOH (organosilicate glass; OSG) as a low-k material to replace SiO_2 (k~4.3). The replacement of Al with Cu was successfully achieved with the release of mass-produced products in 1997 [8], and low-k materials began to be introduced in earnest with the emergence of 90 nm technology [9,10]. Furthermore, over the last 25 years, Cu and low-k materials have served as successful platforms for back-end-of-line (BEOL) interconnects.

However, as the minimum metal pitch approaches the electron mean free path (eMFP) of Cu (39 nm), various interconnection problems have emerged, including patterning restrictions. In particular, while dual-damascene (DD) schemes, which include a metal hard mask (HM), are unaffected, single patterning (SP) is no longer possible at the wavelength of argon fluoride (ArF; 193 nm). Various resolution enhancement techniques (RETs) and immersion lithography have been employed to overcome the limitations of SP, but the use of multiple patterning (MP) is inevitable. If scaling continues at its current rate despite the introduction of extreme ultraviolet (EUV) lithography, MP will be required for future nodes.

Another issue is the limitations of low-k materials. The porous low-k material pSiCOH has been successfully implemented at 45 nm [10], but the development of increasingly low-k materials is hindered by the loss of mechanical strength with an increase in porosity and integration issues arising from plasma-induced damage (PID) [11]. Finally, challenges also arise for Cu metallization as the metal dimensions shrink [12–14]. Cu resistivity increases with higher refractory metal ratios, surface scattering, and grain boundary (GB) scattering in the Cu lines. Various attempts have been reported to overcome this, with the most recent trend being to employ metals with barrier-free advantages and the ability to be used in direct metal etching (DME) [15,16]. From this perspective, ruthenium (Ru) has emerged as the most promising metal to replace Cu [15,17]. This suggests that BEOL interconnects will no longer be confined in Cu damascene structures only.

2. Cu Dual-Damascene Interconnects

Unlike Al, Cu cannot be easily patterned using reactive ion etching (RIE) due to the low volatility of $CuCl_2$ and CuF at low temperatures. Due to this constraint, the damascene process was introduced as a replacement for subtractive etching [18,19]. Cu also has reliability issues due to its high diffusivity into the surrounding dielectric material unless it is perfectly encapsulated [20–22]. Many studies have sought to resolve the diffusion issue, with tantalum nitride (TaN) mostly employed as a diffusion barrier based on comprehensive comparisons of many candidate materials [23]. IBM announced the first product implementation of a Cu damascene structure in 1997 [7], since which this structure has become the standard platform for on-chip interconnects with improved resistivity and reliability compared with Al interconnects.

In the damascene process, the dielectric is first deposited onto the substrate and then etched to form a metal and via profile. Finally, the incoming profile is filled with Cu metal and the excess Cu removed using chemical mechanical polishing (CMP).

Damascene interconnects have two variations: single-damascene (SD) and dual-damascene (DD) structures. Figure 2 presents a schematic comparison of SD and DD interconnects. The SD process produces the via and the trench separately, while the DD process conducts via and trench patterning separately but their metallization together. Due to the economic advantages of reducing the number of steps in a process, the DD process is widely preferred. However, the SD process is still used for particular purposes such as M1 or with thick metal layers used for global interconnects. SD interconnects are employed for M1 because connection with the CNT module is required, and an SD structure is effective if a thick metal such as an inductor is required.

The DD fabrication process can be divided into via-first and trench-first schemes according to the patterning order. In the trench-first scheme, which was the first DD process, there is no fence between the via and trench, so depositing the seed Cu for electroplating is easy. However, this approach suffers from indirect alignment and a narrower depth of focus for via lithography due to the step height, leading to a non-planar resist. These issues place a restriction on scaling. In contrast, the via-first scheme has a broader lithography processing window, though it is susceptible to plasma damage to the dielectric when employing a porous low-k material. Therefore, DD processing with a metal HM has generally been employed since the 32 nm technology generation [19]. This method effectively combines the advantages of the via-first and trench-first schemes. Figure 3 presents the differences in the plasma damage arising from the via-first and trench-first schemes using a titanium nitride (TiN) metal HM.

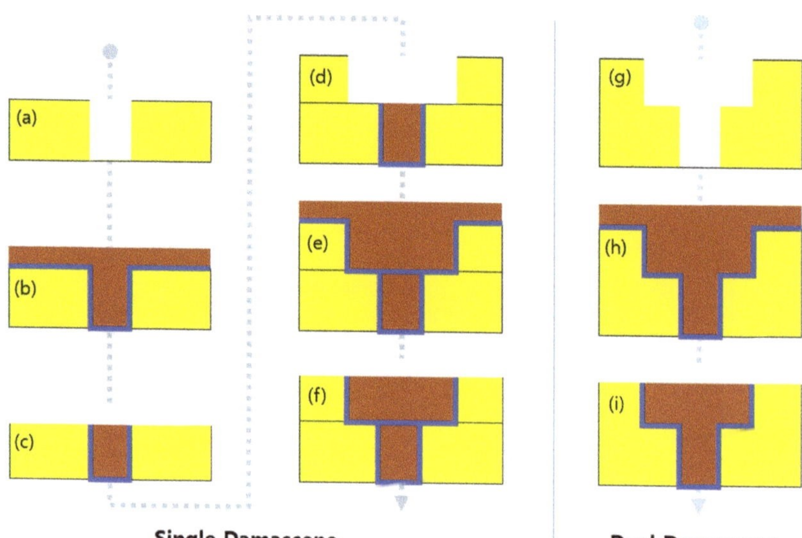

Single Damascene   **Dual Damascene**

Figure 2. Process comparison of the (**a–f**) SD and (**g–i**) DD schemes: (**a**) etching of the via in the deposited ILD, (**b**) filling of the via with Cu, (**c**) CMP of the excess Cu from the via, (**d**) etching of the trench in the deposited ILD, (**e**) filling the trench with Cu, (**f**) CMP of the excess Cu from the trench, (**g**) trench/via etching in the ILD, (**h**) filling with Cu, and (**i**) CMP of the excess Cu.

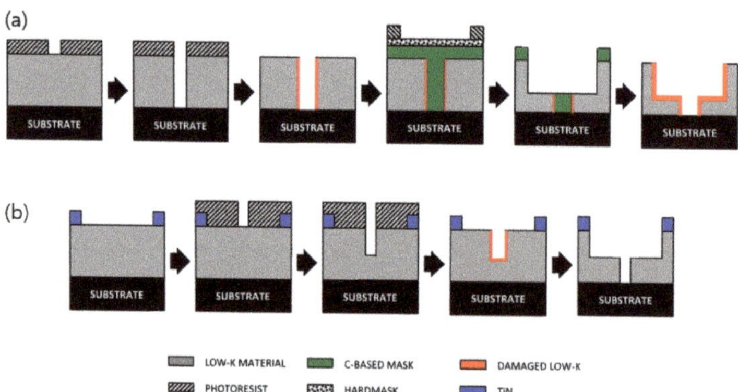

Figure 3. Process flow for (**a**) via-first and (**b**) trench-first approaches with a metal hard mask (HM) [11]. The plasma damage to the low-k material is lower for the trench-first scheme.

Patterned profiles created using the DD process need to be filled using metallization. Cu metallization consists of three steps: barrier/liner deposition, seed Cu deposition, and bulk Cu deposition [12]. Physical vapor deposition (PVD) is employed for the first two steps, while electroplating with additives is used as the standard process for the third. Following metallization, the excess Cu is removed using CMP to produce the final wiring. A dielectric barrier (DB) covers the surface after Cu CMP has been completed. This process is repeated until the final metal is connected to the bonding structure.

The Cu/low-k damascene process has served as a platform for the successful fabrication of BEOL interconnects while satisfying the scaling roadmap over the past 25 years. However, as scaling progresses, patterning, low-k materials, and metallization are all subject to process limitations. The next section summarizes these limitations and describes the approaches that have been proposed to overcome them.

3. Low-k Dielectric Materials

A dielectric material electrically insulates interconnections. When a voltage is applied to an electric wire, the dielectric material induces parasitic capacitance, causing crosstalk noise, power dissipation, and RC delay. These parameters are proportional to the dielectric constant k of the material [8]. However, dielectric materials should not only serve as electrical insulators; they also require reliable mechanical, thermal, chemical, and physical stability under the processing conditions and compatibility with other materials for successful integration [24].

SiO_2 is a dielectric material used for 2 to 0.25 μm CMOS technology generations with a dielectric constant of ~4.2. Though SiO_2 satisfies the dielectric requirements mentioned above, its k value no longer mitigates the RC delay. Therefore, efforts have been made to find materials with a lower k value that also satisfies the other insulator requirements.

The dielectric constant represents the ratio of the permittivity of a material to that of a vacuum and is generally described using Equation (3) [25]:

$$\frac{k-1}{k+2} = \frac{N}{3}\alpha \qquad (3)$$

where $k = \varepsilon/\varepsilon_0$ and ε and ε_0 are the permittivity of the material and vacuum, respectively, N is the number of molecules per unit volume (density), and α is the total polarizability, consisting of electronic (α_e) and distortion (α_d) polarization.

By decreasing the total polarizability (α) and/or density (N), the k value of the dielectric can be reduced. Density has a more substantial effect on the dielectric constant than does polarizability because density can be reduced to 0 (in an air), leading to $k = 1$. In the first-generation low-k material obtained using this method, some Si-O bonds were replaced by less polarized Si-F bonds, resulting in FSG [25,26]. In the second generation, a low dielectric constant was achieved using silsesquioxane-based materials and the chemical vapor deposition (CVD) of OSG (SiCOH) [25]. The low dielectric constant of these materials is partially due to the lower density compared with SiO_2. The density is reduced by breaking the 3D Si-O-Si bonding network via the incorporation of terminating Si-H or Si-R (where R is an organic moiety such as CH_3) [27].

Porosity must also be introduced in order for dense low-k materials to achieve k values below 2.5. This can be accomplished using sacrificial nanoparticles (i.e., porogen) desorbed in a high-temperature baking step, though maintaining a solid matrix structure is essential to avoid weakening the mechanical properties. The resulting material is pSiCOH [8,28]. Table 1 compares the characteristics of low-k materials for each generation.

Table 1. Essential properties of representative dielectric materials [6].

Properties	SiO_2	FSG	Dense Low-k (Osg)	Porous Low-k
density (g/cm³)	2.2	2.2	1.8~1.2	1.0~1.2
dielectric constant (k)	4	3.5~3.8	2.8~3.2	1.9~2.7
modulus (gpa)	55~70	~50	10~20	3~10
hardness (gpa)	3.5	3.36	1.2~2.5	0.3~1.0
cte (ppm/k)	0.6	~0.6	1~5	10~18
thermal conductivity (w/mk)	1.0	1.0	~0.8	0.26
porosity (%)	NA	NA	<10	25~50
average pore size (nm)	NA	NA	<1.0	2.0~10
breakdown field (mv/cm)	>10	>10	8~10	<8

However, low-k materials with high porosity suffer from lower mechanical strength, lower thermal conductivity, poor adhesion to barriers, plasma damage, and moisture absorption [29]. Figure 4 presents the relationship between mechanical strength and changes in the k value. For FSG, the difference from SiO_2 is insignificant in terms of the elastic modulus, but a rapid decrease occurs for dense low-k OSG and pSiCOH.

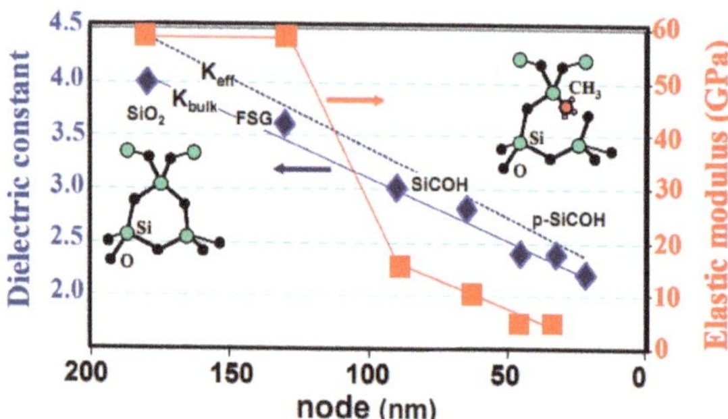

Figure 4. Relationship between the dielectric constant and the elastic modulus for different generation low-k materials. Reprinted/adapted with permission from Ref. [18]. 2009, IEEE.

Unlike low-k inter/intralayer dielectrics (ILDs), low-k DBs offer the integration requirements for downstream manufacturing processes [30]. Low-k DBs are essential components of BEOL interconnects and contribute to their capacitance. This has an increasingly significant impact on technology generations, allowing a decrease in interconnect dimensions. New DBs have a low k value and a low thickness, which reduces the impact on the capacitance of the interconnection and increases reliability concerns. SiN DBs ($k\sim7.0$) were replaced by SiCH and SiCNH ($k\sim5.3$) caps in 90 nm technology, and thus, the contribution of these DBs to total capacitance has been relatively small in several subsequent technical nodes. However, this contribution has since increased because low-k DBs have not scaled in proportion with the ILDs [8]. In other words, though the thickness of ILDs has decreased every time a new technology node is created, the thickness of low-k DBs cannot decrease at the same rate without degrading the DB characteristics [31,32].

Although low-k dielectrics have been introduced, ensuring a reliable interconnect process under $k\sim2.4$ is challenging. However, the use of an air gap is a promising approach to significantly reducing the capacitance because it can achieve an effective dielectric constant below 2.0. Figure 5 presents a cross-sectional image of an air gap.

Although this concept was already known beforehand [33–37], it was first commercially implemented in Intel's 14 nm technology using an 80 nm pitch multilayer (M4 and M6), with RC benefits of up to 17% [38,39]. In addition, IBM recently suggested the possibility of extending the air gap to the thin-wire level by achieving a capacitance reduction of about 20% compared with the baseline process using an ultra-low-k (ULK) dielectric ($k\sim2.4$) [40]. However, air gaps have been limited to selectively applied critical paths.

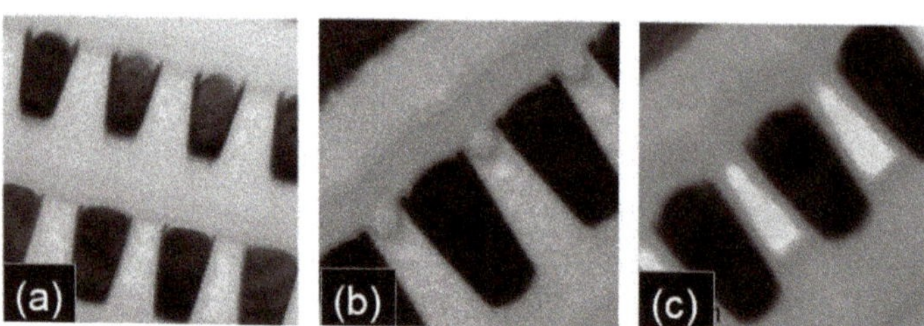

Figure 5. Cross-sectional images of (**a**) a baseline wafer without air gaps, (**b**) with air gaps excluded, and (c) with air gaps Reprinted/adapted with permission from Ref. [40]. 2017, IEEE.

4. Metallization

A typical DD Cu line is surrounded by a barrier, a liner, and a cap (Figure 6) [41]. The barrier promotes metal adhesion to the dielectric, protects Cu from oxidation, and acts as a nucleation layer for the liner metals. The liner facilitates Cu seeding and the plating process and, more importantly, improves the Cu interface for the suppression of

EM. After the emergence of 14 nm technology, a metal cap was applied to the top of the line to improve for EM performance due to the small volume of Cu [42]. A diffusion barrier is required to prevent the Cu from mixing with the surrounding dielectric material when manufacturing a Cu interconnect with a damascene structure. The barrier layer reduces the cross-sectional area of the interconnect, and the effective resistance of the Cu line becomes higher than that of a barrier-free Cu line. The increase in resistivity due to the barrier increases significantly when the width of the interconnect line is reduced. According to Matthiessen's law, line resistivity consists of bulk resistivity, impurity scattering, surface scattering, and GB scattering [43,44]. Before the emergence of 7 nm technology, bulk resistivity and impurity scattering usually determined line resistivity. However, since then, surface scattering and particle boundary scattering have taken on greater importance. For surface scattering, it is essential to minimize the volume fraction of Cu occupied by the barrier and liner, while particle boundary scattering is likely to affect scattering at the interface and the Cu grain size. This section summarizes important studies reported to date on these issues.

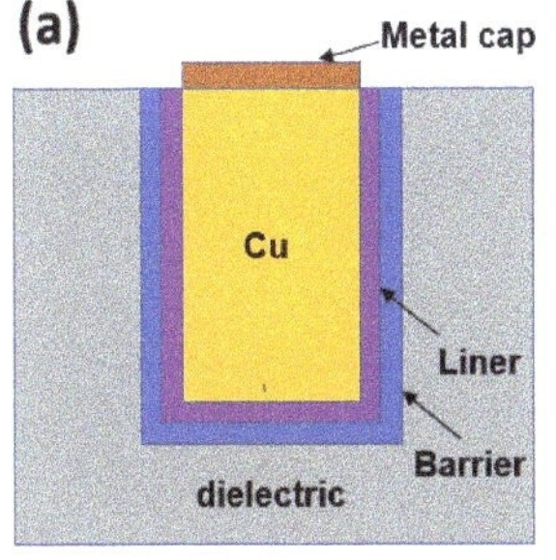

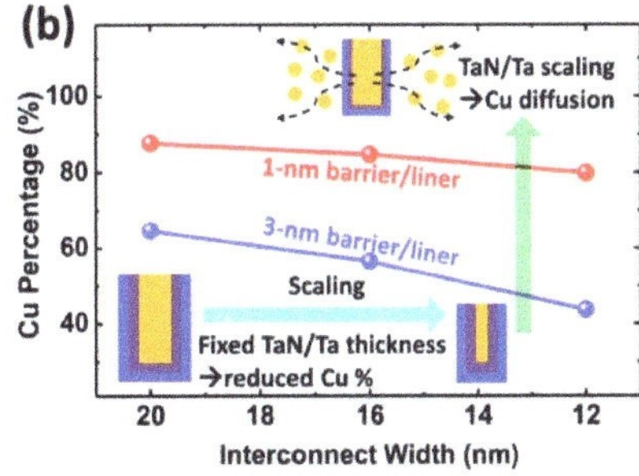

Figure 6. (**a**) Typical metallization of a Cu line with a barrier, liner, and metal cap. (**b**) Scaling challenges for the barrier and liner; in particular, a lower cross-sectional area is occupied by Cu with scaling [41].

4.1. TaN Barrier/Liner Scaling

In order to maximize the Cu volume fraction, the TaN/Ta liner thickness must be reduced. In addition, the thickness of the Cu seed needs to be reduced to prevent the top pinch-off shape during subsequent Cu electroplating. However, the nonconformal step coverage of PVD results in gap-fill problems caused by the overhang and high resistance due to the thick barrier at the via bottom. Therefore, atomic layer deposition (ALD)/CVD is preferred for maximizing the gap-filling window and minimizing the resistance.

Several approaches have been explored for reducing the TaN barrier thickness. First, the minimum thickness can be reduced without sacrificing TaN barrier properties using PVD [45]. In particular, based on time-dependent dielectric breakdown (TDDB) measurements using a planar capacitor structure, TaN barrier characteristics have been shown to be retained even if the thickness is reduced to as low as 0.8 nm. In addition, it has been reported that using a Co/Ru liner instead of a Ta liner significantly improves the overall integrity of the TaN barrier [45]. However, due to the step coverage limit and the overhang problem associated with PVD, it is necessary to switch to ALD as the dimensions decrease. However, TaN thermally deposited using ALD has a higher number of impurities than PVD-fabricated TaN films, a lower film density, a higher resistance, and a lower interface quality, meaning that it cannot emulate the performance of PVD TaN at the same thickness. For this reason, PVD Ta and ALD TaN have been employed as bilayers (1 nm/1 nm) to reduce the thickness while ensuring the barrier properties [46]. Another approach is treating ALD-fabricated TaN in a PVD chamber to transform it into PVD-like film with optimal density and resistivity [47,48]. Using this method, the TaN thickness is 1.2 nm, compared with 1.5 nm for a Co liner and 2.0 nm for a Ru liner [48]. As a result, ALD has been combined with PVD to maintain the barrier properties and a low TaN thickness.

Scaling of the liner has also been actively carried out. The ALD/CVD method is required to overcome the step coverage and overhang problems, but a Ta liner cannot be used. Therefore, replacements for the Ta liner have been considered. Of these, Ru has received significant attention due to its suitability for ALD/CVD and the possibility of direct Cu plating without a PVD Cu seed layer [49–55]. In addition, Co has also been considered as a replacement for the Ta liner [56]. In [41], a Ru liner was found to be superior to a Co liner for Cu filling and electroplating with a liner thickness of 2 nm. However, the resistance of Ru was about 10% higher than that of Co as revealed by the temperature coefficient of resistance (TCR), which was believed to be due to interface and GB scattering. A Co liner has also been shown to outperform a Ru liner in terms of EM characteristics [57].

However, in order to fully exploit the excellent void-free gap-fill performance of Ru liners, approaches to improving their EM deficiencies have been reported. Co caps have been established as a standard process since the emergence of 14 nm technology. As a result, a significant improvement in EM performance has been observed with a Ru liner as the Co cap thickness increases [57]. It has also been demonstrated that the EM problem arises from the diffusion of Co from the cap to the liner due to the Co concentration gradient between the two [57–59]. Thus, a Co-doped Ru liner has been proposed to overcome this problem, and EM improvements have been reported [58,59]. Efforts to scale the TaN barrier and Co/Ru liner continue, but if the metal half-pitch decreases below 10 nm, the barrier/liner scaling limit is reached.

4.2. Selective Barrier Schemes

As advanced technologies emerge, the importance of the resistance and reliability of the vias increases. In particular, via resistance is strongly affected by the bottom thickness of the TaN. Therefore, as scaling progresses, the via resistance rapidly increases due to the thicker TaN. Furthermore, one of the most critical areas of a Cu-based DD structure is the bottom of the via, where the two metal levels meet [60], so selective deposition control of the via bottom barrier is advantageous for improved reliability. In this vein, argon (Ar) sputtering was used as a precleaning PVD barrier/liner [60]. The barrier-first process was conducted in the order of TaN/Ar sputtering/Ta instead of Ar sputtering/TaN/Ta to remove TaN in the via bottom and obtain a Cu/Ta/Cu interface. However, this process disappeared when the precleaning method for removing Cu oxide was changed from Ar sputtering to the oxide-reduction method.

Now that the via size is much smaller, self-assembled monolayers (SAM) have been pursued [61,62]. Figure 7 presents the process for the formation of a selective barrier on the via bottom. The Cu surface is passivated by depositing a SAM on the Cu surface subjected to the CuO_x reduction treatment. The SAM should be highly selective and self-limited and be adsorbed onto metals but not other peripheral dielectrics. However, when the SAM attaches to the metal surface, it interferes with the adsorption of ALD TaN precursors, resulting in significant nucleation delays in ALD TaN growth [61,62]. As a result, ALD TaN does not deposit on the via bottom area where the SAM is located. The SAM is finally removed. The SAM process needs to be developed further before it can be adopted for mass production. Nevertheless, because it can achieve a reduction in the via resistance of about 50% and its TDDB and EM results are equivalent to the process of record (POR), it is likely to be useful in future advanced technologies [61].

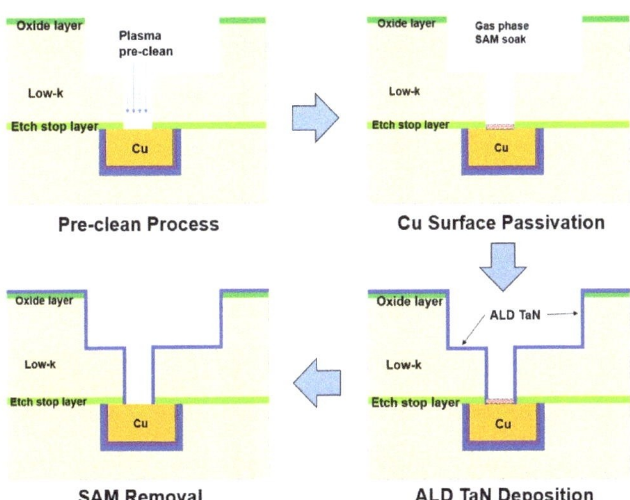

Figure 7. Selective barrier integration process. Reprinted/adapted with permission from Ref. [61]. 2021, IEEE.

4.3. Self-Forming Barriers

Another promising approach to Cu extendibility is the through Co self-forming barrier (tCoSFB), a method in which manganese (Mn) atoms added to the seed Cu diffuse through the thin Co liner layer to form a strong diffusion barrier at the interface between the trench and the dielectric, as seen in Figure 8 [63–66].

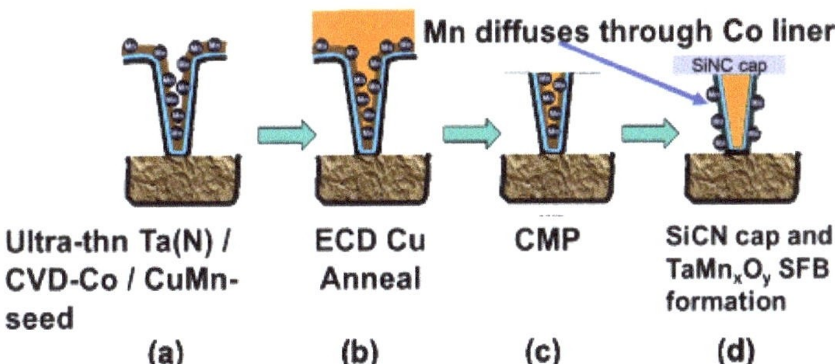

Figure 8. Through Co self-forming barrier (tCoSFB) structure and fabrication process. (**a**) an ultra-thin Ta(N)/CVD-Co film stack is formed on the patterned ULK pSiCOH surfaces, followed by a high-Mn% PVD-Cu(Mn) seed layer. (**b**) Cu electroplating and post-plating anneal. (**c**) CMP. (**d**) a PECVD SiCN(H) dielectric cap is deposited. Reprinted/adapted with permission from Ref. [63]. 2015, IEEE.

Initially, the Mn dopant is employed to form $MnSi_xO_y$ at the dielectric and Cu interface [64,65,67,68]. If CuMn seed deposition occurs without a barrier/liner and Cu plating and annealing are conducted, an SFB is created. This $MnSi_xO_y$ barrier is very thin and uniform, and its EM and TDDB performance is similar to that of POR. However, vertical trench triangular voltage sweep (VT-TVS) and O_2 barrier tests indicate that $MnSi_xO_y$ is inadequate for blocking Cu and O_2 diffusion, which led to the switch from the SFB to tCoSFB containing thin Ta(N) and Co [69].

The advantage of the tCoSFB process is that the cross-sectional area of Cu in the wiring can be maximized, and low line resistance can be obtained due to the 1 nm thickness of the Co liner and Ta barrier. However, because the line resistance and reliability balance are controlled by the Mn concentration of Cu(Mn) PVD, an increase in resistance may occur. Nevertheless, compared with the POR process, the results are encouraging, with improved resistance and EM and TDDB performance [66,70].

4.4. Hybrid Metallization

At the system-on-chip level, vias have become much more important for signal routing, and via resistance has increased significantly when downscaling the bottom contact area of the via. Therefore, introducing barrierless metal to the via is valuable, and metal–metal selective deposition is essential for the high-aspect-ratio (AR) vias. Hybrid metallization is a method of pre-filling a via with a barrierless metal and then filling the remaining metal area using Cu metallization, as seen in Figure 9 [71].

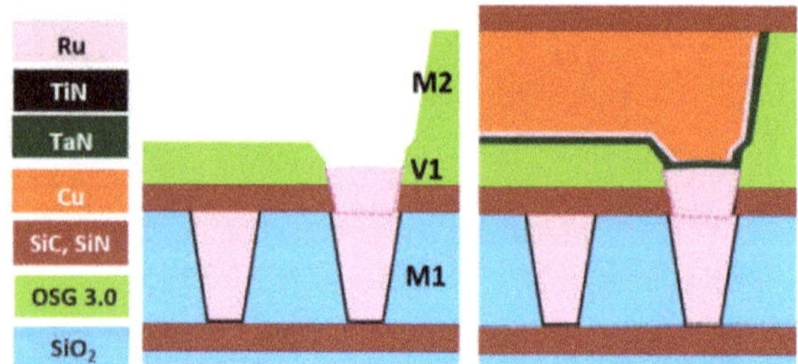

Figure 9. Schematic diagram of Cu hybrid metallization in a two-metal system: (**left**) Ru pre-fill on Ru followed by (**right**) barrier/liner Cu trench metallization. Reprinted/adapted with permission from Ref. [71]. 2020, IEEE.

A barrierless prefill has a number of advantages. First, the prefill moves the barrier position of the Cu DD from the bottom to the top of the via, increasing the tapered via cross-section and reducing the resistance. Second, the gap–fill margin increases because SD trench metallization is employed. Third, because the high-AR vias are excluded, step coverage can be achieved with a thinner barrier/liner [72]. In addition, the optimized barrierless prefilling process may reduce RC delay by increasing the height of the vias and trenches, which can significantly improve circuit performance [73].

The first reported via prefill used Co [73]. However, due to Co ion drift, it was found that barrierless Co was not possible and that TiN barriers are required [72,74]. Ru has been found to be suitable for barrierless prefilling, resulting in a 40% reduction in via resistance and EM results that are similar to those for POR [71].

4.5. Alternative Metals

The thickness of the barrier layer cannot arbitrarily be reduced. If the thickness is less than a certain threshold, it no longer functions as a Cu diffusion barrier. Based on past experimental results, it is unclear whether the barrier/liner combination can have a thickness lower than 2 nm [45–47,75]. However, a barrierless solution is required when the half pitch falls below 10 nm, and thus, new metals to replace Cu have been sought. Some of these metals have a larger bulk resistivity than Cu but do not require a thick barrier/liner and exhibit lower resistance at sufficiently small dimensions because their inelastic average free path is shorter than that of Cu (39 nm) [76,77]. Examples of metals and substitutes proposed under these conditions include Ir, Rh, Mo [78], and W [79], but experimentally verified cases are rare except for Co and Ru [80–83].

Figure 10 displays the line resistance derived from the TCR according to the cross-sectional area of the conductor. Cu has a higher resistance than Ru and Co below ~400 nm^2 at a 16-nm metal width and an AR of 2 [84].

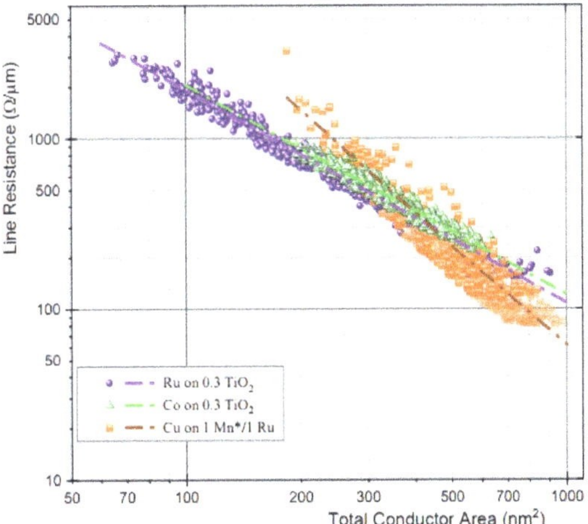

Figure 10. Logarithmic comparison of the damascene line resistance versus the total conductor cross-sectional area for Ru, Co, and Cu nanowires as determined using the TCR. The dotted lines are used to highlight the general trend. Reprinted/adapted with permission from Ref. [84]. 2018, IEEE.

4.6. Ru Semi-Damascene Schemes

Ru can be used in barrierless damascene structures and direct metal etching (DME). Figure 11 shows Ru metal implemented with subtractive etching.

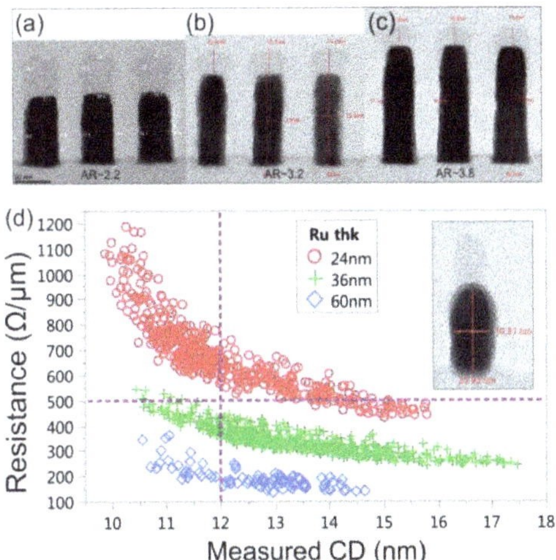

Figure 11. Cross-sectional TEM of patterned Ru films with a thickness of (**a**) 24 nm, (**b**) 36 nm, and (**c**) 60 nm. (**d**) Corresponding R vs. CD for all structures on each wafer. Reprinted/adapted with permission from Ref. [85]. 2018, IEEE.

The use of subtractive etching opens up many possibilities that have not previously been possible for Cu damascene structures. First, because the Ru film is deposited on the entire wafer, the grain size is not limited by the damascene profile width. Therefore, it can significantly restrain the increase in resistance due to GB scattering. Second, the metal thickness is determined by Ru deposition, not by metal CMP. Therefore, if necessary, the resistance can be reduced by increasing the Ru thickness, and there is no AR-related filling problem. Third, because the metal thickness is not affected by the pattern density but is determined by the uniformity of the deposited Ru film, the variation in the metal Rs will be reduced. Finally, a space between Ru metal with an increase in the AR could be introduced by employing an intentional air gap.

The semi-damascene process is an interesting approach that utilizes the advantages of Ru, such as barrierless designs and subtractive etching. This method begins with vias fabricated in a low-k material, and the vias and trench layers are filled using a single deposition step. Subtractive etching is then applied to the trench patterning [14,85–87]. According to the latest result, Ru line resistance outperforms Cu at 270 nm^2 or lower (CD < 12 nm) based on line resistance vs. conducting area plot, and the EM and TDDB results are promising so far [17]. For this reason, the Ru semi-damascene process is the most competitive candidate for use in 3 nm technology and lower.

Figure 12 shows the conceptual summary of the metallization approaches mentioned in this section as they have evolved. As shown in the figure, the main trend of metallization has been to reduce the area ratio occupied by the barrier in the scaled wire. Traditional attempts were to reduce the existing barrier/liner (TaN/Ta) thickness by replacing a new process method (ALD) and a new liner material (Co/Ru). After that, it was gradually developed to implement a barrier-free process using Co or Ru. Moreover, the boundary has recently been extended to the subtraction etch process using Ru instead of the Cu damascene scheme.

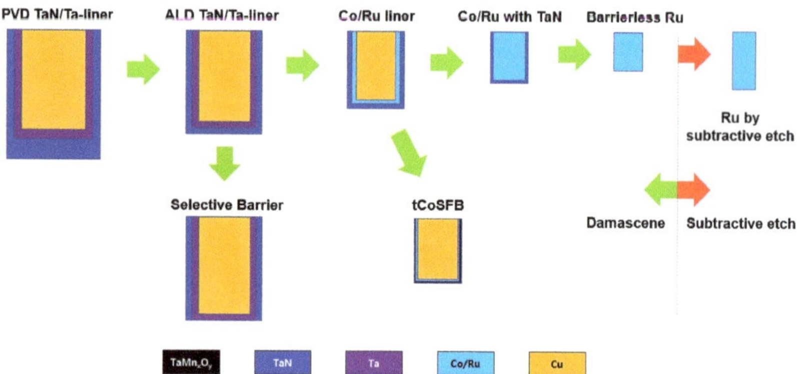

Figure 12. Trends in recent metallization approaches.

In addition to the metallization approaches mentioned here, various studies are being conducted. One of them is the study of a 2D-like barrier. The approach is to minimize the volume fraction of Cu occupied by the barrier by replacing the thick Ta or TaN barrier with 2D-like materials such as graphene [88,89], MoS$_2$ [41,90], TaS$_2$ [41], and WSe$_2$ [41].

In the process of these efforts, the limitation of Cu metallization in CMOS interconnects is becoming more apparent. However, it is interesting that Cu plating is actively applied in other areas, such as solar cells' silicon heterojunction (SHJ) [91,92].

5. Reliability

Recognizing the underlying causes of reliability problems in scaled Cu/low-k interconnects and clearly understanding the degradation mechanisms are essential. EM and TDDB are standard methods for assessing the reliability of metal and dielectrics. Therefore, EM and TDDB are summarized in this section, and countermeasures proposed for improving reliability in the pursuit of scaling are discussed.

5.1. Electromigration

EM is a phenomenon in which metal atoms move in a metal conductor due to a high current density [93]. When EM stress is continuously applied, voids appear on the cathode side of the wire as the metal atoms accumulate on the anode side, resulting in hydrostatic stress. This stress also produces an inverse flux of atoms in the opposite direction to the electron transport flux, which is referred to as the Blech effect or the short-length effect [94,95]. This back-stress force is more evident as the wire length decreases, preventing EM failure because it prevents the formation of a void when it is shorter than the critical threshold length [94,95].

In a Cu damascene structure, the TaN/Ta barrier and the DB act as a boundary for the Cu, and this is where the depletion of metal atoms begins. As a result, Cu EM shifts into the voids through the nucleation, incubation, and growth phases [96]. The dimensions of the vias and metal determine the likelihood that the voids produced in this manner will affect the resistance. Therefore, early failure initiates near the via bottom, and late failure is observed as a void in the metal line.

EM testing is performed under high-current-density and high-temperature conditions to obtain a failure time for when the resistance increases experimentally. The failure time (t) is widely described using Black's equation [97]:

$$t = Aj^{-n} \exp\left(\frac{E_a}{kT}\right) \tag{4}$$

where j is the current density, E_a is the activation energy for diffusion, k is the Boltzmann constant, T is the temperature, A is a constant, and n is the current exponent, whose value is typically between 1 and 2.

The activation energy for diffusion depends on the specific metal atoms involved and the diffusion path. Table 2 summarizes the activation energy for Al and Cu, with the diffusion paths divided into bulk, GB, interface, and surface. It is expected that the lower the activation energy, the faster the failure time detection will be.

Table 2. Activation energy for different diffusion paths for Al, Al/Cu, and Cu metal [12].

Metal	Activation Energy for Different Diffusion Paths (Ev)			
	Bulk	Grain Boundary	Surface	Interface
Al	1.4	0.4~0.5	NA	–
Al/Cu (Alloy)	1.2	0.6~0.7	NA	0.9~1.1
Cu	2.1	1.1~1.2	0.6~0.7	0.8~1.3

GB diffusion is the fastest path for Al, while surface diffusion is the fastest path for Cu due to the low activation energy. Therefore, to improve the EM lifetime, it is necessary to improve the paths with the lowest activation energy. For example, for Cu, the surface and interface should be improved first. For the surface diffusion path, strengthening the adhesion between Cu and the DB is the most important goal, with HN_3 plasma employed to remove Cu oxidation. CoWP [98] is then added, along with a Co cap [42,99] to maintain EM performance and meet the higher current density requirements of advanced technology generations. Similarly, at the interface, there is an adhesion issue between the liner and Cu. Therefore, EM improvement must be verified when evaluating new materials such as Co and Ru liners and reducing barrier/liner thickness [46,47,57–59,75]. For the GBs, the lower the metal width, the more difficult it is to improve the EM because the grain growth is limited by the metal width. However, it has been reported that the grain size increases with a longer annealing time when Cu reflowing is employed, although the result is not yet clear [48]. Finally, there is a method of improving EM by depositing doping impurities together with the seed Cu. Various metals (e.g., Al, Ag, Mn, Mg, and Zr) have been tested in this respect, and Mn is currently the most commonly used. However, the resistance increases as the doping concentration increases, so optimizing EM and resistance are required.

5.2. Time-Dependent Dielectric Breakdown

When long-term stress is applied in a strong electric field, electrical damage occurs to the dielectric material, and insulation is lost, ultimately leading to TDDB [100]. This TDDB has been treated as an essential reliability item in the gate oxide. In the past BEOL interconnects, the size of the electric field applied to the insulator did not become a problem because of the excellent insulation characteristics of SiO_2 and sufficient gaps between wirings. However, as scaling progressed, the size of the electric field across the dielectric gradually increased, and porous low-k materials were introduced, leading to TDDB becoming an essential measure of reliability for BEOL interconnects [101]. The breakdown strength of a porous low-k material decreases as the dielectric constant decreases.

The pores in a porous low-k material shorten the percolation paths, weaken the bonds at the metal−insulator interface, and have a high trap density. The interface between the low-k material and the DB is the most problematic in this respect because of the possibility of a high trap density due to CMP, bond mismatches due to material differences, and the shorter distances arising from a tapered morphology [102]. In addition, a breakdown current may be generated by the defects in the porous low-k material in its as-deposited state or following plasma damage received during the fabrication process such as RIE etch.

In addition to these intrinsic causes, Cu metallization also affects TDDB. Cu diffusion into the dielectric may occur through the metal barrier and DB. Thus, TDDB should be evaluated when applying a change in the characteristics or thickness of a new barrier material. Since the TDDB test is performed at a deliberately high acceleration voltage, a TDDB lifetime model is required for predicting the actual lifetime at the standard operating voltage from the TDDB measurement results. E and $E^{1/2}$ models are widely used as standards for TDDB reliability testing, which is important because they are the most conservative predictors of low-k TDDB [103]. Of these, the $E^{1/2}$ model is most capable of explaining low-k TDDB behavior due to Cu diffusion.

6. Summary

Because introducing a new dielectric material with a lower *k* value to microchip interconnects is likely to prove difficult, efforts to suppress the increase in RC delay due to scaling have concentrated on the metallization process. Furthermore, more innovative approaches to metallization are required when the minimum metal pitch required by the process node is less than 40 nm (10 nm technology or lower).

These innovative approaches, as mentioned earlier, are traditionally moving toward implementing a barrier-less method by introducing new materials from an approach that reduces the thickness of the barrier/liner. Moreover, in recent years, due to the limitations of Cu damascene structures, subtractive etching has been employed as a process option. However, no single metallization approach can satisfy all BEOL interconnect levels simultaneously. Therefore, depending on the purpose of each level, maximum chip performance can be achieved by selectively employing the various metallization options.

As the scaling of interconnects continues, it is doubtful that current metallization approaches will continue to satisfy the requirements for minimizing the RC delay. As in the case of low-k materials, it may no longer be practical to pursue the scaling of metallization. Rather, the innovative approaches proposed to date can contribute to scaling requirements not through direct scaling but with architectural innovations such as super vias and buried power rails.

Funding: This work is supported by the "Ling Yan" Program for Tackling Key Problems in Zhejiang Province under Grant number of No. 2022C01098.

Data Availability Statement: No new data were created or analyzed in this study. Data sharing is not applicable to this article.

Conflicts of Interest: The author declares no conflict of interest.

References

1. Davis, J.A.; Venkatesan, R.; Kaloyeros, A.; Beylansky, M.; Souri, S.; Banerjee, K.; Saraswat, K.; Rahman, A.; Reif, R.; Meindl, J. Interconnect limits on gigascale integration (GSI) in the 21st century. *Proc. IEEE* **2001**, *89*, 305–324. [CrossRef]
2. Bonilla, G.; Lanzillo, N.; Hu, C.-K.; Penny, C.; Kumar, A. Interconnect scaling challenges, and opportunities to enable system-level performance beyond 30 nm pitch. In Proceedings of the 2020 IEEE International Electron Devices Meeting (IEDM), San Francisco, CA, USA, 12–18 December 2020. [CrossRef]
3. Theis, T.N. The future of interconnection technology. *IBM J. Res. Dev.* **2000**, *44*, 379–390. [CrossRef]
4. Havemann, R.; Hutchby, J. High-performance interconnects: An integration overview. *Proc. IEEE* **2001**, *89*, 586–601. [CrossRef]
5. Baccarani, G.; Wordeman, M.; Dennard, R. Generalized scaling theory and its application to a $\frac{1}{4}$ micrometer MOSFET design. *IEEE Trans. Electron Devices* **1984**, *31*, 452–462. [CrossRef]
6. Cheng, Y.-L.; Lee, C.-Y. Porous low-dielectric-constant material for semiconductor microelectronics. In *Nanofluid Flow in Porous Media*; IntechOpen: London, UK, 2018. [CrossRef]
7. Saito, T.; Imai, T.; Noguchi, J.; Kubo, M.; Ito, Y.; Omori, S.; Ohashi, N.; Tamaru, T.; Yamaguchi, H. A high performance liner for copper damascene interconnects. In Proceedings of the IEEE 2001 International Interconnect Technology Conference (Cat. No. 01EX461), Burlingame, CA, USA, 6 June 2001. [CrossRef]
8. Edelstein, D.; Heidenreich, J.; Goldblatt, R.; Cote, W.; Uzoh, C.; Lustig, N.; Roper, P.; McDevitt, T.; Motsiff, W.; Simon, A.; et al. Full copper wiring in a sub-0.25/spl mu/m CMOS ULSI technology. In Proceedings of the International Electron Devices Meeting. IEDM Technical Digest, Washington, DC, USA, 10 December 1997. [CrossRef]
9. Grill, A.; Gates, S.M.; Ryan, T.E.; Nguyen, S.V.; Priyadarshini, D. Progress in the development and understanding of advanced low k and ultralow k dielectrics for very large-scale integrated interconnects—State of the art. *Appl. Phys. Rev.* **2014**, *1*, 011306. [CrossRef]
10. Grill, A. Porous pSiCOH ultralow-k dielectrics for chip interconnects prepared by PECVD. *Annu. Rev. Mater. Res.* **2009**, *39*, 49–69. [CrossRef]
11. Lionti, K.; Volksen, W.; Magbitang, T.; Darnon, M.; Dubois, G.; Lyu, Y.-R.; Hsieh, T.-E. Toward successful integration of porous low-k materials: Strategies addressing plasma damage. *ECS J. Solid State Sci. Technol.* **2014**, *4*, N3071. [CrossRef]
12. Cheng, Y.-L.; Lee, C.-Y.; Huang, Y.-L. Copper metal for semiconductor interconnects. In *Noble and Precious Metals-Properties, Nanoscale Effects and Applications*; IntechOpen: London, UK, 2018; pp. 220–221. [CrossRef]
13. Simon, A.; van der Straten, O.; Lanzillo, N.A.; Yang, C.-C.; Nogami, T.; Edelstein, D.C. Role of high aspect-ratio thin-film metal deposition in Cu back-end-of-line technology. *J. Vac. Sci. Technol. A Vac. Surf. Film.* **2020**, *38*, 053402. [CrossRef]
14. Croes, K.; Adelmann, C.; Wilson, C.J.; Zahedmanesh, H.; Pedreira, O.V.; Wu, C.; Leśniewska, A.; Oprins, H.; Beyne, S.; Ciofi, I.; et al. Interconnect metals beyond copper: Reliability challenges and opportunities. In Proceedings of the 2018 IEEE International Electron Devices Meeting (IEDM), San Francisco, CA, USA, 1–5 December 2018. [CrossRef]
15. Decoster, S.; Camerotto, E.; Murdoch, G.; Kundu, S.; Le, Q.T.; Tőkei, Z.; Jurczak, G.; Lazzarino, F. Patterning challenges for direct metal etch of ruthenium and molybdenum at 32 nm metal pitch and below. *J. Vac. Sci. Technol. B Nanotechnol. Microelectron. Mater. Process. Meas. Phenom.* **2022**, *40*, 032802. [CrossRef]
16. Paolillo, S.; Wan, D.; Lazzarino, F.; Rassoul, N.; Piumi, D.; Tőkei, Z. Direct metal etch of ruthenium for advanced interconnect. *J. Vac. Sci. Technol. B Nanotechnol. Microelectron. Mater. Process. Meas. Phenom.* **2018**, *36*, 03E103. [CrossRef]
17. Murdoch, G.; O'Toole, M.; Marti, G.; Pokhrel, A.; Tsvetanova, D.; Decoster, S.; Kundu, S.; Oniki, Y.; Thiam, A.; Le, Q.; et al. First demonstration of Two Metal Level Semi-damascene Interconnects with Fully Self-aligned Vias at 18MP. In Proceedings of the 2022 IEEE Symposium on VLSI Technology and Circuits (VLSI Technology and Circuits), Honolulu, HI, USA, 13–17 June 2022. [CrossRef]
18. Gambino, J.; Chen, F.; He, J. Copper interconnect technology for the 32 nm node and beyond. In Proceedings of the 2009 IEEE Custom Integrated Circuits Conference, San Jose, CA, USA, 13–16 September 2009. [CrossRef]
19. Kriz, J.; Angelkort, C.; Czekalla, M.; Huth, S.; Meinhold, D.; Pohl, A.; Schulte, S.; Thamm, A.; Wallace, S. Overview of dual damascene integration schemes in Cu BEOL integration. *Microelectron. Eng.* **2008**, *85*, 2128–2132. [CrossRef]
20. Lloyd, J.; Murray, C.; Ponoth, S.; Cohen, S.; Liniger, E. The effect of Cu diffusion on the TDDB behavior in a low-k interlevel dielectrics. *Microelectron. Reliab.* **2006**, *46*, 1643–1647. [CrossRef]
21. Tsu, R.; McPherson, J.W.; McKee, W.R. Leakage and breakdown reliability issues associated with low-k dielectrics in a dual-damascene Cu process. In Proceedings of the 2000 IEEE International Reliability Physics Symposium Proceedings 38th Annual (Cat. No. 00CH37059), San Jose, CA, USA, 10–13 April 2000. [CrossRef]
22. McPherson, J.W. Time dependent dielectric breakdown in copper low-k interconnects: Mechanisms and reliability models. *Materials* **2012**, *5*, 1602–1625. [CrossRef]
23. Edelstein, D.; Uzoh, C.; Cabral, C.; DeHaven, P.; Buchwalter, P.; Simon, A.; Cooney, E.; Malhotra, S.; Klaus, D.; Rathore, H.; et al. A high performance liner for copper damascene interconnects. In Proceedings of the IEEE 2001 International Interconnect Technology Conference (Cat. No. 01EX461), Burlingame, CA, USA, 6 June 2001. [CrossRef]
24. Shamiryan, D.; Abell, T.; Iacopi, F.; Maex, K. Low-k dielectric materials. *Mater. Today* **2004**, *7*, 34–39. [CrossRef]
25. Maex, K.; Baklanov, M.R.; Shamiryan, D.; Iacopi, F.; Brongersma, S.H.; Yanovitskaya, Z.S. Low dielectric constant materials for microelectronics. *J. Appl. Phys.* **2003**, *93*, 8793–8841. [CrossRef]

26. Cheng, Y.; Wang, Y.; Liu, C.; Wu, Y.; Lo, K.; Lan, J. Characterization and reliability of low dielectric constant fluorosilicate glass and silicon rich oxide process for deep sub-micron device application. *Thin Solid Films* **2001**, *398*, 533–538. [CrossRef]

27. Hoofman, R.; Verheijden, G.; Michelon, J.; Iacopi, F.; Travaly, Y.; Baklanov, M.; Tökei, Z.; Beyer, G. Challenges in the implementation of low-k dielectrics in the back-end of line. *Microelectron. Eng.* **2005**, *80*, 337–344. [CrossRef]

28. Sankaran, S.; Arai, S.; Augur, R.; Beck, M.; Biery, G.; Bolom, T.; Bonilla, G.; Bravo, O.; Chanda, K.; Chae, M.; et al. A 45 nm CMOS node Cu/Low-k/Ultra Low-k PECVD SiCOH (k = 2.4) BEOL technology. In Proceedings of the 2006 International Electron Devices Meeting, San Francisco, CA, USA, 11–13 December 2006. [CrossRef]

29. Travaly, Y.; van Aelst, J.; Truffert, V.; Verdonck, P.; Dupont, T.; Camerotto, E.; Richard, O.; Bender, H.; Kroes, C.; de Roest, D.; et al. Key factors to sustain the extension of a MHM-based integration scheme to medium and high porosity PECVD low-k materials. In Proceedings of the 2008 International Interconnect Technology Conference, Burlingame, CA, USA, 1–4 June 2008. [CrossRef]

30. King, S.W. Dielectric barrier, etch stop, and metal capping materials for state of the art and beyond metal interconnects. *ECS J. Solid State Sci. Technol.* **2014**, *4*, N3029. [CrossRef]

31. Cheng, Y.-L.; Chen, S.-A.; Chiu, T.-J.; Wu, J.; Wei, B.-J.; Chang, H.-J. Electrical and reliability performances of nitrogen-incorporated silicon carbide dielectric by chemical vapor deposition. *J. Vac. Sci. Technol. B Nanotechnol. Microelectron. Mater. Process. Meas. Phenom.* **2010**, *28*, 573–576. [CrossRef]

32. Cheng, Y.-L.; Chiu, T.-J.; Wei, B.-J.; Wang, H.-J.; Wu, J.; Wang, Y.-L. Effect of copper barrier dielectric deposition process on characterization of copper interconnect. *J. Vac. Sci. Technol. B Nanotechnol. Microelectron. Mater. Process. Meas. Phenom.* **2010**, *28*, 567–572. [CrossRef]

33. Nitta, S.; Edelstein, D.; Ponoth, S.; Clevenger, L.; Liu, X.; Standaert, T. Performance and reliability of airgaps for advanced BEOL interconnects. In Proceedings of the 2008 International Interconnect Technology Conference, Burlingame, CA, USA, 1–4 June 2008. [CrossRef]

34. Chen, H.-W.; Jeng, S.-P.; Tsai, H.-Y.; Liu, Y.-W.; Yu, C.; Sun, Y. A self-aligned air gap interconnect process. In Proceedings of the 2008 International Interconnect Technology Conference, Burlingame, CA, USA, 1–4 June 2008. [CrossRef]

35. Nakamura, N.; Matsunaga, N.; Kaminatsui, T.; Watanabe, K.; Shibata, H. Cost-effective air-gap interconnects by all-in-one post-removing process. In Proceedings of the 2008 International Interconnect Technology Conference, Burlingame, CA, USA, 1–4 June 2008. [CrossRef]

36. Uno, S.; Noguchi, J.; Ashihara, H.; Oshima, T.; Sato, K.; Konishi, N.; Saito, T.; Hara, K. Dual damascene process for air-gap Cu interconnects using conventional CVD films as sacrificial layers. In Proceedings of the IEEE 2005 International Interconnect Technology Conference, Burlingame, CA, USA, 6–8 June 2005. [CrossRef]

37. Noguchi, J.; Sato, K.; Konishi, N.; Uno, S.; Oshima, T.; Ishikawa, K.; Ashihara, H.; Saito, T.; Kubo, M.; Tamaru, T.; et al. Process and reliability of air-gap Cu interconnect using 90-nm node technology. *IEEE Trans. Electron Devices* **2005**, *52*, 352–359. [CrossRef]

38. Fischer, K.; Agostinelli, M.; Allen, C.; Bahr, D.; Bost, M.; Charvat, P.; Chikarmane, V.; Fu, Q.; Ganpule, C.; Haran, M.; et al. Low-k interconnect stack with multi-layer air gap and tri-metal-insulator-metal capacitors for 14nm high volume manufacturing. In Proceedings of the 2015 IEEE International Interconnect Technology Conference and 2015 IEEE Materials for Advanced Metallization Conference (IITC/MAM), Grenoble, France, 18–21 May 2015. [CrossRef]

39. Fischer, K.; Chang, H.; Ingerly, D.; Jin, I.; Kilambi, H.; Longun, J.; Patel, R.; Pelto, C.; Petersburg, C.; Plekhanov, P.; et al. Performance enhancement for 14nm high volume manufacturing microprocessor and system on a chip processes. In Proceedings of the 2016 IEEE International Interconnect Technology Conference/Advanced Metallization Conference (IITC/AMC), San Jose, CA, USA, 23–26 May 2016. [CrossRef]

40. Penny, C.; Gates, S.; Peethala, B.; Lee, J.; Priyadarshini, D.; Nguyen, S.; McLaughlin, P.; Liniger, E.; Hu, C.-K.; Clevenger, L.; et al. Reliable airgap BEOL technology in advanced 48 nm pitch copper/ULK interconnects for substantial power and performance benefits. In Proceedings of the 2017 IEEE International Interconnect Technology Conference (IITC), Hsinchu, Taiwan, 16–18 May 2017. [CrossRef]

41. Lo, C.-L.; Helfrecht, B.A.; He, Y.; Guzman, D.M.; Onofrio, N.; Zhang, S.; Weinstein, D.; Strachan, A.; Chen, Z. Opportunities and challenges of 2D materials in back-end-of-line interconnect scaling. *J. Appl. Phys.* **2020**, *128*, 080903. [CrossRef]

42. Chang, S.Y.; Wan, C.C.; Wang, Y.Y. Selectivity enhancement of electroless Co deposition for Cu capping process via spontaneous diazonium ion reduction. *Electrochem. Solid-State Lett.* **2007**, *10*, D43. [CrossRef]

43. Shimada, M.; Moriyama, M.; Ito, K.; Tsukimoto, S.; Murakami, M. Electrical resistivity of polycrystalline Cu interconnects with nano-scale linewidth. *J. Vac. Sci. Technol. B Microelectron. Nanometer Struct. Process. Meas. Phenom.* **2006**, *24*, 190–194. [CrossRef]

44. Steinhögl, W.; Schindler, G.; Steinlesberger, G.; Engelhardt, M. Size-dependent resistivity of metallic wires in the mesoscopic range. *Phys. Rev. B* **2002**, *66*, 075414. [CrossRef]

45. Witt, C.; Yeap, K.; Lesniewska, A.; Wan, D.; Jordan, N.; Ciofi, I.; Wu, C.; Tokei, Z. Testing the limits of TaN barrier scaling. In Proceedings of the 2018 IEEE International Interconnect Technology Conference (IITC), Santa Clara, CA, USA, 4–7 June 2018. [CrossRef]

46. Motoyama, K.; van der Straten, O.; Maniscalco, J.; Huang, H.; Kim, Y.; Choi, J.; Lee, J.; Hu, C.-K.; McLaughlin, P.; Standaert, T.; et al. Ru liner scaling with ALD TaN barrier process for low resistance 7 nm Cu interconnects and beyond. In Proceedings of the 2018 IEEE International Interconnect Technology Conference (IITC), Santa Clara, CA, USA, 4–7 June 2018. [CrossRef]

47. Wu, Z.; Li, R.; Xie, X.; Suen, W.; Tseng, J.; Bekiaris, N.; Vinnakota, R.; Kashefizadeh, K.; Naik, M. PVD-treated ALD TaN for Cu interconnect extension to 5 nm node and beyond. In Proceedings of the 2018 IEEE International Interconnect Technology Conference (IITC), Santa Clara, CA, USA, 4–7 June 2018. [CrossRef]
48. Bhosale, P.S.; Maniscalco, J.; Lanzillo, N.; Nogami, T.; Canaperi, D.; Motoyama, K.; Huang, H.; McLaughlin, P.; Shaviv, R.; Stolfi, M.; et al. Modified ALD TaN barrier with Ru liner and dynamic Cu reflow for 36nm pitch interconnect integration. In Proceedings of the 2018 IEEE International Interconnect Technology Conference (IITC), Santa Clara, CA, USA, 4–7 June 2018. [CrossRef]
49. Yang, C.-C.; Spooner, T.; Ponoth, S.; Chanda, K.; Simon, A.; Lavoie, C.; Lane, M.; Hu, C.-K.; Liniger, E.; Gignac, L.; et al. Physical, electrical, and reliability characterization of Ru for Cu interconnects. In Proceedings of the 2006 International Interconnect Technology Conference, Burlingame, CA, USA, 5–7 June 2006. [CrossRef]
50. Abe, M.; Ueki, M.; Tada, M.; Onodera, T.; Furutake, N.; Shimura, K.; Saito, S.; Hayashi, Y. Highly-oriented PVD ruthenium liner for low-resistance direct-plated Cu interconnects. In Proceedings of the 2007 IEEE International Interconnect Technology Conference, Burlingame, CA, USA, 4–6 June 2007. [CrossRef]
51. Mori, K.; Ohmori, K.; Torazawa, N.; Hirao, S.; Kaneyama, S.; Korogi, H.; Maekawa, K.; Fukui, S.; Tomita, K.; Inoue, M.; et al. Effects of Ru-Ta alloy barrier on Cu filling and reliability for Cu interconnects. In Proceedings of the 2008 International Interconnect Technology Conference, Burlingame, CA, USA, 1–4 June 2008. [CrossRef]
52. Huang, H.Y.; Hsieh, C.H.; Jeng, S.M.; Tao, H.J.; Cao, M.; Mii, Y.J. A new enhancement layer to improve copper interconnect performance. In Proceedings of the 2010 IEEE International Interconnect Technology Conference, Burlingame, CA, USA, 6–9 June 2010. [CrossRef]
53. Rullan, J.; Ishizaka, T.; Cerio, F.; Mizuno, S.; Mizusawa, Y.; Ponnuswamy, T.; Reid, J.; McKerrow, A.; Yang, C.-C. Low resistance wiring and 2Xnm void free fill with CVD Ruthenium liner and DirectSeed TM copper. In Proceedings of the 2010 IEEE International Interconnect Technology Conference, Burlingame, CA, USA, 6–9 June 2010. [CrossRef]
54. Yang, C.-C.; Cohen, S.; Shaw, T.; Wang, P.-C.; Nogami, T.; Edelstein, D. Characterization of "Ultrathin-Cu"/Ru (Ta)/TaN liner stack for copper interconnects. *IEEE Electron Device Lett.* **2010**, *31*, 722–724. [CrossRef]
55. Swerts, J.; Armini, S.; Carbonell, L.; Delabie, A.; Franquet, A.; Mertens, S.; Popovici, M.; Schaekers, M.; Witters, T.; Tökei, Z.; et al. Scalability of plasma enhanced atomic layer deposited ruthenium films for interconnect applications. *J. Vac. Sci. Technol. A Vac. Surf. Film.* **2012**, *30*, 01A103. [CrossRef]
56. He, M.; Zhang, X.; Nogami, T.; Lin, X.; Kelly, J.; Kim, H.; Spooner, T.; Edelstein, D.; Zhao, L. Mechanism of Co liner as enhancement layer for Cu interconnect gap-fill. *J. Electrochem. Soc.* **2013**, *160*, D3040. [CrossRef]
57. Wu, Z.; Chen, F.; Shen, G.; Hu, Y.; Pethe, S.; Lee, J.J.; Tseng, J.; Suen, W.; Vinnakota, R.; Kashefizadeh, K.; et al. Pathfinding of Ru-Liner/Cu-Reflow Interconnect Reliability Solution. In Proceedings of the 2018 IEEE International Interconnect Technology Conference (IITC), Santa Clara, CA, USA, 4–7 June 2018. [CrossRef]
58. Motoyama, K.; van der Straten, O.; Maniscalco, J.; Cheng, K.; DeVries, S.; Huang, H.; Shen, T.; Lanzillo, N.; Hosadurga, S.; Park, K.; et al. Co-doped Ru liners for highly reliable Cu interconnects with selective Co cap. In Proceedings of the 2020 IEEE International Interconnect Technology Conference (IITC), San Jose, CA, USA, 5–8 October 2020. [CrossRef]
59. Motoyama, K. EM performance improvements for Cu interconnects with Ru-based liner and Co cap in advanced nodes. In Proceedings of the 2021 IEEE International Interconnect Technology Conference (IITC), Kyoto, Japan, 6–9 July 2021. [CrossRef]
60. Alers, G.; Rozbicki, R.T.; Harm, G.J.; Kailasam, S.; Ray, G.W.; Danek, M. Barrier-first integration for improved reliability in copper dual damascene interconnects. In Proceedings of the IEEE 2003 International Interconnect Technology Conference (Cat. No. 03TH8695), Burlingame, CA, USA, 4 June 2003. [CrossRef]
61. You, S.; Ren, H.; Naik, M.; Chen, L.; Chen, F.; Cervantes, C.L.; Xie, X.; Kashefizadeh, K. Selective Barrier for Cu Interconnect Extension in 3nm Node and Beyond. In Proceedings of the 2021 IEEE International Interconnect Technology Conference (IITC), Kyoto, Japan, 6–9 July 2021. [CrossRef]
62. Kawasaki, H.; Iwashita, M.; Warashina, H.; Nagai, H.; Iwai, K.; Komatsu, H.; Ozaki, Y.; Pattanaik, G. Advanced Damascene integration using selective deposition of barrier metal with Self Assemble Monolayer. In Proceedings of the 2021 IEEE International Interconnect Technology Conference (IITC), Kyoto, Japan, 6–9 July 2021. [CrossRef]
63. Nogami, T.; Briggs, B.D.; Korkmaz, S.; Chae, M.; Penny, C.; Li, J.; Wang, W.; McLaughlin, P.S.; Kane, T.; Parks, C.; et al. Through-Cobalt Self Forming Barrier (tCoSFB) for Cu/ULK BEOL: A novel concept for advanced technology nodes. In Proceedings of the 2015 IEEE International Electron Devices Meeting (IEDM), Washington, DC, USA, 7–9 December 2015. [CrossRef]
64. Koike, J.; Haneda, M.; Iijima, J.; Wada, M. Cu alloy metallization for self-forming barrier process. In Proceedings of the 2006 International Interconnect Technology Conference, Burlingame, CA, USA, 5–7 June 2006. [CrossRef]
65. Siew, Y.K.; Jourdan, N.; Ciofi, I.; Croes, K.; Wilson, C.J.; Tang, B.J.; Demuynck, S.; Wu, Z.; Ai, H.; Cellier, D.; et al. Cu wire resistance improvement using mn-based self-formed barriers. In Proceedings of the IEEE International Interconnect Technology Conference, San Jose, CA, USA, 20–23 May 2014. [CrossRef]
66. Nogami, T.; Zhang, X.; Kelly, J.; Briggs, B.; You, H.; Patlolla, R.; Huang, H.; McLaughlin, P.; Lee, J.; Shobha, H.; et al. Comparison of key fine-line BEOL metallization schemes for beyond 7 nm node. In Proceedings of the 2017 Symposium on VLSI Technology, Kyoto, Japan, 5–8 June 2017. [CrossRef]
67. Usui, T.; Nasu, H.; Koike, J.; Wada, M.; Takahashi, S.; Shimizu, N.; Nishikawa, T.; Yoshimaru, A.; Shibata, H. Low resistive and highly reliable Cu dual-damascene interconnect technology using self-formed MnSi/sub x/O/sub y/barrier layer. In Proceedings of the IEEE 2005 International Interconnect Technology Conference, Burlingame, CA, USA, 6–8 June 2005. [CrossRef]

68. Usui, T.; Tsumura, K.; Nasu, H.; Hayashi, Y.; Minamihaba, G.; Toyoda, H.; Sawada, H.; Ito, S.; Miyajima, H.; Watanabe, K.; et al. High performance ultra low-k (k = 2.0/keff = 2.4)/Cu dual-damascene interconnect technology with self-formed MnSixOy barrier layer for 32 nm-node. In Proceedings of the 2006 International Interconnect Technology Conference, Burlingame, CA, USA, 5–7 June 2006. [CrossRef]

69. Nogami, T.; Chae, M.; Penny, C.; Shaw, T.; Shobha, H.; Li, J.; Cohen, S.; Hu, C.-K.; Zhang, X.; He, M.; et al. Performance of ultrathin alternative diffusion barrier metals for next-Generation BEOL technologies, and their effects on reliability. In Proceedings of the IEEE International Interconnect Technology Conference., San Jose, CA, USA, 20–23 May 2014. [CrossRef]

70. Nogami, T.; Huang, H.; Shobha, H.; Patlolla, R.; Kelly, J.; Penny, C.; Hu, C.-K.; Sil, D.; Devries, S.; Lee, J.; et al. Technology challenges and enablers to extend Cu metallization to beyond 7 nm node. In Proceedings of the 2019 Symposium on VLSI Technology, Kyoto, Japan, 9–14 June 2019. [CrossRef]

71. van der Veen, M.H.; Soethoudt, J.; Delabie, A.; Pedreira, O.V.; Gonzalez, V.V.; Lariviere, S.; Teugels, L.; Jourdan, N.; Decoster, S.; Struyf, H.; et al. Hybrid Metallization with Cu in sub 30nm Interconnects. In Proceedings of the 2020 IEEE International Interconnect Technology Conference (IITC), San Jose, CA, USA, 5–8 October 2020. [CrossRef]

72. van der Veen, M.H.; Vandersmissen, K.; Dictus, D.; Demuynck, S.; Liu, R.; Bin, X.; Nalla, P.; Lesniewska, A.; Hall, L.; Croes, K.; et al. Cobalt bottom-up contact and via prefill enabling advanced logic and DRAM technologies. In Proceedings of the 2015 IEEE International Interconnect Technology Conference and 2015 IEEE Materials for Advanced Metallization Conference (IITC/MAM), Grenoble, France, 18–21 May 2015. [CrossRef]

73. Zheng, J.-F.; Chen, P.; Baum, T.H.; Lieten, R.R.; Hunks, W.; Lippy, S.; Frye, A.; Li, W.; O'Neill, J.; Xu, J.; et al. Selective Co growth on Cu for void-free via fill. In Proceedings of the 2015 IEEE International Interconnect Technology Conference and 2015 IEEE Materials for Advanced Metallization Conference (IITC/MAM), Grenoble, France, 18–21 May 2015. [CrossRef]

74. Steinhögl, W.; Schindler, G.; Steinlesberger, G.; Traving, M.; Engelhardt, M. Comprehensive study of the resistivity of copper wires with lateral dimensions of 100 nm and smaller. *J. Appl. Phys.* **2005**, *97*, 023706. [CrossRef]

75. Lanzillo, N.A.; Yang, C.-C.; Motoyama, K.; Huang, H.; Cheng, K.; Maniscalco, J.; Van Der Straten, O.; Penny, C.; Standaert, T.; Choi, K. Exploring the limits of cobalt liner thickness in advanced copper interconnects. *IEEE Electron Device Lett.* **2019**, *40*, 1804–1807. [CrossRef]

76. Hu, C.-K.; Kelly, J.; Chen, J.H.-C.; Huang, H.; Ostrovski, Y.; Patlolla, R.; Peethala, B.; Adusumilli, P.; Spooner, T.; Gignac, L.M.; et al. Electromigration and resistivity in on-chip Cu, Co and Ru damascene nanowires. In Proceedings of the 2017 IEEE International Interconnect Technology Conference (IITC), Hsinchu, Taiwan, 16–18 May 2017; pp. 1–3. [CrossRef]

77. Bekiaris, N.; Wu, Z.; Ren, H.; Naik, M.; Park, J.H.; Lee, M.; Ha, T.H.; Hou, W.; Bakke, J.R.; Gage, M.; et al. Cobalt fill for advanced interconnects. In Proceedings of the 2017 IEEE International Interconnect Technology Conference (IITC), Hsinchu, Taiwan, 16–18 May 2017. [CrossRef]

78. Tierno, D.; Hosseini, M.; van der Veen, M.; Dangol, A.; Croes, K.; Demuynck, S.; Tokei, Z.; Litta, E.; Horiguchi, N. Reliability of Barrierless PVD Mo. In Proceedings of the 2021 IEEE International Interconnect Technology Conference (IITC), Kyoto, Japan, 6–9 July 2021. [CrossRef]

79. Choi, D.; Barmak, K. On the potential of tungsten as next-generation semiconductor interconnects. *Electron. Mater. Lett.* **2017**, *13*, 449–456. [CrossRef]

80. Wen, L.G.; Roussel, P.; Pedreira, O.V.; Briggs, B.; Groven, B.; Dutta, S.; Popovici, M.I.; Heylen, N.; Ciofi, I.; Vanstreels, K.; et al. Atomic layer deposition of ruthenium with TiN interface for sub-10 nm advanced interconnects beyond copper. *ACS Appl. Mater. Interfaces* **2016**, *8*, 26119–26125. [CrossRef] [PubMed]

81. Zhang, X.; Huang, H.; Patlolla, R.; Wang, W.; Mont, F.W.; Li, J.; Hu, C.-K.; Liniger, E.G.; McLaughlin, P.S.; Labelle, C.; et al. Ruthenium interconnect resistivity and reliability at 48 nm pitch. In Proceedings of the 2016 IEEE International Interconnect Technology Conference/Advanced Metallization Conference (IITC/AMC). San Jose, CA, USA, 23–26 May 2016. [CrossRef]

82. Nogami, T.; Patlolla, R.; Kelly, J.; Briggs, B.; Huang, H.; Demarest, J.; Li, J.; Hengstebeck, R.; Zhang, X.; Lian, G.; et al. Cobalt/copper composite interconnects for line resistance reduction in both fine and wide lines. In Proceedings of the 2017 IEEE International Interconnect Technology Conference (IITC), Hsinchu, Taiwan, 16–18 May 2017. [CrossRef]

83. Dutta, S.; Kundu, S.; Gupta, A.; Jamieson, G.; Granados, J.F.G.; Bommels, J.; Wilson, C.J.; Tokei, Z.; Adelmann, C. Highly scaled ruthenium interconnects. *IEEE Electron Device Lett.* **2017**, *38*, 949–951. [CrossRef]

84. van der Veen, M.H.; Heyler, N.; Pedreira, O.V.; Ciofi, I.; Decoster, S.; Gonzalez, V.V.; Jourdan, N.; Struyf, H.; Croes, K.; Wilson, C.J.; et al. Damascene benchmark of Ru, Co and Cu in scaled dimensions. In Proceedings of the 2018 IEEE International Interconnect Technology Conference (IITC), Santa Clara, CA, USA, 4–7 June 2018. [CrossRef]

85. Wan, D.; Paolillo, S.; Rassoul, N.; Kotowska, B.K.; Blanco, V.; Adelmann, C.; Lazzarino, F.; Ercken, M.; Murdoch, G.; Bommels, J.; et al. Subtractive etch of ruthenium for sub-5nm interconnect. In Proceedings of the 2018 IEEE International Interconnect Technology Conference (IITC), Santa Clara, CA, USA, 4–7 June 2018. [CrossRef]

86. Murdoch, G.; Tokei, Z.; Paolillo, S.; Pedreira, O.V.; Vanstreels, K.; Wilson, C.J. Semidamascene Interconnects for 2nm node and Beyond. In Proceedings of the 2020 IEEE International Interconnect Technology Conference (IITC), San Jose, CA, USA, 5–8 October 2020. [CrossRef]

87. Lesniewska, A.; Roussel, P.; Tierno, D.; Gonzalez, V.V.; van der Veen, M.H.; Verdonck, P.; Jourdan, N.; Wilson, C.; Tokei, Z.; Croes, K. Dielectric reliability study of 21 nm pitch interconnects with barrierless Ru fill. In Proceedings of the 2020 IEEE International Reliability Physics Symposium (IRPS), Dallas, TX, USA, 28 April–30 May 2020. [CrossRef]

88. Son, M.; Jang, J.; Lee, Y.; Nam, J.; Hwang, J.Y.; Kim, I.S.; Lee, B.H.; Ham, M.-H.; Chee, S.-S. Copper-graphene heterostructure for back-end-of-line compatible high-performance interconnects. *NPJ 2D Mater. Appl.* **2021**, *5*, 41. [CrossRef]
89. Nogami, T.; Nguyen, S.; Huang, H.; Lanzillo, N.; Shobha, H.; Li, J.; Peethela, B.; Parbatani, A.; van Schravendijk, B.; Varadarajan, B.; et al. Electromigration and line R of graphene capped Cu dual Damascene interconnect. In Proceedings of the 2021 IEEE International Electron Devices Meeting (IEDM), San Francisco, CA, USA, 11–16 December 2021. [CrossRef]
90. Lo, C.-L.; Zhang, K.; Smith, R.S.; Shah, K.; Robinson, J.A.; Chen, Z. Large-area, single-layer molybdenum disulfide synthesized at BEOL compatible temperature as Cu diffusion barrier. *IEEE Electron Device Lett.* **2018**, *39*, 873–876. [CrossRef]
91. Yu, J.; Li, J.; Zhao, Y.; Lambertz, A.; Chen, T.; Duan, W.; Liu, W.; Yang, X.; Huang, Y.; Ding, K. Copper metallization of electrodes for silicon heterojunction solar cells: Process, reliability and challenges. *Sol. Energy Mater. Sol. Cells* **2021**, *224*, 110993. [CrossRef]
92. Yang, L.; Zhong, S.; Zhang, W.; Li, X.; Li, Z.; Zhuang, Y.; Wang, X.; Zhao, L.; Cao, X.; Deng, X.; et al. Study and development of rear-emitter Si heterojunction solar cells and application of direct copper metallization. *Prog. Photovolt. Res. Appl.* **2018**, *26*, 385–396. [CrossRef]
93. Blech, I.A.; Herring, C. Stress generation by electromigration. *Appl. Phys. Lett.* **1976**, *29*, 131–133. [CrossRef]
94. Christiansen, C.; Li, B.; Gill, J. Blech effect and lifetime projection for cu/low-k interconnects. In Proceedings of the 2008 International Interconnect Technology Conference, Burlingame, CA, USA, 1–4 June 2008. [CrossRef]
95. Cheng, Y.-L.; Lee, S.Y.; Chiu, C.; Wu, K. Back stress model on electromigration lifetime prediction in short length copper interconnects. In Proceedings of the 2008 IEEE International Reliability Physics Symposium, Phoenix, AZ, USA, 27 April–1 May 2008. [CrossRef]
96. Zhao, W.-S.; Zhang, R.; Wang, D.-W. Recent Progress in Physics-Based Modeling of Electromigration in Integrated Circuit Interconnects. *Micromachines* **2022**, *13*, 883. [CrossRef]
97. Lloyd, J.R. Black's law revisited-Nucleation and growth in electromigration failure. *Microelectron. Reliab.* **2007**, *47*, 1468–1472. [CrossRef]
98. Hu, C.-K.; Gignac, L.; Rosenberg, R.; Liniger, E.; Rubino, J.; Sambucetti, C.; Stamper, A.; Domenicucci, A.; Chen, X. Reduced Cu interface diffusion by CoWP surface coating. *Microelectron. Eng.* **2003**, *70*, 406–411. [CrossRef]
99. Hu, C.-K.; Gignac, L.; Lian, G.; Cabral, C.; Motoyama, K.; Shobha, H.; Demarest, J.; Ostrovski, Y.; Breslin, C.M.; Ali, M.; et al. Mechanisms of electromigration damage in Cu interconnects. In Proceedings of the 2018 IEEE International Electron Devices Meeting (IEDM), San Francisco, CA, USA, 1–5 December 2018. [CrossRef]
100. Kimura, M. Oxide breakdown mechanism and quantum physical chemistry for time-dependent dielectric breakdown. In Proceedings of the 1997 IEEE International Reliability Physics Symposium Proceedings. 35th Annual, Denver, CO, USA, 8–10 April 1997. [CrossRef]
101. Ogawa, E.; Kim, J.; Haase, G.; Mogul, H.; McPherson, J. Leakage, breakdown, and TDDB characteristics of porous low-k silica-based interconnect dielectrics. In Proceedings of the 2003 IEEE International Reliability Physics Symposium Proceedings, 2003. 41st Annual, Dallas, TX, USA, 30 March–4 April 2003. [CrossRef]
102. Oshida, D.; Takewaki, T.; Iguchi, M.; Taiji, T.; Morita, T.; Tsuchiya, Y.; Tsuchiya, H.; Yokogawa, S.; Kunishima, H.; Aizawa, H.; et al. Quantitative analysis of correlation between insulator surface copper contamination and TDDB lifetime based on actual measurement. In Proceedings of the 2008 International Interconnect Technology Conference, Burlingame, CA, USA, 1–4 June 2008. [CrossRef]
103. Noguchi, J. Dominant factors in TDDB degradation of Cu interconnects. *IEEE Trans. Electron Devices* **2005**, *52*, 1743–1750. [CrossRef]

Article

Design of a Capacitorless DRAM Based on Storage Layer Separated Using Separation Oxide and Polycrystalline Silicon

Geon Uk Kim [1], Young Jun Yoon [2], Jae Hwa Seo [3], Sang Ho Lee [1], Jin Park [1], Ga Eon Kang [1], Jun Hyeok Heo [1], Jaewon Jang [1], Jin-Hyuk Bae [1], Sin-Hyung Lee [1] and In Man Kang [1,*]

[1] School of Electronic and Electrical Engineering, Kyungpook National University, Daegu 41566, Korea
[2] Korea Multi-Purpose Accelerator Complex, Korea Atomic Energy Research Institute, Gyeongju 38180, Korea
[3] Power Semiconductor Research Center, Korea Electrotechnology Research Institute, Changwon 51543, Korea
* Correspondence: imkang@ee.knu.ac.kr

Abstract: In this study, a capacitorless one-transistor dynamic random-access memory (1T-DRAM) based on a polycrystalline silicon (Poly-Si) metal-oxide-semiconductor field-effect transistor (MOS-FET) with a storage layer separated using a separation oxide was designed and analyzed using technology computer-aided design (TCAD). The channel and storage layers were separated using a separation oxide to improve the inferior retention time of the conventional 1T-DRAM, and we adopted the underlap structure to reduce Shockley-Read-Hall recombination. In addition, poly-Si, which has several advantages, including low manufacturing cost and availability of high-density three-dimensional (3D) memory arrays, is used to easily fabricate silicon-on-insulator (SOI)-like structures. Accordingly, we extracted memory performance by analyzing the effect of grain boundary (GB). The proposed 1T-DRAM achieved a sensing margin of 14.10 μA/μm and a retention time of 251 ms at T = 358 K, even in the existence of a GB.

Keywords: one-transistor dynamic random-access memory; metal-oxide-semiconductor field-effect transistor; polycrystalline silicon

Citation: Kim, G.U.; Yoon, Y.J.; Seo, J.H.; Lee, S.H.; Park, J.; Kang, G.E.; Heo, J.H.; Jang, J.; Bae, J.-H.; Lee, S.-H.; et al. Design of a Capacitorless DRAM Based on Storage Layer Separated Using Separation Oxide and Polycrystalline Silicon. *Electronics* **2022**, *11*, 3365. https://doi.org/10.3390/electronics11203365

Academic Editors: Yi Zhao and ChoongHyun Lee

Received: 13 September 2022
Accepted: 15 October 2022
Published: 18 October 2022

Publisher's Note: MDPI stays neutral with regard to jurisdictional claims in published maps and institutional affiliations.

Copyright: © 2022 by the authors. Licensee MDPI, Basel, Switzerland. This article is an open access article distributed under the terms and conditions of the Creative Commons Attribution (CC BY) license (https://creativecommons.org/licenses/by/4.0/).

1. Introduction

Conventional one-transistor (1T) one-capacitor (1C) dynamic random-access memory (DRAM) is a key product of the semiconductor industry. It comprises one transistor for reading and writing operations and one capacitor for storing charges. Recently, the transistor size has continued to shrink for improved performance, such as higher speed and lower power consumption, and severe difficulties have arisen in capacitor fabrication. Therefore, many researchers have proposed 1T-DRAM without capacitors to replace conventional DRAM [1–10]. The 1T-DRAM utilizes the floating body effect of a partially depleted silicon-on-insulator (SOI) to retain memory performance [11]. 1T-DRAM program methods include impact ionization, gate-induced barrier lowering, and band-to-band tunneling (BTBT), and excess holes are stored in the floating body [12–14]. However, SOI wafers are significantly costly and difficult to fabricate. A low-cost method for forming SOI-like structures using polycrystalline silicon (poly-Si) has been proposed [15]. In addition, poly-Si can replace conventional silicon fabrication processes and be applied to three-dimensional (3D) memory arrays. However, even if the manufacturing problem of SOI structures is addressed, miniaturization of the device results in an insufficient charge storage region. This increases the recombination/generation rate owing to the strong electric field between the body and source/drain junction, which degrades the retention time.

In this study, we present a 1T-DRAM based on poly-Si with a storage layer separated using a separation oxide. The proposed 1T-DRAM uses an underlap structure and separation oxide to improve the retention time. In addition, because the poly-Si used to fabricate SOI-like structures in a simple manner is an aggregate of grains [16,17], a grain boundary (GB) is formed where different grains contact each other. Therefore, the memory characteristics were analyzed assuming that a single GB existed at the center of the channel and storage layers. The memory performance of the proposed 1T-DRAM was investigated through a two-dimensional simulation and evaluated in terms of sensing margin and retention time.

2. Device Structure and Simulation

Figure 1 shows a cross-sectional view of the poly-Si-based 1T-DRAM with a storage layer separated using a separation oxide. The program, erase, and hold operations are performed by the top gate, while conventional metal-oxide-semiconductor field-effect transistor (MOSFET) operations are performed by the bottom gate. The bottom gate uses a heavily doped poly-Si to minimize the difficulty in metal-based gate fabrication. The geometric parameters are summarized in Table 1. The underlap structure and separation oxide are considered to be the main parameters that influence the memory performance of 1T-DRAM. The underlap structure reduces the electric field between the storage layer and source/drain. The separation oxide separates the channel and storage layers, enabling the excess holes to be stored more effectively.

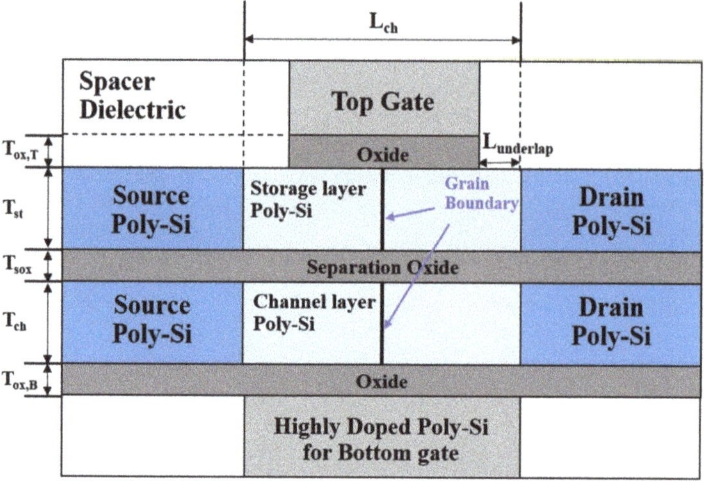

Figure 1. Poly-Si based 1T-DRAM with storage layer separated using separation oxide.

Table 1. Parameters for the proposed 1T-DRAM.

Parameter	Value
Chanel Length (L_{ch})	70 nm
Underlap Length ($L_{underlap}$)	0–20 nm
Top Gate Oxide (HfO_2) Thickness ($T_{ox,T}$)	8 nm
Channel Layer Thickness (T_{ch})	8 nm
Separation Oxide (SiO_2) Thickness (T_{sox})	1–30 nm
Storage Layer Thickness (T_{st})	8 nm
Bottom Gate Oxide (SiO_2) Thickness ($T_{ox,B}$)	3 nm
n^+-Source Doping Concentration	1×10^{20} cm^{-3}
n^+-Drain Doping Concentration	1×10^{20} cm^{-3}
p-Channel Layer Doping Concentration	1×10^{16} cm^{-3}
p^+-Storage Layer Doping Concentration	1×10^{18} cm^{-3}
p^+-Poly-Si Bottom gate doping concentration	1×10^{18} cm^{-3}
Top Gate Work function	5.20 eV

Figure 2 shows the main fabrication steps of a proposed device crystallized via excimer laser crystallization (ELC) [18]. First, SiO_2 is deposited and the bottom gate is etched. Second, amorphous silicon (a-Si) is deposited by LPCVD and activated by excimer laser irradiation to form a poly-Si bottom gate. Third, SiO_2 is deposited to form a bottom gate oxide. Fourth, after deposition a-Si through LPCVD, a channel layer is formed through ELC. Fifth, a hard mask is formed through SiO_2 deposition and etching. Sixth, ion implantation is performed to form the n-type source and drain of the channel layer. Seventh, SiO_2 is deposited to form a separation oxide. Eighth, after deposition a-Si by LPCVD, a storage layer is formed through ELC. Ninth, the top gate oxide is formed through HfO_2 deposition and etching. Tenth, the metal top gate was deposited. Finally, ion implantation was performed to form the n-type source and drain of the storage layer. The device can be fabricated through the above process. In addition, through silicon via process (TSV) is performed to connect the channel layer and the storage layer. The stack structure requires high temperature processing, which can threaten dopant diffusion. These manufacturing challenges can be overcome using ELC.

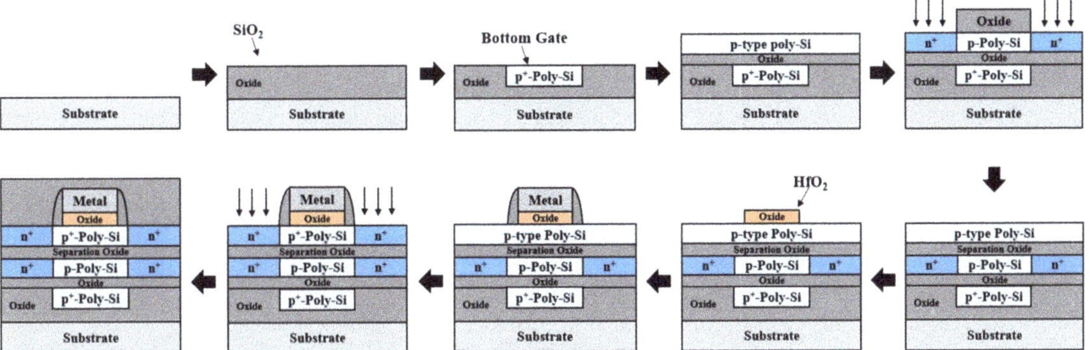

Figure 2. Key fabrication steps of the proposed 1T-DRAM crystallized via excimer laser crystallization [18].

The work function of the top gate is 5.20 eV. The top gate work function increases the energy band of the storage layer, thereby generating a potential well that can store additional holes. The proposed 1T-DRAM consists of an 8 nm thick storage layer and the channel layer. The doping concentrations of the storage and channel layers are 1×10^{18} and 1×10^{16} cm^{-3}, respectively, and that of the source/drain regions is 1×10^{20} cm^{-3}. The channel length is set to 70 nm. A long channel is used because the longer the channel, the wider energy band of the hole storage area, which is advantageous for hole storage [19]. The top gate dielectric is designed with 3 nm thick HfO_2. This is because the high dielectric constant of 22 increases the effect of the top gate controllability. Accordingly, the effect of the top gate work function on the storage layer is increased, so that the hole can be stored more effectively. For high accuracy, physical models such as Fermi-Dirac statistical, Shockley-Read-Hall (SRH) recombination, Auger recombination, nonlocal BTBT, and trap-assisted tunneling (TAT) models, doping-dependent and field-dependent mobility models, bandgap narrowing model, and quantum confinement effect were used in the simulations [20]. The design and analysis of the device were performed using Sentaurus technology computer-aided design (TCAD) simulation tool. Furthermore, the GB trap distribution of poly-Si was used as the calibrated data, which is reported in [15] (Figure 3).

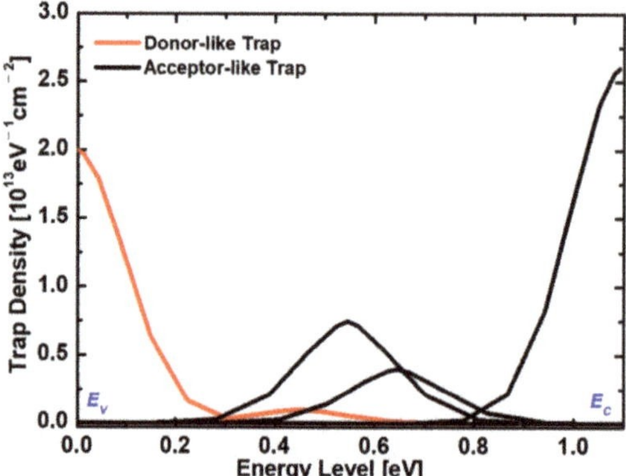

Figure 3. GB trap distribution calibrated using experimental data reported in [15].

3. Results

Figure 4a shows the drain current (I_{ds}) versus gate voltage ($V_{gs,B}$) transfer characteristics of the proposed device. At different temperatures of 300 K and 358 K, the threshold voltage (V_{th}) is 1.00 V and 1.12 V, respectively. The heavily doped poly-Si bottom gate performs the conventional MOSFET operation. Although the gate controllability is reduced compared to that of the metal-based gate, this gate structure can be easily fabricated in a 3D stacked manner.

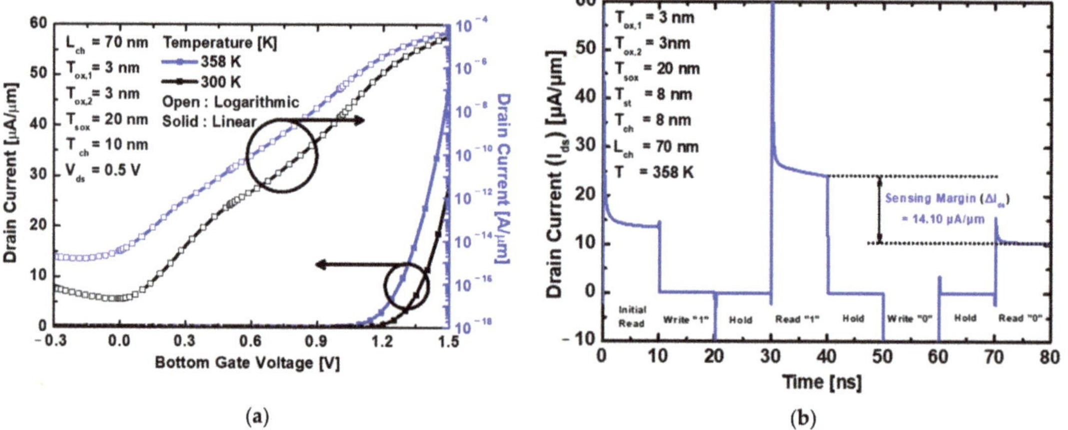

(a)

(b)

Figure 4. (**a**) Linear (solid symbol) and logarithmic (open symbol) I_{ds}-$V_{gs,B}$ transfer characteristics for the proposed 1T-DRAM at V_{ds} = 0.5 V. (**b**) Transient characteristic of the proposed 1T-DRAM.

Figure 4b shows the transient characteristics of the proposed device. Table 2 summarizes the bias conditions for the memory performance of 1T-DRAM. The program operation uses the BTBT mechanism. Generated excess holes are accumulated in the storage layer. The erase operation reduces the potential barrier of the storage layer and a negative voltage is applied to drain. Thus, accumulated excess holes are removed by drifting towards the drain. As shown in Figure 4b, the proposed device extracted a sensing margin of 14.10 μA/μm at a temperature of 358 K.

Table 2. Operating bias scheme for memory performance.

	Write "1" (Program)	Write "0" (Erase)	Read	Hold
Top Gate Voltage ($V_{gs,T}$)	−3.2 V	0.5 V	0.0 V	−0.3 V
Bottom Gate Voltage ($V_{gs,B}$)	0.0 V	0.0 V	1.3 V	0.0 V
Drain Voltage (V_{ds})	3.0 V	−0.7 V	0.0 V	0.0 V

Figure 5a shows the BTBT rate of the proposed device during the program operation. The program operation is performed using BTBT between the storage layer and drain. The program operation of the tunneling method is advantageous because it consumes less power than that of the impact ionization method. As shown in Figure 5b, the top gate voltage ($V_{gs,T}$) = −3.2 V and V_d = 3.0 V are applied to form a thin tunneling barrier between the storage layer and drain. BTBT occurs through a thin tunneling barrier, and the generated excess holes accumulate below the valence band of the top gate oxide. In addition, the stored excess holes are maintained owing to the negative voltage and high work function.

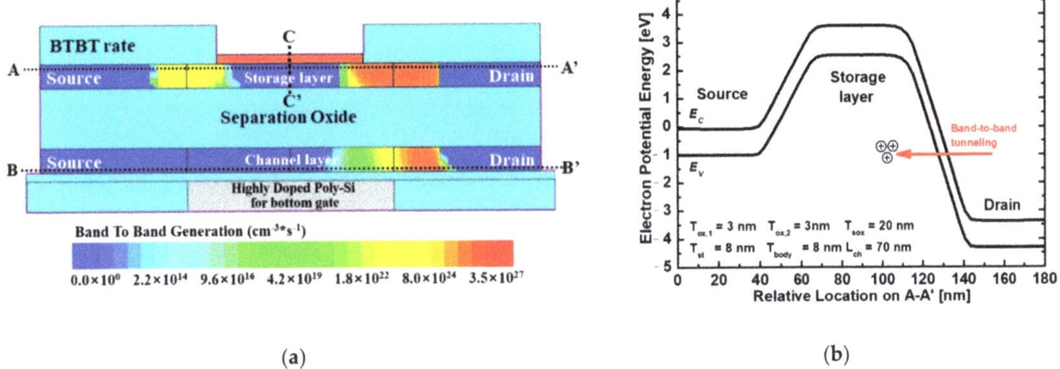

Figure 5. (**a**) Simulation profiles of the proposed 1T-DRAM during the program operation showing BTBT rate. (**b**) Energy band diagram of the storage layer. The energy band is extracted at 2 nm below the top gate oxide.

Figure 6a shows the hole density of the proposed device during the hold operation. The hold "1" state is defined after a program operation, and the hold "0" state is defined after an erase operation. As shown in Figure 6a, the hole density is different between hold "1" and hold "0" states. Figure 6b shows the energy band diagram of the storage layer during hold "1" and hole "0" operations in the storage layer. In the hold "1" state, an additional positive voltage is applied to the storage layer by the stored excess holes, reducing the potential barrier than in the hold "0" state. Moreover, a hold voltage of $V_{gs,T}$ = −0.3 V is applied during hold operation to store excess holes more effectively.

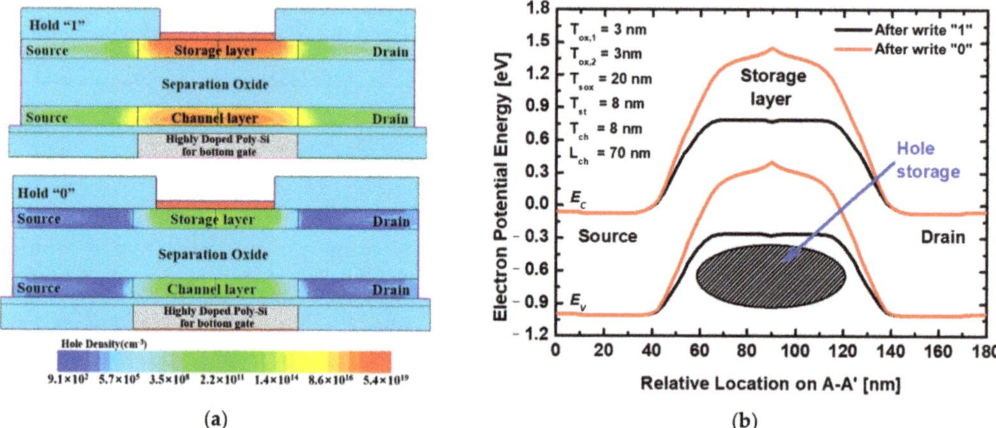

Figure 6. (**a**) Simulation profile of the proposed 1T-DRAM during the hold operation showing hole density. (**b**) Energy band diagram of the storage layer during the hold operation. The energy band is extracted at 2 nm below the top gate oxide.

Figure 7a shows the electron density of the proposed device during the read operations. During the read operation, the current flows in the channel layer when $V_{gs,B} = 1.3$ V and $V_{ds} = 0.5$ V. Figure 7a shows the difference in electron density in the channel layer. As shown in Figure 7b, when excess holes are stored in the storage layer, they affect the potential barrier of the channel layer and have the same effect as that of lowering of V_{th} by applying an additional voltage. However, after the stored holes are erased, the potential barrier of the channel layer increases owing to the low hole density of the storage layer. This causes a drain current difference between read "1" and read "0" states, defined as the sensing margin.

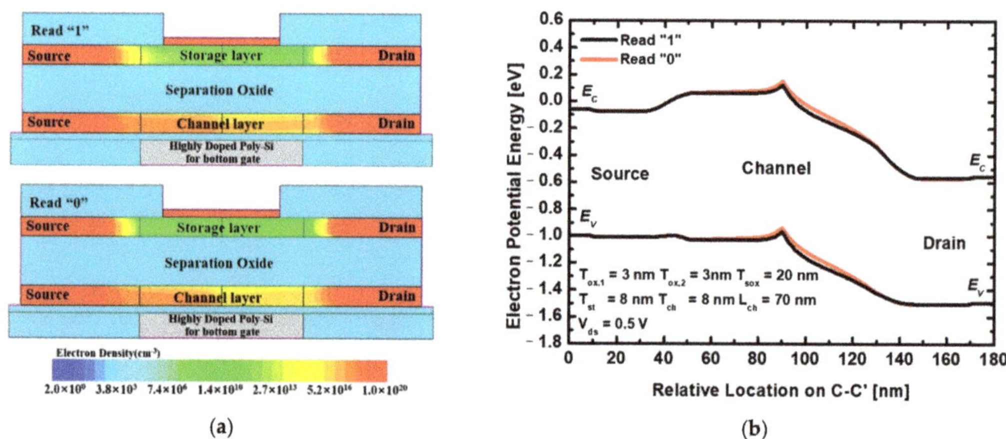

Figure 7. (**a**) Simulation profile of the proposed 1T-DRAM during the read operation showing electron density. (**b**) Energy band diagram of the channel layer during the read operation. The energy band is extracted at 2 nm above the bottom gate oxide.

3.1. Effect of $L_{underlap}$ Variation

It is important to analyze the retention time among the memory performance of 1T-DRAM. The retention time is the time it takes for accumulated excess holes by recombination/generation during hold operation to return to a steady state after program/erase operation. The retention time of 1T-DRAM is defined as the hold time, which is 50% of the sensing margin. Figure 8a shows the hole density of the proposed device depending on the $L_{underlap}$ in the hold "1" state. As $L_{underlap}$ increases, the influence of the top gate on the storage layer and the source/drain junction decreases. Therefore, the recombination rate of the storage layer is reduced and the retention time is increased. However, when the effect of the top gate is reduced, it is difficult to form a sufficient tunneling barrier for BTBT to occur at the storage layer and drain junction during program operation, as shown in Figure 8b. This eventually decreases the hole density in the storage layer, as shown in Figure 8a. Moreover, the sensing margin decreases because the hole density in the storage layer is insufficient. On the other hand, when $L_{underlap}$ decreases, the tunneling barrier between the storage layer and drain becomes very thin and excess holes are generated, resulting in a decreased potential barrier of the storage layer, as shown in Figure 9a. This increases the rate of recombination between the storage layer and the source/drain so that the retention time is reduced. Figure 9b shows the sensing margin and retention time depending on the $L_{underlap}$ of the proposed device. At a temperature of 358 K, the sensing margin reached the peak value at $L_{underlap}$ = 0 nm, and the retention time reached the peak value of 251 ms at $L_{underlap}$ = 10 nm. In addition, this performance adheres to the international roadmap for devices and systems (IRDS) [21].

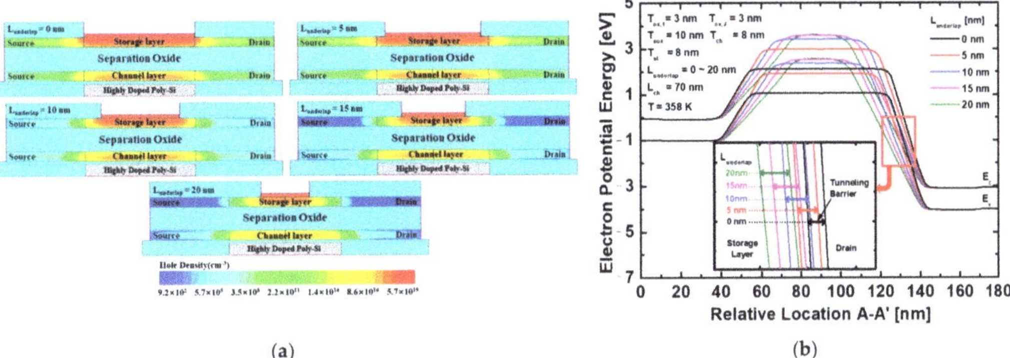

(a)   (b)

Figure 8. (**a**) Simulation profile of the proposed 1T-DRAM with different Lunderlap during hold "1" operation showing hole density. (**b**) Energy band diagram of the storage layer with different Lunderlap during program operation. The energy band is extracted 2 nm below top gate oxide.

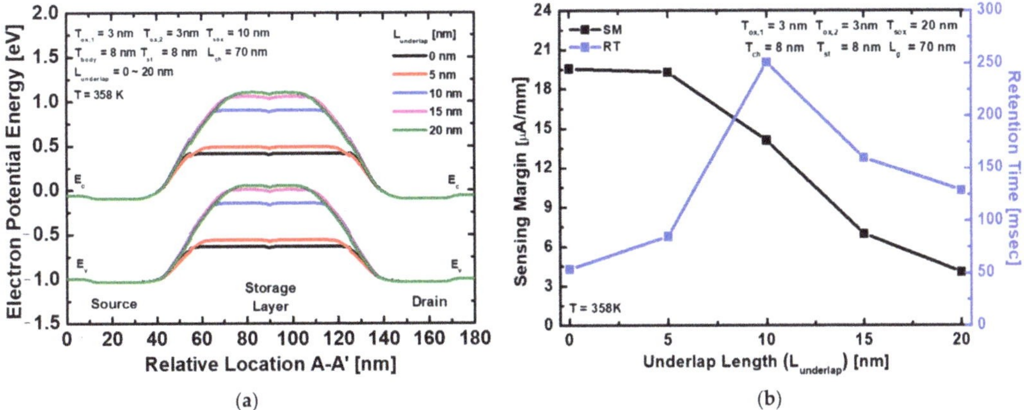

(a)   (b)

Figure 9. (**a**) Energy band diagram of the storage layer with different $L_{underlap}$ during hold "1" operation. The energy band is extracted at 2 nm below the top gate oxide. (**b**) Variations in sensing margin and retention time with $L_{underlap}$ of proposed 1T-DRAM.

3.2. Effect of T_{sox} Variation

The conventional 1T-DRAM stores excess holes in a floating body. However, the stored holes tend to decrease rapidly during the hold and read operations, limiting the retention time. Therefore, the proposed 1T-DRAM effectively stores excess holes and reduces the SRH recombination rate by separating the channel and storage layers with a separation oxide to improve the retention time.

Figure 10a shows the SRH recombination rate of the proposed device depending on the T_{sox} in the hold "1" state. As T_{sox} decreases, the recombination rate increases because the source/drain junction of the storage layer is influenced by the electric field of the bottom gate. As shown in Figure 10b, when $T_{sox} = 1$ nm, the SRH recombination rate attains a peak value and decreases gradually. However, when T_{sox} exceeds 20 nm, the effect between the storage and channel layers and the retention time decreases. In terms of the sensing margin, as T_{sox} increases, the effect of the top gate work function on the channel layer gradually decreases, and the potential barrier of the channel layer decreases, as shown in Figure 11a. Consequently, the read current and sensing margin increase. However, when $T_{sox} > 15$ nm, the relationship between the storage and channel layers weakens. Therefore, the excess holes stored in the storage layer cannot be applied as an additional positive voltage. Therefore, the difference between the read "1" and "0" currents decrease. Figure 11b shows the sensing margin and retention time depending on T_{sox} of the proposed device. At a temperature of 358 K, the sensing margin achieved a peak value of 15.60 µA/µm at $T_{sox} = 15$ nm, and the retention time achieved a peak value of 251 ms at $T_{sox} = 20$ nm. Additionally, when $T_{sox} = 20$ nm, a sufficient sensing margin of 14.10 µA/µm is extracted. Accordingly, optimal memory performance is obtained at $T_{sox} = 20$ nm.

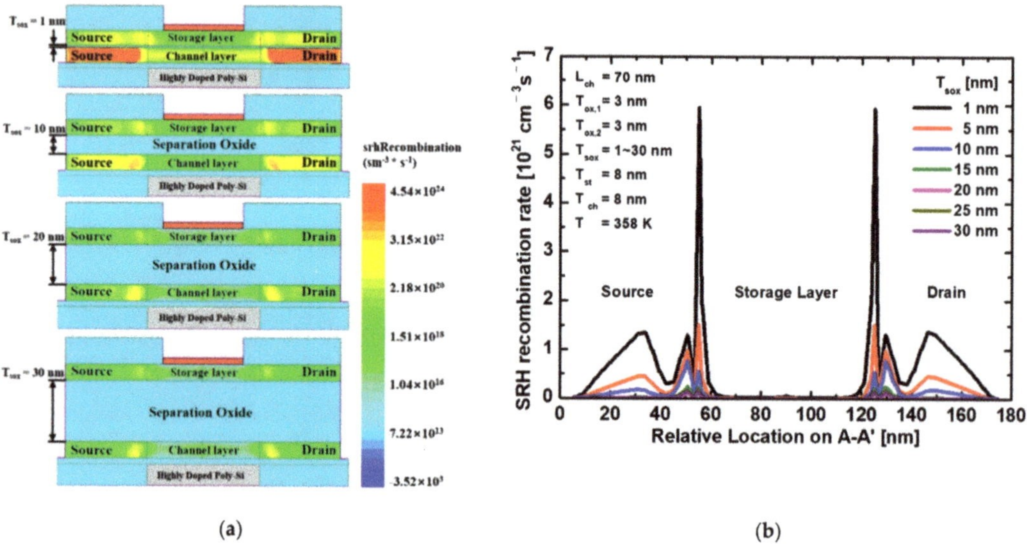

Figure 10. (**a**) Simulation profile of the 1T-DRAM with different T_{sox} during hold "1" operation showing SRH recombination. (**b**) SRH recombination rate of storage layer with different T_{sox} during the read operation. The energy band is extracted at 2 nm below the top gate oxide.

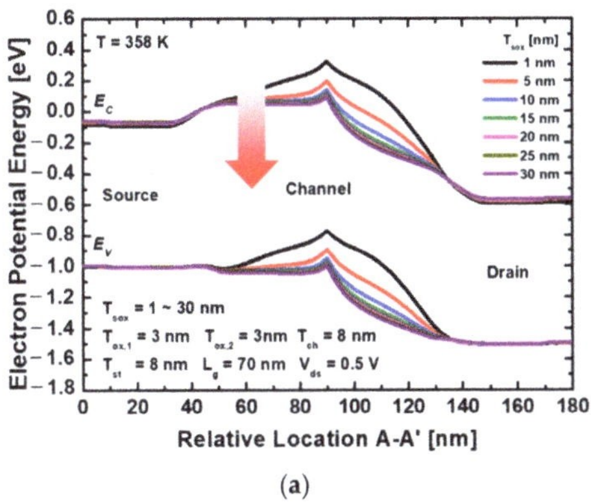

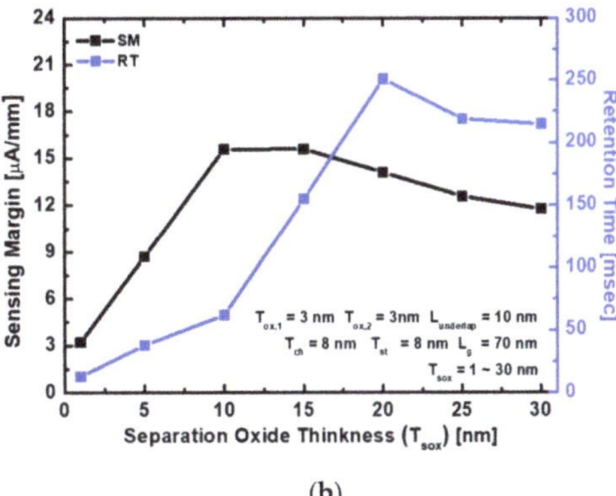

(a) (b)

Figure 11. (**a**) Energy band diagram of the channel layer with different T_{sox} during the read operation. The energy band is extracted at 2 nm above the bottom gate oxide. (**b**) Variations in sensing margin and retention time with T_{sox} of proposed 1T-DRAM.

3.3. Effect of GB Location

Because the proposed device uses poly-Si to fabricate an SOI-like structure easily, it is necessary to analyze its effect on the GB. The grain size of poly-Si can be adjusted according to the growth and annealing temperatures. Therefore, it can be significantly larger than that of a recently scaled-down device. When investigating the impact of the GB, it is assumed that a single GB is located at the center of the storage and channel layers, as shown in Figure 12a. Figure 12b shows the memory characteristics based on GB location. The sensing margins and retention times at GB locations are summarized in Table 3.

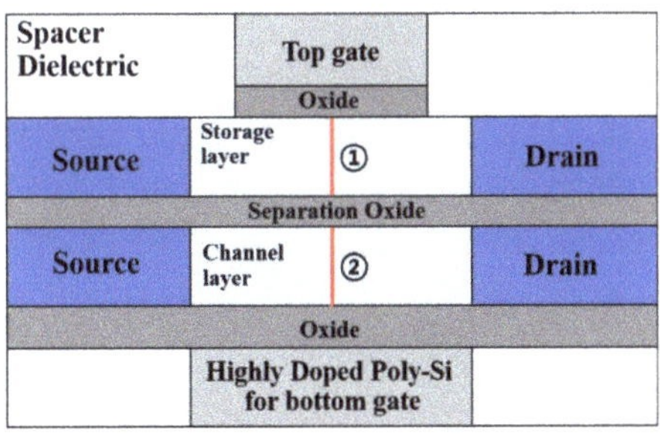

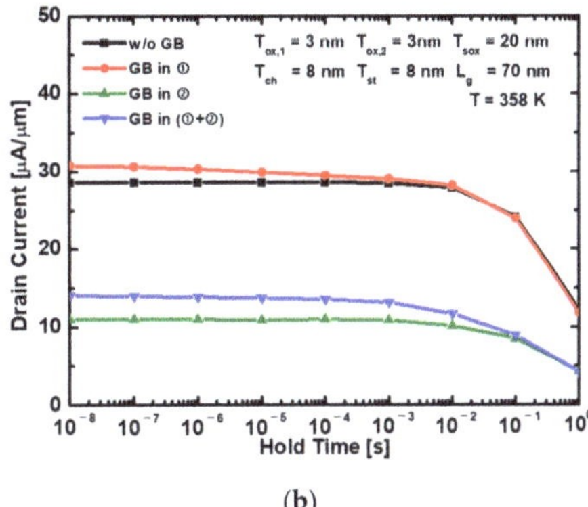

(a) (b)

Figure 12. (**a**) Cross-section depending on the GB location of the proposed 1T-DRAM. (**b**) Variation of currents in read "1" and read "0" status depending on the GB position.

Table 3. Memory performance depending on the GB location of the proposed 1T-DRAM.

	Sensing Margin [μA/μm]	Retention Time [ms]
w/o GB	28.58	648
GB in ①	30.79	507
GB in ②	11.09	484
GB in (① + ②)	14.10	251

When a GB exists in the storage layer, the sensing margin is larger than that without a GB in the storage layer. This is because the difference between read "1" and read

"0" increases owing to the hole/electrons captured by the GB trap of the storage layer. However, the retention time is reduced by the TAT mechanism through GB traps, as shown in Figure 13a. This is because the recombination/generation rates of hole are increased by the TAT mechanism. When a GB exists in the channel layer, electrons are captured in the region where the GB is located; therefore, potential barriers increase, as shown in Figure 13b. The increased potential barrier disturbs the current flowing from the source to the drain, significantly decreasing the sensing margin. In addition, the reduced sensing margin degrades the retention time. As a result, the memory performance of the proposed 1T-DRAM exhibits a superior performance of 28.58 $\mu A/\mu m$ sensing margin and 648 ms retention time without a GB. Furthermore, when the GB exists in the storage and channel layers, the sensing margin and retention time are 14.10 $\mu A/\mu m$ and 251 ms, respectively, which correspond to an excellent memory performance.

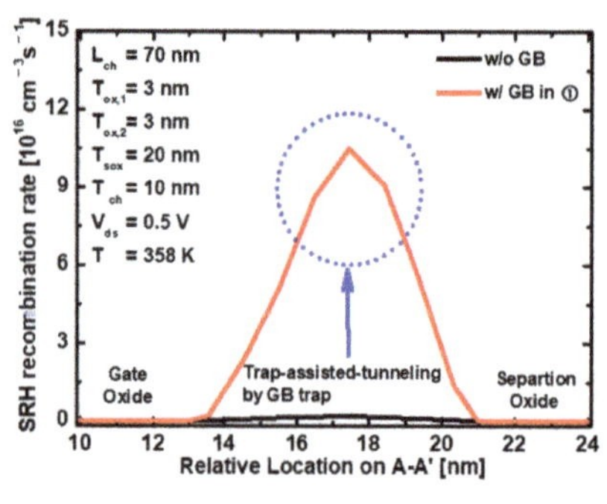

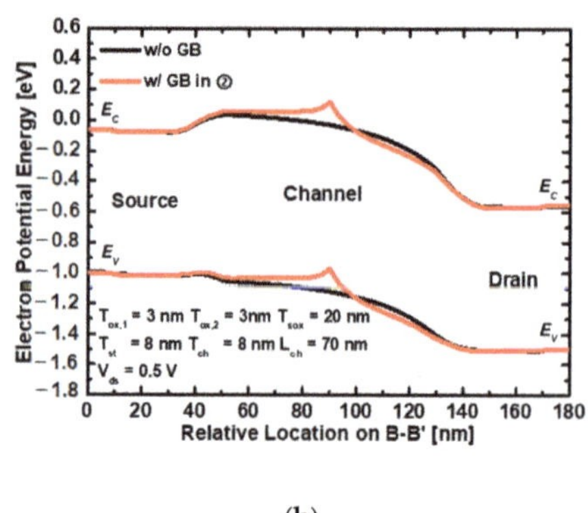

(a)   (b)

Figure 13. (a) SRH recombination of the hold state with and without GB in ①. The energy band is extracted at the center of the storage region. (b) Energy band diagram of the read operation with and without GB in ②. The energy band is extracted at 2 nm above the bottom gate oxide.

Table 4 presents the previously reported memory performance of the 1T-DRAM. The 1T-DRAM developed in this study exhibits a superior memory performance at T = 358 K compared with other devices.

Table 4. Memory performances of various 1T-DRAM reported in the literature.

No	Structure	Sensing Margin [$\mu A/\mu m$]	Retention Time [ms]
1.	Junctionless FinFET-Based 1T-DRAM (Vertical GB) [22]	11.7	64.27
2.	Junctionless FinFET-Based 1T-DRAM (Horizontal GB) [22]	11.3	148
3.	SiGe GAA JLFET-Based 1T-DRAM [18]	0.39	10
4.	Double-gate Si/SiGe 1T-DRAM [23]	6.16	131
5.	Dopingless 1T-DRAM [24]	0.012	170
6.	Dual Gate 1T-DRAM (w/GB) [25]	6.58	340.1
7.	Nanotube 1T-DRAM (w/GB) [26]	422	120
8.	Ge/GaAs TFET-Based 1T-DRAM [27]	1.11	120
9.	This study (w/o GB)	28.58	648
10.	This study (w/GB)	14.10	251

4. Conclusions

In this study, a poly-Si-based 1T-DRAM with a storage layer separated using a separation oxide was designed and analyzed using TCAD simulation. In addition to analyzing the effect of GB on memory performance, the important parameters $L_{underlap}$ and T_{sox} were evaluated. The proposed 1T-DRAM obtained high retention time despite the effect of GB because the storage efficiency was improved by separating the channel and the storage layers with separation oxide. As a result, the optimized parameters are $L_{ch} = 70$ nm, $L_{underlap} = 10$ nm, and $T_{sox} = 20$ nm, achieving a sensing margin of 14.10 μA/μm and a high retention time of 251 ms at T = 358 K. Therefore, the proposed 1T-DRAM has the potential to replace the conventional 1T-1C DRAM.

Author Contributions: Conceptualization, G.U.K.; investigation, G.U.K., data curation, G.U.K., S.H.L. and J.P.; validation, Y.J.Y., J.H.S., G.E.K., J.H.H., J.J., J.-H.B., S.-H.L. and I.M.K.; writing—original draft preparation, G.U.K.; writing—review and editing, I.M.K. All authors have read and agreed to the published version of the manuscript.

Funding: This work was supported by the National Research Foundation of Korea (NRF) grant funded by the Korea government (MSIT) (No. NRF-2020R1A2C1005087). This study was supported by the BK21 FOUR project funded by the Ministry of Education, Korea (4199990113966). This research was supported by Basic Science Research Program through the National Research Foundation of Korea (NRF) funded by the Ministry of Education (NRF-2021R1A6A3A13039927). This research was supported by National R&D Program through the National Research Foundation of Korea (NRF) funded by Ministry of Science and ICT (2021M3F3A2A03017764). This research was supported by National R&D Program through the National Research Foundation of Korea (NRF) funded by Ministry of Science and ICT (NRF-2022M3I7A1078936). This investigation was financially supported by Semiconductor Industry Collaborative Project between Kyungpook National University and Samsung Electronics Co. Ltd. The EDA tool was supported by the IC Design Education Center (IDEC), Korea.

Institutional Review Board Statement: Not applicable.

Informed Consent Statement: Not applicable.

Data Availability Statement: Not applicable.

Conflicts of Interest: The authors declare no conflict of interest.

References

1. Ieong, M.; Narayanan, V.; Singh, D.; Topol, A.; Chan, V.; Ren, Z. Transistor scaling with novel materials. *Mater. Today* **2006**, *9*, 26–31. [CrossRef]
2. Bawedin, M.; Cristoloveanu, S.; Flandre, D. A capacitorless 1T-DRAM on SOI based on dynamic coupling and double-gate operation. *IEEE Electron Device Lett.* **2008**, *29*, 795–798. [CrossRef]
3. Okhonin, S.; Nagoga, M.; Sallese, J.M.; Fazan, P. A capacitor-less 1T-DRAM cell. *IEEE Electron Device Lett.* **2002**, *23*, 85–87. [CrossRef]
4. Mandelman, J.A.; Dennard, R.H.; Bronner, G.B.; DeBrosse, J.K.; Divakaruni, R.; Li, Y.; Radens, C.J. Challenges and future directions for the scaling of dynamic random-access memory (DRAM). *IBM J. Res. Dev.* **2002**, *46*, 187–212. [CrossRef]
5. Giusi, G.; Alam, M.A.; Crupi, F.; Pierro, S. Bipolar mode operation and scalability of double-gate capacitorless 1T-DRAM cells. *IEEE Trans. Electron Devices* **2010**, *57*, 1743–1750. [CrossRef]
6. Giusi, G.; Iannaccone, G. Junction engineering of 1T-DRAMs. *IEEE Electron Device Lett.* **2013**, *34*, 408–410. [CrossRef]
7. Ertosun, M.G.; Cho, H.; Kapur, P.; Saraswat, K.C. A nanoscale vertical double-gate single-transistor capacitorless DRAM. *IEEE Electron Device Lett.* **2008**, *29*, 615–617. [CrossRef]
8. Moon, D.I.; Choi, S.J.; Han, J.W.; Kim, S.; Choi, Y.K. Fin-width dependence of BJT-based 1T-DRAM implemented on FinFET. *IEEE Electron Device Lett.* **2010**, *31*, 909–911. [CrossRef]
9. Han, J.W.; Ryu, S.W.; Kim, D.H.; Kim, C.J.; Kim, S.; Moon, D.I.; Choi, S.J.; Choi, Y.K. Fully depleted polysilicon TFTs for capacitorless 1T-DRAM. *IEEE Electron Device Lett.* **2009**, *30*, 742–744. [CrossRef]
10. Kuo, C.; King, T.J.; Hu, C. A capacitorless double gate DRAM technology for sub-100-nm embedded and stand-alone memory applications. *IEEE Trans. Electron Devices* **2003**, *50*, 2408–2416. [CrossRef]
11. Ouisse, T.; Ghibaudo, G.; Brini, J.; Cristoloveanu, S.; Borel, G. Investigation of floating body effects in silicon-on-insulator metal-oxide-semiconductor field-effect transistors. *J. Appl. Phys.* **1991**, *70*, 3912–3919. [CrossRef]

12. Okhonin, S.; Nagoga, M.; Sallese, J.M.; Fazan, P. A SOI capacitor-less 1T-DRAM concept. In Proceedings of the 2001 IEEE International SOI Conference, Proceedings (Cat. No. 01CH37207), Durango, CO, USA, 1–4 October 2001; pp. 153–154. [CrossRef]

13. Guegan, G.; Touret, P.; Molas, G.; Raynaud, C.; Pretext, J. A novel capacitor-less 1T-DRAM on partially depleted SOI pMOSFET based on direct-tunneling current in the partial n+ poly gate. In Proceedings of the 2009 International Conference on Solid State Devices and Materials, Miyagi, Japan, 7–9 October 2009; pp. 446–447. [CrossRef]

14. Yoshida, E.; Tanaka, T. A capacitorless 1T-DRAM technology using gate-induced drain-leakage (GIDL) current for low-power and high-speed embedded memory. *IEEE Trans. Electron Devices* **2006**, *53*, 692–697. [CrossRef]

15. Seo, J.H.; Yoon, Y.J.; Yu, E.; Sun, W.; Shin, H.; Kang, I.M.; Lee, J.H.; Cho, S. Fabrication and characterization of a thin-body poly-Si 1T DRAM with charge-trap effect. *IEEE Electron Device Lett.* **2019**, *40*, 566–569. [CrossRef]

16. Koyanagi, M.; Baba, Y.; Hata, K.; Wu, I.W.; Lewis, A.G.; Fuse, M.; Bruce, R. The charge-pumping technique for grain boundary trap evaluation in polysilicon TFTs. *IEEE Electron Device Lett.* **1992**, *13*, 152–154. [CrossRef]

17. Yu, E.; Kim, Y.; Lee, J.; Cho, Y.; Lee, W.J.; Cho, S. Processing and characterization of ultra-thin poly-crystalline silicon for memory and logic applications. *J. Semicond. Technol. Sci.* **2018**, *18*, 172–179. [CrossRef]

18. Lee, I.C.; Tsai, T.C.; Tsai, C.C.; Yang, P.Y.; Wang, C.L.; Cheng, H.C. High-performance vertically stacked bottom-gate and top-gate polycrystalline silicon thin-film transistors for three-dimensional integrated circuits. *Solid-State Electron.* **2012**, *77*, 26–30. [CrossRef]

19. Yoon, Y.J.; Lee, J.S.; Kim, D.S.; Lee, S.H.; Kang, I.M. One-transistor dynamic random-access memory based on gate-all-around junction-less field-effect transistor with a Si/SiGe heterostructure. *Electronics* **2020**, *9*, 2134. [CrossRef]

20. *User's Manual*; Version L-2016.03; Synopsys: Mountain View, CA, USA, 2016.

21. More Moore, 2021 International Roadmap for Devices and Systems (IRDS™) Edition. 2021. Available online: https://irds.ieee.org/editions/2021 (accessed on 1 September 2022).

22. Cho, M.S.; Mun, H.J.; Lee, S.H.; Jang, J.; Bae, J.H.; Kang, I.M. Simulation of capacitorless dynamic random access memory based on junctionless FinFETs using grain boundary of polycrystalline silicon. *Appl. Phys. A* **2020**, *126*, 943. [CrossRef]

23. Yoon, Y.J.; Cho, M.S.; Kim, B.G.; Seo, J.H.; Kang, I.M. Capacitorless one-transistor dynamic random-access memory based on double-gate metal-oxide-semiconductor field-effect transistor with Si/SiGe heterojunction and underlap structure for improvement of sensing margin and retention time. *J. Nanosci. Nanotechnol.* **2019**, *19*, 6023–6030. [CrossRef]

24. James, A.; Saurabh, S. Dopingless 1T DRAM: Proposal, design, and analysis. *IEEE Access* **2019**, *7*, 88960–88969. [CrossRef]

25. Lee, S.H.; Jang, W.D.; Yoon, Y.J.; Seo, J.H.; Mun, H.J.; Cho, M.S.; Jang, J.; Bae, J.H.; Kang, I.M. Polycrystalline-silicon-MOSFET-based capacitorless DRAM with grain boundaries and its performances. *IEEE Access* **2021**, *9*, 50281–50290. [CrossRef]

26. Park, J.; Cho, M.S.; Lee, S.H.; An, H.D.; Min, S.R.; Kim, G.U.; Yoon, Y.J.; Seo, J.H.; Jang, J.; Bae, J.H.; et al. Design of Capacitorless DRAM Based on Polycrystalline Silicon Nanotube Structure. *IEEE Access* **2021**, *9*, 163675–163685. [CrossRef]

27. Yoon, Y.J.; Seo, J.H.; Kang, I.M. Capacitorless one-transistor dynamic random-access memory based on asymmetric double-gate Ge/GaAs-heterojunction tunneling field-effect transistor with n-doped boosting layer and drain-underlap structure. *Jpn. J. Appl. Phys.* **2018**, *57*, 04FG03. [CrossRef]

 *electronics*

Article

Simulation Study of Low Turn-Off Loss and Snapback-Free SA-IGBT with Injection-Enhanced *p*-Floating Layer

Xiaodong Zhang [1], Ming Gong [2], Junfeng Pan [3], Mingxin Song [4,*], Hang Zhang [5] and Linlin Zhang [6,*]

[1] School of Computer Science and Technology, Hainan University, Haikou 570228, China; zhangxiaodong@hainanu.edu.cn

[2] School of Information and Communication Engineer, Hainan University, Haikou 570228, China; 20213003584@hainanu.edu.cn

[3] College of Foreign Languages, Hainan University, Haikou 570228, China; panjunfeng1@126.com

[4] School of Applied Science and Technology, Hainan University, Haikou 570228, China

[5] Ecological Environmental Information and Emergency Protection Center, Ecological Environment Bureau of Heihe, Heihe 164399, China; hangzhang991417617@163.com

[6] Key Laboratory of Agro-Forestry Environmental Processes and Ecological Regulation of Hainan Province, School of Ecological and Environmental Sciences, Hainan University, Haikou 570228, China

* Correspondence: songmingxin@126.com (M.S.); 6962zhanglinlin@163.com (L.Z.)

Abstract: In this study, a shorted-anode IGBT with an injection-enhanced *p*-floating layer (IEPF-IGBT) under the N-buffer layer is proposed. Compared to conventional shorted-anode IGBT (SA-IGBT), the IEPF-IGBT has the structural characteristics of an injection-enhanced P-floating (IEPF) layer inserted into the N-buffer layer and the P+ collector region. The IEPF layer and P+ collector region pinch off the electron path during the turn-on period to suppress the snapback effect with a half-cell pitch of 10 μm. In addition, the IEPF layer acts as an injection-enhanced layer that influences the current injection of the holes. There is 56.3% reduction in the turn-off loss of the IEPF-IGBT at the same forward voltage drop.

Keywords: injection-enhanced *p*-floating; shorted anode; snapback; turn-off loss

Citation: Zhang, X.; Gong, M.; Pan, J.; Song, M.; Zhang, H.; Zhang, L. Simulation Study of Low Turn-Off Loss and Snapback-Free SA-IGBT with Injection-Enhanced *p*-Floating Layer. *Electronics* **2022**, *11*, 2351. https://doi.org/10.3390/electronics11152351

Academic Editor: Francesco Giuseppe Della Corte

Received: 23 June 2022
Accepted: 26 July 2022
Published: 28 July 2022

Publisher's Note: MDPI stays neutral with regard to jurisdictional claims in published maps and institutional affiliations.

Copyright: © 2022 by the authors. Licensee MDPI, Basel, Switzerland. This article is an open access article distributed under the terms and conditions of the Creative Commons Attribution (CC BY) license (https://creativecommons.org/licenses/by/4.0/).

1. Introduction

IGBTs have received increasing amounts of attention in theoretical research due to their uncomplicated driving circuits and low driving power. An IGBT is a voltage-controlled device consisting of MOS and BJT parts [1,2]. In high-voltage, high-current, fast-switching, and high-power applications, IGBTs show low resistance, so they have become a promising power semiconductor device [3].

To achieve low turn-off loss and high frequency, IGBT anode engineering is a versatile solution for extracting excess carriers. Examples of IGBT anode engineering include dual-gate structures [4,5], striped anode structures [6], the Schottky anode method [7], shorted-anode structures [8–14], the super junction method [15–17], and other structures [18]. However, a dual-gate structure is not a suitable solution for conventional drive control. A striped anode structure is an effective way to reduce turn-off loss; however, this leads to increased turn-on loss [6]. The Schottky anode method decreases the current density in the holes during the on-state period [7]. The super junction method is a method that suppresses or eliminates the snapback effect, but the super junction process leads to increased manufacturing costs [15–17]. It should be pointed out that the charge balance conditions must be met [19]. For a vertical IGBT structure, a shorted-anode IGBT is a promising device. However, this structure demonstrates a snapback effect during the turn-on period. In addition, because the MOS cells are in a parallel connection, currents with an uneven distribution can lead to unstable operation [20]. An SA-IGBT with an anode trench can suppress or eliminate the snapback effect, but for vertical IGBTs, the etching process for anode trenches is difficult and costly [21]. To improve the performance of IGBT devices, the active region on the front side of the device and the drift region in the middle have been improved by researchers. However, improving the back side of the device will also contribute improvements in device performance [22,23].

In this study, a 1.2 kV shorted-anode IGBT developed using Silvaco TCAD that experiences no spring back phenomenon and low turn-off loss is proposed and is achieved by injecting an enhanced IEPF layer. The IEPF-IGBT is snapback-free and has a small cell pitch and is shown to have improved turn-off loss compared to conventional SA-IGBTs.

2. Device Structure and Mechanism

Figure 1a depicts the device structure of the IEPF-IGBT. There is an IEPF layer that has been inserted into the N-buffer layer and into the N-neutral region. The N-neutral region is lightly doped. Furthermore, during the turn-on period, there is a gap (L_G) between each P-floating and N-buffer layer that conducts the electron current. By observing Figure 1b, we can see that conventional SA-IGBTs are composed of multiple MOS cells that are connected in parallel. In order to eliminate the snapback effect, conventional SA-IGBTs must have a device cell with a large width. Figure 1c is the equivalent circuit diagram of the IEPF-IGBT, and the existence of path resistance (R_{neu}) between N and P is emphasized.

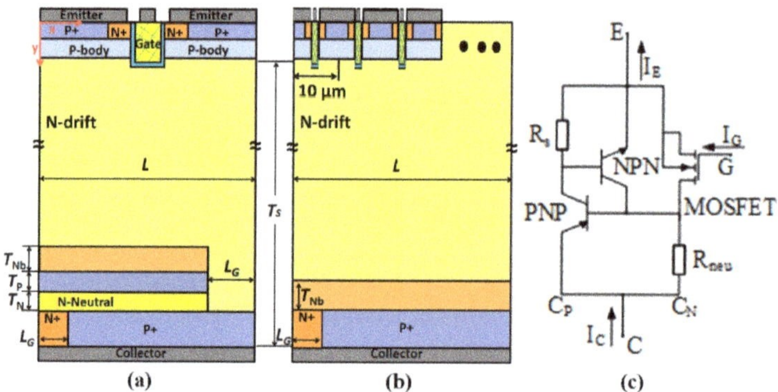

Figure 1. Schematic cross section of the devices under study. (**a**) IEPF-IGBT; (**b**) conventional SA-IGBT; (**c**) equivalent circuit diagram of the IEPF-IGBT.

The snapback *I–V* curves of the IEPF-IGBT and the conventional SA-IGBT are shown in Figure 2. The green box line in the figure marks the enlarged portion of Figure 2. It can be seen that the SA-IGBT experiences the snapback effect, but the IEPF-IGBT does not. The forward conduction voltage drop (at 100 A/cm^2 current density) is lower than that observed for the conventional SA-IGBT, with a half collector length of 10 μm and 360 μm, respectively. Because of the enlarged resistance above the P-collector region, the hole conductivity in the IEPF-IGBT was modulated earlier than in the conventional SA-IGBT [24].

Figure 3 shows the current flow paths of the electrons and holes at the beginning of conduction mode [25]. Figure 3a,b show the electron current flow into the N-collector around the IEPF layer in the IEPF-IGBT in unipolar mode and demonstrates that the IEPF layer enlarged the path resistance effectively.

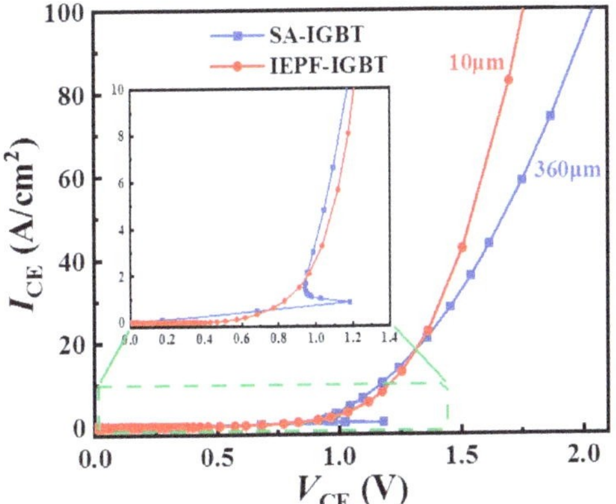

Figure 2. The *I–V* characteristics of IEPF-IGBT and SA-IGBT.

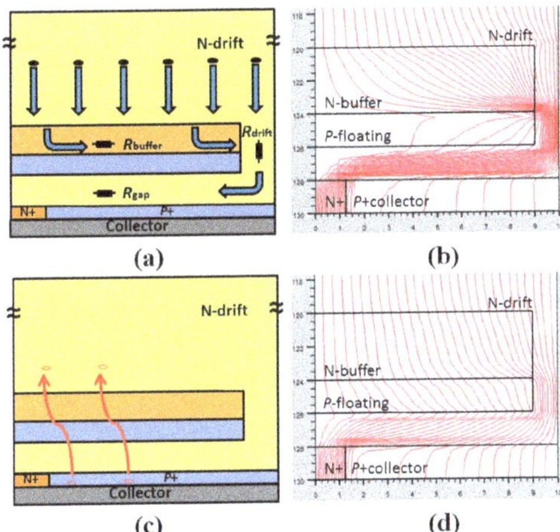

Figure 3. (a) Electron flow path; (b) current flow lines at the collector side of the IEPF-IGBT in low current density mode; (c) bipolar conduction path; (d) current flow lines at the collector side of the IEPF-IGBT in high-current-density bipolar mode.

The simple snapback model [26–28] used for Figure 1a can be described as follows:

$$V_{SB} = \left[1 + \frac{R_{drift} + R_{channel}}{(R_{neu} + R_{buf}) \cdot (L - L_G)} \right] \cdot V_{critical} \tag{1}$$

For IEPF-IGBT, the IEPF layer prevents carriers from rapidly flowing out of the device and suppresses V_{SB} by increasing $(L - L_G)$.

Figure 3c,d show the current flow of the holes through the IEPF layer in bipolar mode. In the bipolar mode, the holes can pass through the IEPF layer.

3. Results and Discussion

The devices used in this article have a voltage that can withstand the rating of 1.2 kV, and the corresponding key parameters of the devices are listed in Table 1.

Table 1. Device specifications.

Parameters	IEPF-IGBT	SA-IGBT
Gate oxide thickness (nm), T_{ox}	50	50
Gate trench depth (µm), D_{G}	6	6
Drift region doping (cm^{-3}), N_{d}	5×10^{13}	5×10^{13}
N-buffer doping (cm^{-3}), N_{Nb}	5×10^{16}	5×10^{16}
N-buffer thickness (µm), T_{Nb}	4	4
N-neutral doping (cm^{-3}), N_{N}	5×10^{13}	5×10^{13}
P-collector thickness (µm), T_{Pb}	2	2
N-collector length (µm), L_{N}	1.25	45
Half collector length (µm), L	10	360
Wafer thickness (µm), T_{S}	130	130
P-base doping (cm^{-3}), N_{base}	7×10^{17}	7×10^{17}
P-floating doping (cm^{-3}), N_{Pb}	1×10^{18}	1×10^{18}
P-floating thickness (µm), T_{pf}	2	
P-floating gap (µm), L_{G}	1~9	
Carrier lifetime (µs), τ	1	1

The parameters used in this paper refer to previous reports [29–31]. The gate oxide thickness (T_{ox}) is 50 nm. The MOS cell pitch is 10 µm [30,32]. To enhance the conductivity modulation effect, the thickness of these P collector regions is 2 µm.

Figure 4 shows the characteristic curves of the forward off-state breakdown voltage. The breakdown voltages of the IEPF-IGBT and the conventional SA-IGBT are 1437 V and 1498 V, respectively. Because there is a gap between the N-buffer layers, the N-drift of the IEPF-IGBT is less thick than the N-drift of the SA-IGBT. This leads to the IEPF-IGBT demonstrating a lower breakdown voltage than the SA-IGBT, and this small drawback of the IEPF-IGBT seems acceptable.

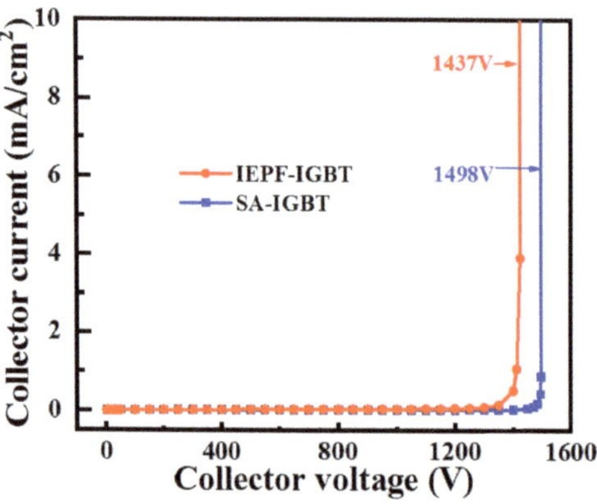

Figure 4. BV curves of the IEPF-IGBT and SA-IGBT, respectively. The gate voltage is zero.

Figure 5 shows the forward conduction characteristics of the IEPF-IGBT with different N-neutral doping. At $N_{\text{N}} = 5 \times 10^{13}$ cm^{-3} and $N_{\text{N}} = 1 \times 10^{14}$ cm^{-3}, the characteristic curves of forward conduction show no snapback effect, and their curves are highly coincidental. However, when the doping concentration is beyond $N_{\text{N}} = 3 \times 10^{14}$ cm^{-3}, the snapback effect can be observed. As N_{N} is lower, the snapback effect is weaker. According to

Equation (1), because the doping of the N-neutral region influences the resistance of the electron flow path (R_{neu}), the snapback effect can be modulated by N_N.

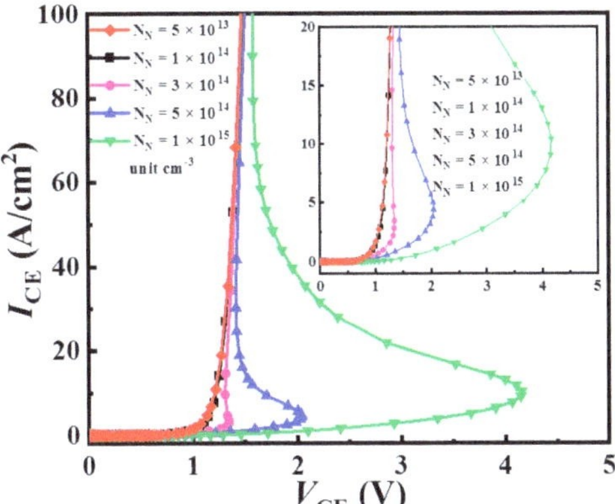

Figure 5. Forward conduction characteristics of the IEPF-IGBT with different N_N. The gate voltage is 15 V.

Figure 6a shows the doping of the IEPF layer during turn on. Obviously, the higher doping of the IEPF layer does not affect the snapback effect. This is because the lower doping concentration enables the IEPF layer to not pinch off the resistance path with the P-collector region. At a low V_{CE}, the N-neutral region exhibits a high resistance path because it is depleted by the PN junction of p-floating/N-neutral and N-neutral/p-collector. As the doping concentration of the IEPF layer increases, the V_{CE} decreases. This is due to the IEPF layer acting as holes during the current amplification stage. Figure 6b explains this phenomenon by analyzing the current densities of the holes along the IEPF layer (y = 125 μm in Figure 3c) at a current density of 100 A/cm². Figure 6b shows that the hole currents can be amplified at the IEPF layer, which, in turn, reduces the on-state voltage drop.

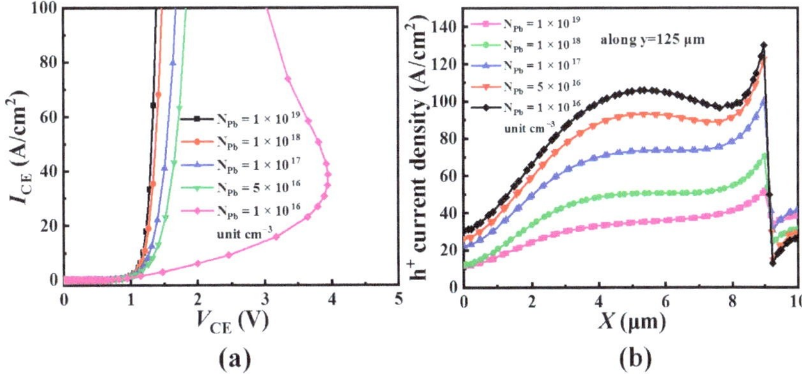

Figure 6. (**a**) Dependence of *I–V* curves on the doping concentration of the IEPF layer; (**b**) dependence of hole current density along y = 125 μm during on-state (100 A/cm² current density).

The different thickness of the N-neutral region (T_N) in the IEPF-IGBT shows the tradeoff between V_{CE} and E_{OFF} in Figure 7. Because the larger T_N enables the IEPF layer and the P-collector region to pinch off the electron flow path, as the T_N increases, the tradeoff charactcristics are promoted.

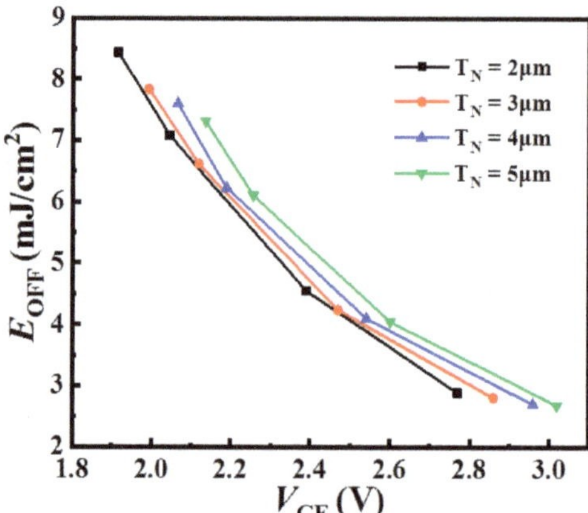

Figure 7. Dependence of T_N on the E_{OFF}–V_{CE} relationships of the IEPF-IGBT. The gate voltage is 15 V and then falls down to −5 V.

Figure 8 shows the influence of the gap of the IEPF layer (L_G). As L_G increases, V_{CE} and V_{SB} increase. As L_G increases to beyond 3 µm, the snapback effect appears. Additionally, when L_G is less than 4 µm, the snapback effect does not appear. This is due to the increased length and the resistance of the electron path because the entire electron current passes through the gap between the IEPF layers compared to the conventional SA-IGBT.

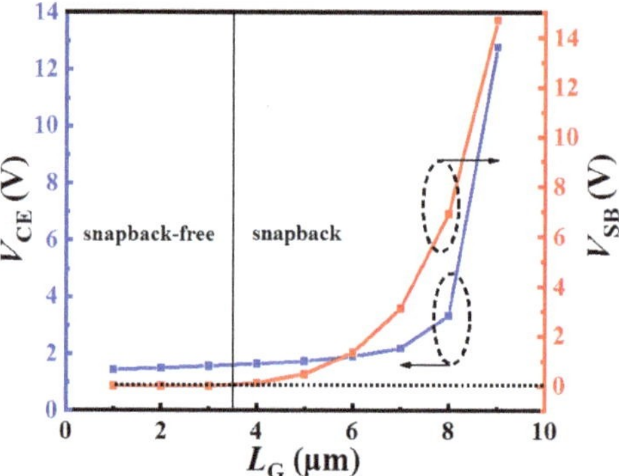

Figure 8. Dependence of V_{CE} on the gap of the IEPF layer (L_G). The shorter L_G contributed to the lower V_{CE} and V_{SB}.

Figure 9 shows the current and voltage variation curves during the turn-off process of the IEPF-IGBT and the SA-IGBT at a current density of 100 A/cm². Half collector lengths of 10 µm and 360 µm are used for the IEPF-IGBT and conventional SA-IGBT, respectively. When the turn-off loss is 12 mJ/cm², the turn-off transient times of the turn-off current curves are 32 ns and 603 ns for the IEPF-IGBT and the SA-IGBT, respectively. As such, the IEPF-IGBT shows a shorter time than the SA-IGBT. The IEPF-IGBT extracts the remaining electrons quickly due to the shorter length of collector [33,34].

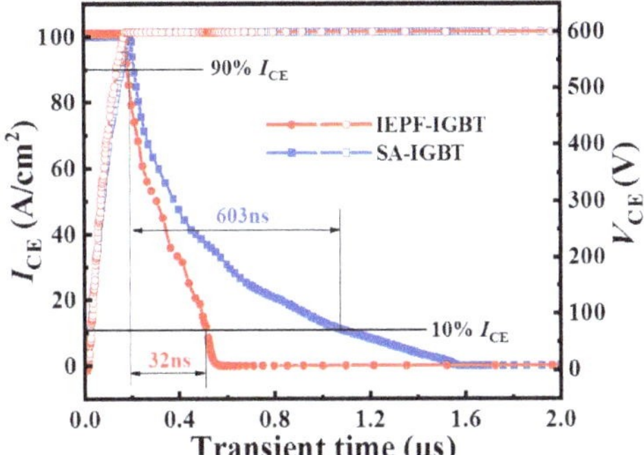

Figure 9. Turn-off voltage and current waveforms of the IEPF-IGBT and SA-IGBT, respectively.

Figure 10 depicts the key trade-off curve for the device: V_{CE} vs. E_{OFF}, for the IEPF-IGBT and the SA-IGBT. By observing Figure 10, we can observe that the trade-off characteristics of the IEPF-IGBT are better than those of the conventional SA-IGBT. At V_{CE} = 1.58 V, the E_{OFF} of the IEPF-IGBT and the SA-IGBT are 25.2 and 11 mJ/cm^2, respectively. The IEPF-IGBT shows a 56.3% E_{OFF} reduction compared to the SA-IGBT.

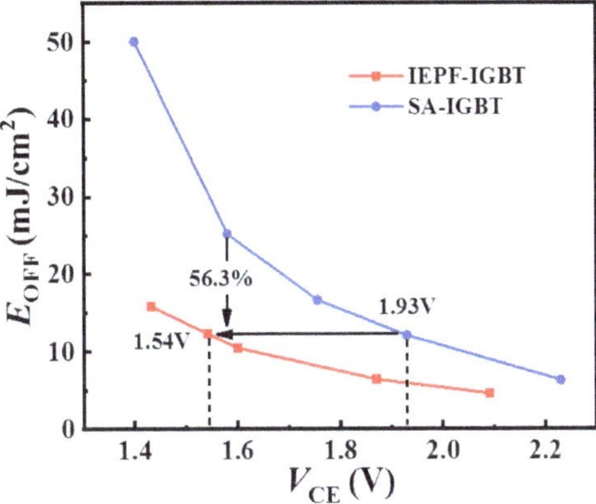

Figure 10. E_{OFF}–V_{CE} relationships of the IEPF-IGBT and the SA-IGBT.

A flow diagram of the device process is shown in Figure 11. Since the front-side process of the device is basic MOS trench technology, only the key process flow for the back side is provided. Figure 11a shows the first step, where two ion implantations of the N-type substrate are performed to form the N-buffer region and the low-doped N-neural region. Figure 11b shows the first step, which continues with ion implantation to form the IEPF layer. Figure 11c shows the third step, which continues with the formation of P+ and N+ collector regions by means of ion implantation. Last, the metal electrode is formed, as shown in Figure 11d.

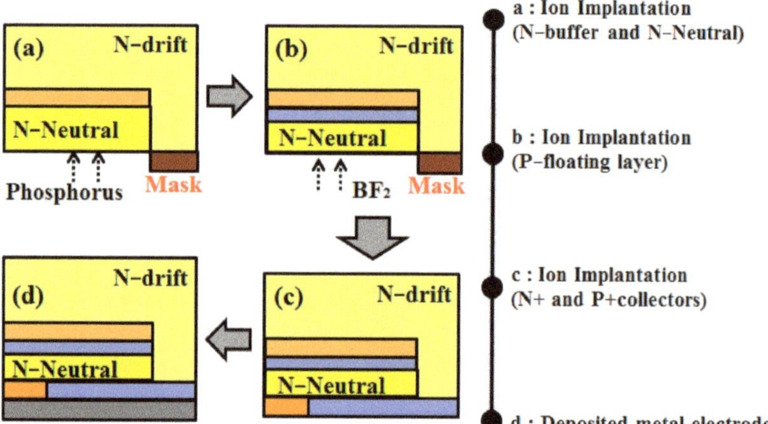

Figure 11. The flow diagram of the device process for the IEPF-IGBT.

4. Conclusions

In this study, a 1.2 kV IGBT with an IEPF layer that act as a current amplification stage and that suppresses the snapback effect during the turn-on period is proposed. The comparative study results show that the IEPF-IGBT can reduce the turn-off energy loss. The goal of eliminating the snapback effect is achieved. The IEPF-IGBT shows a 56.3% E_{OFF} reduction compared to the SA-IGBT.

Author Contributions: Conceptualization, software and writing—original draft preparation, X.Z.; methodology and validation, M.G.; resources, data curation and software, H.Z.; writing—review and editing, J.P.; supervision, L.Z.; project administration and funding acquisition, M.S. All authors have read and agreed to the published version of the manuscript.

Funding: This research was funded in part by Hainan Provincial Natural Science Foundation of China (No. 622QN285, 620RC553, 521QN210), in part by Youth Fund Project of Hainan University (No. HDQN202113), in part by Research initiation fund of Hainan University (No. KYQD(ZR)20062).

Data Availability Statement: Not applicable.

Conflicts of Interest: The authors declare no conflict of interest.

References

1. Rahimo, M.; Kopta, A.; Schlapbach, U.; Vobecky, J.; Schnell, R.; Klaka, S. The Bi-mode Insulated Gate Transistor (BIGT) a potential technology for higher power applications. In Proceedings of the 2009 21st International Symposium on Power Semiconductor Devices & IC's, Barcelona, Spain, 14–18 June 2009; pp. 283–286. [CrossRef]
2. Rahimo, M.; Schlapbach, U.; Kopta, A.; Vobecky, J.; Schneider, D.; Baschnagel, A. A High Current 3300V Module Employing Reverse Conducting IGBTs Setting a New Benchmark in Output Power Capability. In Proceedings of the 2008 20th International Symposium on Power Semiconductor Devices and IC's, Orlando, FL, USA, 18–22 May 2008; pp. 68–71. [CrossRef]
3. Shan, Y.; Tseng, K.J.; Simanjorang, R.; Tu, P. Experimental Comparison of High-Speed Gate Driver Design for 1.2-kV/120-A Si IGBT and SiC MOSFET Modules. *IET Power Electron.* **2017**, *10*, 979–986.
4. Zhu, L.; Chen, X. An Investigation of a Novel Snapback-Free Reverse-Conducting IGBT and with Dual Gates. *IEEE Trans. Electron Devices* **2012**, *59*, 3048–3053. [CrossRef]
5. Wei, J.; Luo, X.; Huang, L.; Zhang, B. Simulation Study of a Novel Snapback-Free and Low Turn-Off Loss Reverse-Conducting IGBT with Controllable Trench Gate. *IEEE Electron Device Lett.* **2018**, *39*, 252–255. [CrossRef]
6. Luther-King, N.; Sweet, M.; Spulber, O.; Vershinin, K.; De Souza, M.; Narayanan, E.S. Striped anode engineering: A concept for fast switching power devices. *Solid-State Electron.* **2002**, *46*, 903–909. [CrossRef]
7. Sun, L.; Duan, B.; Wang, Y.; Yang, Y. Analysis of the Fast-Switching LIGBT with Double Gates and Integrated Schottky Barrier Diode. *IEEE Trans. Electron Devices* **2019**, *66*, 2675–2680. [CrossRef]
8. Simpson, M.R. Analysis of negative differential resistance in the I-V characteristics of shorted-anode LIGBT's. *IEEE Trans. Electron Devices* **1991**, *38*, 1633–1640. [CrossRef]

9. Zhang, L.; Zhu, J.; Sun, W.; Du, Y.; Yu, H.; Huang, K.; Shi, L. A high current density SOI-LIGBT with Segmented Trenches in the Anode region for suppressing negative differential resistance regime. In Proceedings of the 2015 IEEE 27th International Symposium on Power Semiconductor Devices & IC's (ISPSD), Hong Kong, 10–14 May 2015; pp. 49–52. [CrossRef]
10. Huang, L.; Luo, X.; Wei, J.; Zhou, K.; Deng, G.; Sun, T.; Ouyang, D.; Fan, D.; Zhang, B. A Snapback-Free Fast-Switching SOI LIGBT with Polysilicon Regulative Resistance and Trench Cathode. *IEEE Trans. Electron Devices* **2017**, *64*, 3961–3966. [CrossRef]
11. Zhang, L.; Zhu, J.; Sun, W.; Chen, M.; Zhao, M.; Huang, X.; Chen, J.; Qian, Y.; Shi, L. Novel Snapback-Free Reverse-Conducting SOI-LIGBT with Dual Embedded Diodes. *IEEE Trans. Electron Devices* **2017**, *64*, 1187–1192. [CrossRef]
12. Chen, W.; Huang, Y.; He, L.; Han, Z.; Huang, Y. A snapback-free TOL-RC-LIGBT with vertical P-collector and N-buffer design. *Chin. Phys. B* **2018**, *27*, 088501. [CrossRef]
13. Duan, B.; Sun, L.; Yang, Y. Analysis of the Novel Snapback-Free LIGBT with Fast-Switching and Improved Latch-Up Immunity by TCAD Simulation. *IEEE Electron Device Lett.* **2019**, *40*, 63–66. [CrossRef]
14. Wu, J.; Kong, M.; Yi, B.; Chen, X.B. An Ultralow Turn-Off Loss SOI-LIGBT with a High-Voltage p-i-n Diode Integrated on Field Oxide. *IEEE Trans. Electron Devices* **2019**, *66*, 1831–1836. [CrossRef]
15. Antoniou, M.; Udrea, F.; Bauer, F.; Nistor, I. A new way to alleviate the RC IGBT snapback phenomenon: The Super Junction solution. In Proceedings of the 2010 22nd International Symposium on Power Semiconductor Devices & IC's (ISPSD), Hiroshima, Japan, 6–10 June 2010; pp. 153–156.
16. Antoniou, M.; Lophitis, N.; Udrea, F.; Bauer, F.; Vemulapati, U.R.; Badstuebner, U. On the Investigation of the "Anode Side" SuperJunction IGBT Design Concept. *IEEE Electron Device Lett.* **2017**, *38*, 1063–1066. [CrossRef]
17. Findlay, E.M.; Udrea, F.; Antoniou, M. Investigation of the Dual Implant Reverse-Conducting SuperJunction Insulated-Gate Bipolar Transistor. *IEEE Electron Device Lett.* **2019**, *40*, 862–865. [CrossRef]
18. Green, D.W.; Vershinin, K.V.; Sweet, M.; Narayanan, E.M.S. Anode Engineering for the Insulated Gate Bipolar Transistor—A Comparative Review. *IEEE Trans. Power Electron.* **2007**, *22*, 1857–1866. [CrossRef]
19. Zhang, L.; Zhu, J.; Ma, J.; Cao, S.; Li, A.; Li, S.; Ye, R.; Sun, W.; Zhao, J.; Shi, L. Turn-Off Transient of Superjunction SOI Lateral IGBTs: Mechanism and Optimization Strategy. *IEEE Trans. Electron Devices* **2019**, *66*, 1409–1415. [CrossRef]
20. Findlay, E.M.; Udrea, F. Reverse-Conducting Insulated Gate Bipolar Transistor: A Review of Current Technologies. *IEEE Trans. Electron Devices* **2019**, *66*, 219–231. [CrossRef]
21. Jiang, H.; Zhang, B.; Chen, W.; Li, Z.; Liu, C.; Rao, Z.; Dong, B. A Snapback Suppressed Reverse-Conducting IGBT with a Floating p-Region in Trench Collector. *IEEE Electron Device Lett.* **2012**, *33*, 417–419. [CrossRef]
22. Zhu, L.; Chen, X. A novel double trench reverse conducting IGBT with robust freewheeling switch. *J. Semicond.* **2014**, *35*, 084004. [CrossRef]
23. Chen, X.; Cheng, J.; Teng, G.; Guo, H. Novel trench gate field stop IGBT with trench shorted anode. *J. Semicond.* **2016**, *37*, 54008. [CrossRef]
24. Bhojani, R.; Lutz, J.; Baburske, R.; Schulze, H.-J.; Niedemostheide, F.-J. A novel Injection Enhanced Floating Emitter (IEFE) IGBT structure improving the ruggedness against short-circuit and thermal destruction. In Proceedings of the 2017 29th International Symposium on Power Semiconductor Devices and IC's (ISPSD), Sapporo, Japan, 28 May–1 June 2017; pp. 113–116. [CrossRef]
25. Long, H.; Sweet, M.R.; Luther-King, N.; Narayanan, E. Current Distribution Analysis of IGBT Cells. In Proceedings of the 2009 International Conference on Solid State Devices and Materials, Sendai, Japan, 7–9 October 2009.
26. Chen, W.-Z.; Li, Z.-H.; Zhang, B.; Ren, M.; Zhang, J.-P.; Liu, Y.; Li, Z.-J. A snapback suppressed reverse-conducting IGBT with uniform temperature distribution. *Chin. Phys. B* **2014**, *23*, 018505. [CrossRef]
27. Yamamoto, T.; Miyake, M.; Kato, H.; Feldmann, U.; Mattausch, H.J.; Miura-Mattausch, M. Compact modeling of injection-enhanced insulated-gate bipolar transistor for accurate circuit switching prediction. *Jpn. J. Appl. Phys.* **2014**, *53*, 04EP13. [CrossRef]
28. Vemulapati, U.; Kaminski, N.; Silber, D.; Storasta, L.; Rahimo, M. Reverse conducting–IGBTs initial snapback phenomenon and its analytical modelling. *IET Circuits Devices Syst.* **2014**, *8*, 168–175. [CrossRef]
29. Jiang, M.; Shen, Z.J. Simulation Study of an Injection Enhanced Insulated-Gate Bipolar Transistor with p-Base Schottky Contact. *IEEE Trans. Electron Devices* **2016**, *63*, 1991–1995. [CrossRef]
30. Deng, G.Q.; Luo, X.R.; Wei, J.; Zhou, K.; Huang, L.; Sun, T.; Liu, Q.; Zhang, B. A Snapback-Free Reverse Conducting Insulated-Gate Bipolar Transistor with Discontinuous Field-Stop Layer. *IEEE Trans. Electron Devices* **2018**, *65*, 1856–1861. [CrossRef]
31. Zhang, X.-D.; Wang, Y.; Wu, X.; Bao, M.-T.; Yu, C.-H.; Cao, F. An Improved VCE–EOFF Tradeoff and Snapback-Free RC-IGBT with P⁺ Pillars. *IEEE Trans. Electron Devices* **2020**, *67*, 2859–2864. [CrossRef]
32. Deng, G.; Luo, X.; Zhou, K.; He, Q.; Ruan, X.; Liu, Q.; Sun, T.; Zhang, B. A snapback-free RC-IGBT with Alternating N/P buffers. In Proceedings of the 2017 29th International Symposium on Power Semiconductor Devices and IC's (ISPSD), Sapporo, Japan, 28 May–1 June 2017; pp. 127–130. [CrossRef]
33. Chen, W.; Zhang, B.; Li, Z.; Ren, M.; Li, Z. A short-contacted double anodes IGBT. In Proceedings of the 2012 IEEE 11th International Conference on Solid-State and Integrated Circuit Technology, Xi'an, China, 29 October–1 November 2012; pp. 1–3.
34. Jiang, H.; Zhang, B.; Chen, W.; Qiao, M.; Li, Z.; Liu, C.; Rao, Z.; Dong, B. Low turnoff loss reverse-conducting IGBT with double n-p-n electron extraction paths. *Electron. Lett.* **2012**, *48*, 457–458. [CrossRef]

MDPI
St. Alban-Anlage 66
4052 Basel
Switzerland
www.mdpi.com

Electronics Editorial Office
E-mail: electronics@mdpi.com
www.mdpi.com/journal/electronics

Disclaimer/Publisher's Note: The statements, opinions and data contained in all publications are solely those of the individual author(s) and contributor(s) and not of MDPI and/or the editor(s). MDPI and/or the editor(s) disclaim responsibility for any injury to people or property resulting from any ideas, methods, instructions or products referred to in the content.

www.ingramcontent.com/pod-product-compliance
Lightning Source LLC
LaVergne TN
LVHW071021170726
843515LV00006B/1177